Advanced Transport Systems

Milan Janić

Advanced Transport Systems

Analysis, Modeling, and Evaluation of Performances

Springer

Milan Janić
Transport and Planning Department
Faculty of Civil Engineering
and Geosciences
Delft University of Technology
Delft
The Netherlands

ISBN 978-1-4471-7045-7 ISBN 978-1-4471-6287-2 (eBook)
DOI 10.1007/978-1-4471-6287-2
Springer London Heidelberg New York Dordrecht

Softcover reprint of the hardcover 1st edition 2014

Printed on acid-free paper

Springer is part of Springer Science+Business Media (www.springer.com)

To my wife Vesna at the 30th anniversary of our marriage

Preface

The transport system has, is, and will continue to be a foundation of the economy of each country/nation, as well as that of the world. In particular, in the twenty-first century, it will further strengthen its role in integrating and globalizing economic activities, and will thus also influence the quality of people's lives. In the past, transport demand in terms of both the number of passengers and the volumes of freight/goods shipments has been constantly growing in the medium- to long-term period(s) despite being affected, from time to time, by the local and global economic and political crises. This demand has been satisfied by the capacity of the transport system generally consisting of the transport infrastructure, transport means/vehicles, and workforce. Material, energy, and labor has been consumed in order to provide transport services according to the specified internal organization containing operating rules and procedures, and under given external regulation and constraints. On the one hand, such developments have produced the above-mentioned positive contributions to the national economies and social welfare. On the other, they have affected the environment and society in terms of land use/take for expanding the transport infrastructure, energy consumption from non-renewable sources (coal, crude oil, and natural gas) and related emissions of Green House Gases (GHG), local noise, congestion, and safety (traffic incidents and accidents), Since both passenger and freight transport demand are predicted to double over the next 20 and triple over the next 50 years, solutions for serving them more efficiently and effectively while mitigating impacts on the environment and society need to be provided. Therefore, in addition to creating transport policies and monitoring schemes aiming to reduce physical transport demand (i.e., telecommuting) and implementing advanced transport planning and operating tools and techniques, potential solutions also lie in developing advanced technologies individually and/or in combination with advanced operational concepts. Generally, this implies providing: (i) sufficiently capacitated and environmentally friendlier, i.e., more energy/fuel efficient, cleaner, quieter, and safer, technologies based on an increased use of renewable energy/fuel sources (such as, for example, biomass fuels (liquid) hydrogen, wind and solar energy), nanotechnologies, and information technologies; and (ii) the advanced organizational and operational forms and concepts of using transport infrastructure, transport means/vehicles, and accompanied resources.

Experience so far indicates that commercialization, i.e., development and implementation, of the advanced components—technologies and related operational concepts—of the transport system has been an evolutionary rather than a revolutionary process. The main reasons include: (i) a rather long time for maturing up to full commercialization; (ii) an inherent threat from confronting existing and forthcoming even stricter institutional/policy regulation/constraints; (iii) relatively high development costs; (iv) frequently uncertain long-term overall commercial and social feasibility; and (v) a relatively long path for obtaining operational certification implying full environmental and societal/policy acceptance. Under such circumstances, most such transport technologies and operational concepts, except a couple of futuristic ones, have been mostly gradually updated and improved, usually based on the closest previous counterparts. In the given context, this justifies deeming them "innovative" or "advanced" rather than completely "new". In this book, the attribute "advanced" is used for all such technologies and operational concepts.

The book describes analysis, modeling, and evaluation of performances of the selected advanced transport systems. Some of them have already been commercialized, i.e., implemented and operationalized, and/or are planned to be so, while others are still at the conceptual level waiting for further elaboration. Their performances are considered as derived from the technical/technological design and solutions of the infrastructure, transport means/vehicles, and supporting facilities and equipment used according to the specified operational rules and procedures, and economic, environmental, social, and policy conditions/constraints.

Analysis and modeling implies examination of their infrastructural, technical/technological, operational, economic, environmental, social, and policy performances. Evaluation based on a Strengths, Weaknesses, Opportunities and Threats (SWOT)-like analysis implies assessment of the advantages and disadvantages of these systems. In such context, Strengths and Opportunities are considered as advantages, while Weaknesses and Threats are considered disadvantages. Both are considered from the aspects of academics/researchers, but also from those of particular actors/stakeholders involved such as users of transport services–passengers and freight/goods shippers/receivers, transport infrastructure and service providers/operators, investors, policy makers at different institutional levels (local, national, international), and members of the local community/society.

Particular advanced transport systems have been selected according to the following criteria: (i) the level of advancement of particular performances; (ii) representativeness through transport modes (rail, road, air, water/sea, intermodal); (iii) their spatial scale (area) of operation (urban and inter-urban); (iv) category of demand served (passengers, freight/goods); (v) availability/accessibility of relevant information (from science-based and publically-accessible relevant sources); and (vi) the level of systematic scientific elaboration as compared to that used in this book.

The selected advanced transport systems are clustered in the book's chapters respecting the type and number of their advanced performances independently of the transport mode, spatial/geographical scale of operation, and type of transport demand they serve.

The widely dispersed and in some cases scarce material collected from the various available sources such as research (including my own), literature (books and papers in scientific and professional journals), and websites is presented from the traffic and transport engineering and planning and design perspective. Most facts and issues are scientifically supported and accurate regarding the fundamental relationships between particular variables (parameters). Nevertheless, some of them, particularly those related to futuristic concepts, contain a level of fuzziness in the absolute terms, which, however, does not compromise their relevance in the given context. As such, the book aims to be informative as much as possible but by no means exhaustive—to the contrary, it intends to provide academics, researchers, consultants, policy/decision makers, and professionals from the transport industry and related fields with material for current and future research and development of the transport system.

October 2013 Milan Janić

Acknowledgments

The author gratefully acknowledges the support of organizations and individuals in getting this book into publishable form.

Firstly, I would like to thank very much the Transport Institute of Delft University of Technology (Prof. Bart van Arem—Director, and Dr. Arjan van Binsbergen—Secretary) (TU Delft, Delft, The Netherlands) for some financial support in finalizing the work on the book. Secondly, I benefited greatly from the valuable discussions, materials, exhibits, and data provided by Mr. John P. Christy, Lead Engineer—Operations Airport Technology, Boeing and Ms. Mary E. Kane, Trademark and Copyright Licensing, Boeing Intellectual Property Management (Boeing Company/Commercial Airplanes, Seattle, Washington, USA).

Thirdly, I give my special gratitude to Mr. Andrej Grah Whatmough for his excellent help polishing the language.

On the personal side, the great effort of writing this book was continuously supported and inspired by my wife Vesna and my son Miodrag, doctor of medicine, who was at the same time also working enthusiastically on his Ph.D. thesis.

Milan Janić

Contents

Abbreviations

ABD	Additional breaking device
ACN	Aircaft classification number
AGV	Automotice grande vitesse
AMT	Automatic manual transmission
ANA	Air Nippon Airways
APT	Air passenger transport
APU	Auxiliary power unit
ASM	Available seat mile
ATA	Air Transport Association
ATAG	Air Transport Action Group
ATC	Air traffic control
ATMS	Automated manual system
AVL	Automatic vehicle location
atm	Atmosphere
ATMS	Automated manual system
BAU	Business as usual
BEV	Battery electric vehicle
BR	Bypass ratio
BRT	Bus rapid transit
CAD	Computer aided dispatching
CDA	Continuous descent approach
CDB	Central business district
CEN	Comité Européen de Normalisation
CIA	Central Intelligence Agency
CIFT	Commercial intermodal freight train
CNG	Compressed natural gas
CO	Carbon monoxide
CSS	Carbon capture and storage
DC	Direct-current
DM	Decision making (maker)
DPF	Diesel particulate filter

DWT	Deadweight tonnage
EADS	European Aeronautic Defense and Space Company
EAT	Economic analysis technique
EBHA	Electrical backup hydrostatic actuators
EC	European Commission
ECB	European Central Bank
EDS	Electro dynamic suspension
EEC	Electronic engine controller
EEDI	Energy efficiency design index
EEOI	Energy efficiency operational indicator
EET	Evacuated tube transport
EFB	Electronic flight bag
EGR	Exhaust gas recirculation
EHA	Electro-hydrostatic actuators
EMS	Electromagnetic suspension
EPNL	Equivalent perceived noise level
EPS	Enhanced permissible speed
ETCS	European train control system
ETOPS	Extended range twin-engine operational performances
EU	European Union
FAA	Federal Aviation Administration
FAME	Fatty acid methyl ester
FL	Flight level
FMS	Flight management system
g	Gravitational acceleration
GAO	Government Accountability Office (US)
GDP	Gross domestic product
GHG	Green house gases
GIS	Geographic information system
GS	Glide slope
GPS	Global positioning satellite
Gt	Giga ton
HFCV	Hydrogen fuel cell vehicle
HFO	Heavy fuel oil
hp	Horse power
HPC	High pressure compressor
HPT	High pressure turbine
HS	High speed
HSs	Hub-and spoke(s) (network)
HSR	High speed rail
HYV	Hybrid vehicle

HV	Hydrogen vehicle
IATA	International Air Transport Association
ICAO	International Civil Aviation Organization
ICE	Inter-city-express
ICEV	Internal combustion engine vehicle
ICT	Information communication technologies
IFR	Instrument flight rules
ILS	Instrument landing system
IMA	Integrated modular avionics
IMC	Instrument metrological conditions
IMF	International monetary fund
IMO	International Maritime Organization
INA	Integrated noise area
IPC	Intermediate pressure compressor
IPT	Intermediate pressure turbine
ITS	Intelligent transport systems
JAXA	Japan Aerospace eXploration Agency
km	Kilometer
Kn	Kilo-Newton
kts	Knot
kW	Kilowatt
kWh	Kilowatt-hour
l	Liter
LAPCAT	Long-Term Advanced Propulsion Concepts and Technologies
lb	Pound-mass
LCA	Life cycle analysis
LCC	Low cost carrier
LEM	Linear electric motor
LH_2	Liquid hydrogen
LIFT	Long intermodal freight train
LIM	Linear induction motor
LNG	Liquefied natural gas
LPP	Lean premixed pre-vaporized (concept)
L/R	Line or ring (network)
LSM	Linear synchronous motor
LU	Loading unit
m	Meter
M	Mixed (network)
MAGLEV	MAGnetic levitation
MCA	Multi criteria analysis
MCDM	Multi-criteria decision-making (method)

MEPC	The Marine Environment Protection Committee
MFD	Multi-functional display
MJ	Mega joule
MLS	Microwave landing system
MLW	Maximum landing weight
MS	Manual system
MTOW	Maximum take-off weight
MW	Mega watt
MWh	Mega watt hour
MZFW	Maximum zero fuel weight
NASA	National Aeronautics and Space Administration
NextGen	Next generation (air transport system)
nm	Nautical mile
NSS	Network systems server
NO_x	Nitrogen oxide
OMs	Overall emission(s)
OEW	Operating empty weight
Pa	Pascal
PDE	Pulse detonated engine
PEM	Polymer electrolyte membrane
P–P	Point-to-point (network)
PS	Permissible speed
RFID	Radio frequency identification
PR	Priority
PRT	Personal rapid transit
RAT	Ram air turbine
RNAV	aRea navigation
ROL	Rich-burn/quick-quench/lean-burn
RPK	Revenue passenger kilometer
rpm	Rotations per minute
RTK	Revenue ton-kilometer
SAW	Simple additive weighting (method)
SBSP	Space-based solar power
SCMR	Specific maximum continuous rating
SCR	Selective catalytic reduction
SESAR	Single European Sky ATM Research
SEEMP	Ship energy efficiency management plan
SFC	Specific fuel consumption
SN	Specific noise
SRM	Steam methane reforming
SSP	Space solar power

STA	Supersonic Transport Aircraft
TCD	Trunk line with collecting/distribution forks (network)
TEN	Trans-European Transport Network
TEU	Twenty foot equivalent unit
TGV	Train à grande vitesse
TOPSIS	Technique for order preference by similarity to ideal solution (method)
TOW	Take-off weight
TRM	Transrapid maglev
TSFC	Thrust specific fuel consumption
TTW	Tank-to-wheel
TU	Transport unit
TVM	Transmission voie-machine (transmission track-*machine)*
UFT	Underground freight transport
UIC	International Union of Railways
ULD	Unit load device
U.S.	United States
VFR	Visual flight rules
VMC	Visual meteorological conditions
VOCs	Volatile organic compound(s)
WSC	World Shipping Council
WHRS	Waste heat recovery system
WIF	Water in fuel
WTT	Well-to-tank
WTW	Well-to-wheel

Chapter 1
Advanced Transport Systems: General

1.1 Definition

The transport system can be considered as a physical entity for the mobility of persons and physical movements of freight/goods shipments between their (ultimate) origins and destinations. The entity consists of infrastructure, transport means/vehicles, supporting facilities and equipment, workforce, and organizational forms of their use. Energy/fuel is consumed to build/manufacture and operate the infrastructure, transport means/vehicles, and facilities and equipment. The transport system includes different forms/modes such as rail, road, water, air, and their sensible/wise combinations operating as intermodal or multimodal transport service networks. Depending on the volumes and intensity of passenger and freight/goods demand, each mode has different self-contained components distinguished mainly with respect to the type of technologies, resources used, and concepts of providing transport services. Consequently, in the remaining text, the term "systems" is used for these rather complex components of the transport system.

The above-mentioned systems operated by different transport modes provide services in urban, suburban, and interurban regions, thus covering different spatial/geographical scales implying short, medium, and long transport distances, respectively. These systems include both conventional and advanced elements. In the remaining text, those with predominantly advanced elements as compared to their preceding counterparts are referred to as "advanced systems." The attribute "advanced" implies that the given system is superior compared to its closest preceding counterpart(s) in the same or different transport mode(s), with respect to one, a few, and/or all infrastructural, technical/technological,[1] and/or operational

---

[1] A specific advancement in technical/technological performances is made by use of new materials (composites) based on the elements of nanotechnology. This is the science and engineering of examining, monitoring, and modifying materials at nanoscale (atomic and/or molecular level). By changing the structure of materials in terms of their physical, mechanical, electrical, magnetic properties, heat conduction, and light reflection, this approach will also be able to produce improved and/or new generation of concrete, steel, aluminum, etc., materials currently widely used in construction of transport infrastructure and transport means/vehicles (Khan 2011).

M. Janić, *Advanced Transport Systems*,
DOI: 10.1007/978-1-4471-6287-2_1, © Springer-Verlag London 2014

performances. In many cases, economic, environmental, and social/policy performances are also taken into account to refer to such systems as "advanced."

Similar to their conventional counterparts, advanced transport systems consist of physical infrastructure, transport means/vehicles, workforce/labor, and supporting facilities and equipment. An important part of the latter is ITS (Intelligent Transport Systems),[2] which, with components such as sensors and microchips, have already become and will increasingly continue to be an unavoidable part of the transport system. In addition, advanced transport systems consume energy/fuel to perform their primary function of transporting persons and freight/goods shipments according to the specified organization of transport services based on the given operational rules and procedures. In such context, they are designed to provide safe/secure, efficient, effective, environmentally, and socially friendlier services then their conventional ("non-advanced") counterparts. The first implies the lack of incidents and accidents due to known reasons. The second refers to the lower total, average, and marginal costs of services offered to users/passengers and freight/goods shippers/receivers. In particular, lower average costs per unit of output (p-km and/or t-km) (p-km—passenger kilometer; t-km—ton-kilometer) can make these systems commercially more feasible then their conventional counterparts. The third implies the quality of transport services provided to users/passengers and freight/goods shippers/receivers—attributed to the improved accessibility, regularity, punctuality, reliability, and shorter travel/service time, higher riding comfort, etc. The last includes, on the one hand, lower absolute and relative impacts on the environment and society in terms of land use (take), energy consumption and related emissions of GHG (Green House Gases),[3] local noise, congestion, and waste, and on the other, greater contribution to social welfare such as employment and GDP (Gross Domestic Product) on the local, regional, national, and global scale.

1.2 Classification

Advanced transport systems can be classified with respect to different attributes/criteria. Some relate to their advanced components and some to the level of their commercialization.

---

[2] The ITS enable collecting, processing, and distribution of information about the system states and operations thanks to tracking and telematics applications, scheduling services/operations, informing/notification of users/passengers and freight/goods shippers/receivers, monitoring security, and detecting potential all kind of threats.

[3] The U.S. (United Sates) and EU (European Union) have set up the targets for reducing emissions of GHG (Green House Gases) from transport sector for about 20 and 50 % by the year 2020 and 2050, respectively, as compared to the year 1990. In addition, some research suggests that if the scenario of the economic and social development continues as BAU (Business As Usual), the world's total emissions of GHG in terms of CO_2 from transport sector will reach about 5 Gt/y (Giga tons/year) by the year 2050. However, if the Green Growth and CSS (Carbon Capture and Storage) scenario is going to be implemented, this amount will likely not be greater than 1 Gt/y (Gt—Giga tons) (EC 2010; Hawksworth 2008).

1.2.1 Attributes/Criteria Related to Advanced Components

Advanced transport systems can be classified depending on a single and/or combination of dominant (prevailing) advanced components and related performances of their infrastructure, technics/technologies of transport means/vehicles and supporting facilities and equipment, energy/fuel used, pattern of operations, economic/business model, and impacts/effects on the environment and society. Consequently, in this book, advanced transport systems are ultimately distinguished and elaborated, independently on the transport mode, as follows:

- Systems with advanced technics/technologies of transport means/vehicles often implying modified operations and in some specific cases infrastructure such as high-speed tilting passenger trains, road mega trucks, large commercial freight aircraft, and advanced commercial aircraft (for example, the most recent Boeing B787-8 and the forthcoming Airbus A350);
- Systems with sometimes slightly modified technics/technologies of transport means/vehicles and advanced operations aiming at improving the efficiency and effectiveness of transport services such as the BRT (Bus Rapid Transit) System systems, advanced freight collection/distribution networks, LIFTs (Long Intermodal Freight Train(s)) (in Europe), and the APT (Air Passenger Transport) system;
- Systems with advanced technics/technologies of transport means/vehicles and energy/fuel contributing to the consequent environmental effects/impacts, including advanced passenger cars, large advanced container ships, LH_2 (Liquid Hydrogen)-fuelled commercial subsonic aircraft and advanced STA (Supersonic Transport Aircraft); the latter two are alternatives to their current counterparts using crude-oil derivatives-petrol/diesel and kerosene, respectively; and
- Systems with advanced infrastructure, technics/technologies of transport means/vehicles, and consequently operations and business model such as HSR (High-Speed Rail), TRM (TransRapid MAGLEV (MAGnetic LEVitation)) system, PRT (Personal Rapid Transit) and UFT (Underground Urban Freight) systems in urban areas, and the long-distance ETT (Evacuated Tube Transport) system; additionally, the advanced technologies and procedures in the ATC (Air Traffic Control) system for increasing the airport runway capacity can be categorized in this category.

At present, except the ETT system, none of the above-mentioned existing and/or forthcoming systems possesses all six—infrastructural, technical/technological, operational, economic, environmental, and social/policy—advanced elements. On the one hand, this indicates a lack of completely new systems in the medium- to long-term future, and on the other, their present and prospective mainly evolutionary rather than revolutionary development and commercialization.

1.2.2 *Attributes/Criteria Related to Level of Commercialization*

Advanced transport systems are usually developed in five phases reflecting the level of their commercialization as follows:

- Exploratory research delivering ideas and concepts;
- Applied research resulting in understanding and further elaboration of the particular ideas and concepts;
- Pre-industrial development resulting in prototypes and carrying out pilot operational trials;
- Industrialization resulting in production/manufacturing; and
- Commercialization implying physical implementation and operationalization.

Consequently, advanced transport systems can be categorized into four categories as follows:

- *Category I* includes systems that have passed all five phases and are fully commercialized;
- *Category II* includes systems that have passed all five phases but have been commercialized on a very limited scope and scale;
- *Category III* includes systems that have passed two or at most three of the above-mentioned (five) phases, implying that they are still waiting for or just undergoing pilot operational trials and industrialization; and
- *Category IV* includes systems in the exploratory phase waiting for the "green light" in order to pass to subsequent phase(s).

This book considers advanced transport systems categorized according to the level of their commercialization as given in Table 1.1.

1.3 Performances

Dealing with advanced transport systems usually raises the question of their performances, i.e., their ability to satisfy current and prospective needs and expectations of particular actors/stakeholders involved. Such an approach requires analyzing, modeling, and evaluating particular performances.

1.3.1 *Definition*

Advanced transport systems are generally characterized by infrastructural, technical/technological, operational, economic, environmental, social, and policy performances.

Table 1.1 Classification of advanced transport systems respecting their level of commercialization

Category	Systems
I	BRT (Bus Rapid Transit)
	High-Speed Tilting Trains
	HSR (High-Speed Rails)
	APT (Air Passenger Transport)[a]
	Advanced Commercial Aircraft
	Large Commercial Freight Aircraft
II	Advanced Freight Collection/Distribution Networks (in Europe)
	LIFTs (Long Intermodal Freight Trains) (in Europe)
	Road Mega-Trucks (in Europe)
	Large Advanced Container Ships
	PRT (Personal Rapid Transit)
	TRM (TransRapid MAGLEV)
III	Super HSR (High-Speed Rail) (in Europe)
	Advanced Passenger Cars
IV	UFT (Underground Freight Transport)
	ETT (Evacuated Tube Transport)
	Advanced ATC (Air Traffic Control) Technologies
	LH_2 (Liquid Hydrogen)-Fuelled Commercial Air Transportation
	Advanced STA (Supersonic Transport Aircraft)

[a] "The Wright Brothers created the single greatest cultural force since the invention of writing. The airplane became the first World Wide Web, bringing people, languages, ideas, and values together."—Bill Gates, CEO, Microsoft Corporation

- *Infrastructural and technical/technological performances* mainly reflect physical, constructive, technical, and technological features of infrastructure, transport means/vehicles, and supporting facilities and equipment, respectively, enabling them to carry out the specified transport operations serving the specified volumes of passenger and freight/goods demand under given conditions.
- *Operational performances* imply quantitative and qualitative capabilities to serve given volumes of passenger and freight/goods demand.
- *Economic performances* reflect the efficiency of serving given volumes of passenger and freight/goods demand expressed by costs of services covered by the relevant charges (prices).
- *Environmental and social performances* reflect the intensity and scale of physical impacts on the environment and society. If monetized, these impacts are considered as externalities.
- *Policy performances* reflect compliance with current and future medium- to long-term transport policy regulations and specified targets.

Although these performances are usually considered independently, they are inherently strongly dependent and interactive with each other as shown in Fig. 1.1.

In the "top-down" consideration, infrastructural performances influence technical/technological performances and consequently create mutual influence

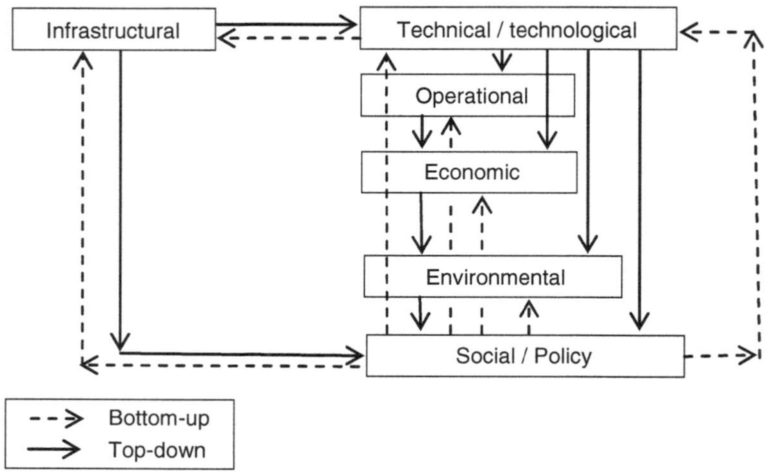

Fig. 1.1 Potential interaction between advanced transport system performances

between these and all other performances. In the "bottom-up" consideration, social/policy performances influence infrastructural and technical/technological performances and consequently also create mutual influence of these and all other performances.

For example, the technical/technological performances of transport means/vehicles and supporting facilities and equipment can require completely new infrastructure and consequently pattern of operations, which can influence the economic, environmental, and social performances. Some examples include HSR (High-Speed Rail), TRM (TransRapid MAGLEV (MAGnetic LEVitation)), and ETT (Evacuated Tube Transport). In some other cases, the economic performances can strongly influence the operational performances. Examples of this include advanced freight collection/distribution networks, long intermodal freight trains, and road mega trucks in Europe, as well as large commercial freight aircraft. In addition, the technical/technological performances may directly influence the operational and indirectly the economic performances of particular advanced transport systems using the existing infrastructure. Examples include the BRT (Bus Rapid Transit) System system in urban areas, high-speed tilting passenger trains, large advanced container ships, advanced commercial subsonic and supersonic passenger aircraft, and advanced ATC (Air Traffic Control) technologies aimed at increasing the airport runway capacity. Last but not least, the required environmental and/or social performances can speed up development and commercialization of completely technically/technologically new systems such as advanced passenger cars (that use electricity and/or LH_2 (Liquid Hydrogen) instead of the currently used crude oil-based petrol/diesel), the PRT (Personal Rapid Transit) and UFT (Underground Freight Transport) system (that uses electricity), and advanced STA (Supersonic Transport Aircraft) (that uses LH_2 instead of the crude-oil derivative kerosene (JP-1)).

The main actors/stakeholders involved in dealing with advanced transport systems include:

- Investors, constructors/manufacturers of infrastructure, transport means/vehicles, supportive facilities and equipment, and suppliers of raw material energy/fuel;
- Providers and operators of transport infrastructure and services, respectively;
- Users of transport services (passengers and freight/goods shippers/receivers);
- Policy/decision makers at local, regional, national, and international level and related associations; and
- The local population both benefiting and being affected by the given systems.

Their interests and individual objectives can coincide or be in conflict with each other. For example, investors generally prefer to see a return on their investments over the specified/planned period of time. Suppliers of raw material and energy, and all related manufacturers prefer growth of these systems bringing them economic benefits. In this case, the objectives and interests coincide and influence each other downstream the chain of commercialization of the given advanced transport system(s). Providers and operators of transport infrastructure prefer its utilization at least at the level of covering operational and maintenance costs under the given pricing policy. Transport operators prefer efficient, effective, and safe transport means/vehicles providing services attractive to their users. Users/passengers and freight/goods shippers/receivers have the same preferences as transport operators, but from the perspective of experiencing the expected quality, safety, and security of the consumed services at reasonable/acceptable prices. Policy/decision makers and related associations at the particular institutional/organizational levels prefer efficient, effective, and safe advanced transport systems fully satisfying the overall user, social, and policy needs. In certain respects, these preferences and objectives, particularly those related to the policy, may be in direct conflict with those of transport infrastructure and service providers, and in indirect conflict with those of raw materials' suppliers and system manufacturers. Local communities usually have two sets of conflicting preferences. On the one hand, acting as prospective users they prefer advanced transport systems with the maximal availability in space (as close as possible and easily physically accessible) and time (frequent services). On the other, due to their proximity, the same people often complain about the impacts of these systems such as local air pollution, noise, induced road congestion, and compromised/demolished landscape. Consequently, these last mentioned preferences are essentially conflicting with the preferences of all other actors/stakeholders involved including those they themselves experience as users of the systems.

The above-mentioned approach in dealing with performances of advanced transport systems indicates that it is very difficult if not even impossible to simultaneously satisfy (conflicting) objectives and interests of all actors/stakeholders involved. However, as it will be shown, it is possible to come very close to achieving some balance between them.

1.3.2 Analyzing, Modeling, and Evaluation

Analyzing the performances of advanced transport systems implies gaining insight into their characteristics and the main influencing factors.

Modeling the performances of advanced transport systems implies defining the indicators and measures of performances and establishing the analytical/quantitative relationships between them and the main influencing factors. This enables sensitivity analysis to be carried out systematically by providing a range of inputs for planning and designing the considered system(s). In addition, modeling provides the opportunity to check the quality of particular models/methodologies by using the inputs from the considered cases and comparing their outputs/results with their real-life counterparts. Last but not least, modeling provides input for planning and design of particular system's performances according to the "what-if" scenario approach. Consequently, this enables their further evaluation according to the given set of attributes/criteria, chosen according to their relevance to the particular actors/stakeholders involved.

Evaluation of performances of advanced transport systems can generally be qualitative and quantitative. Qualitative evaluation implies identification of advantages (Strengths and Opportunities) and disadvantages (Weaknesses and Threats) of the particular advanced system perceived by the current and prospective actors/stakeholders, i.e., applying a simplified SWOT analysis. Quantitative evaluation implies choosing the preferable among the specified set of alternatives with respect to the specified attributes/criteria reflecting their performances relevant for the DM (Decision Maker) by using one of the multicriteria evaluation methods. This enables ranking and calculating the scores of the available alternatives and then choosing the one with the highest score as the preferred option.

1.4 Composition of the Book

In addition to this introductory chapter, the book consists of six chapters, each consisting of sections (subchapters) elaborating on a particular advanced transport system. At the beginning of each section, bullet-like historical milestones in development of the given system are provided. At the end of each section, a qualitative evaluation of this system is presented by emphasizing its presumed advantages and disadvantages viewed by the particular actors/stakeholders involved.

Chapter 2 elaborates the advanced operational and technological performances of the BRT (Bus Rapid Transit) Systems high-speed tilting passenger train(s), and advanced commercial subsonic aircraft.

Chapter 3 deals with the operational and economic performances of advanced freight collection/distribution networks, road mega trucks and LIFTs (Long

Intermodal Freight Trains) (in Europe), as well as large commercial freight aircraft.

Chapter 4 elaborates the technical/technological and environmental performances of advanced passenger cars, large advanced container ships, and the LH_2 (Liquid Hydrogen)-fuelled commercial air transport system.

Chapter 5 deals with the multicriteria ranking of different HS (High-Speed) passenger transport systems—HSR (High-Speed Rail), APT (Air Passenger Transport), and TRM (TransRapid Maglev)—with respect to their infrastructural, technical/technological, operational, economic, and environmental, and social/policy performances.

Chapter 6 elaborates performances of the future systems such as: PRT (Personal Rapid Transit) and UFT (Underground Freight Transport) as urban and/or suburban passenger and freight transport systems, respectively; ETT (Evacuated Tube Transport) as a very high-speed long-distance intercontinental transport system for both passengers and freight/goods; advanced Air Traffic Control (ATC) technologies and operations aimed at increasing the airport runway capacity; and advanced long-haul STA (Supersonic Transport Aircraft).

The last Chap. 7 summarizes the potential contribution of the advanced transport systems to sustainability, i.e., greening, of the transport sector abovementioned.

References

EC. (2010). *EU Transport GHG: Routes to 2050?—Towards the Decarbonisation of the EU's Transport Sector by 2050*. Brussels: European Commission DG Environment.

Hawksworth, J. (2008). *The word in 2050: Can rapid global growth be reconciled by moving to a low carbon economy?*. London, UK: Pricewaterhose Coopers LLP, Economics.

Khan, S. M. (2011). Nanotechnology in transportation: Evaluation of a revolutionary technology. *TR NEWS* No. 277, Transportation Research Board of The National Academies, Washington D.C., USA, pp. 3–8.

Chapter 2
Advanced Transport Systems: Operations and Technologies

2.1 Introduction

This chapter describes BRT (Bus Rapid Transit) as an advanced mature public transport system operating in many urban and suburban areas round the world, high-speed tilting passenger trains operating along medium- to long-distance passenger corridors/markets in many countries worldwide, and an advanced subsonic commercial aircraft—the Boeing B787-8, which has recently started commercial operation.

The BRT systems are considered as advanced compared to the conventional urban bus systems mainly thanks to advanced operations. A BRT system can be defined as a "rapid modes of transportation that combines the quality of rail transit and flexibility of buses" (Thomson 2001).

High-speed tilting trains transport users/passengers along the curved segments of the conventional rail lines/tracks at higher speeds than their conventional counterparts thanks to advanced technology—the tilting mechanism. This compensates increased centrifugal and centripetal forces due to higher speed in the curved segments of the line by tilting on the opposite side from the direction of the force, i.e., if the force is directed to the left, the train tilts to the right, and vice versa. Such a tilting mechanism makes these trains advanced transport means/vehicles in terms of technology, despite the fact that some other components influencing their performances remain similar to those of their conventional counterparts.

The subsonic commercial aircraft (Boeing B787-8) is considered advanced thanks to its innovative design and the new materials used in its construction, and perceived superior economic and environmental performances as compared to those of its conventional counterparts.

M. Janić, *Advanced Transport Systems*,
DOI: 10.1007/978-1-4471-6287-2_2,

2.2 Bus Rapid Transit Systems

1974	The first BRT (Bus Rapid Transit) system in the world—the "Integrated Transportation Network"—begins operations in Curitiba (Brazil)
1999/2000	The world's largest BRT system—Transmilenio—begins operations in Bogota (Columbia)

2.2.1 *Definition, Development, and Use*

The BRT (Bus Rapid Transit) systems are considered as a flexible rubber-tired rapid transit mode that combines stations, vehicles, services, running ways, and ITS (Intelligent Transport System) into an integrated system with a strong positive image and identity. Flexibility implies that these systems can be incrementally implemented as permanently integrated systems of facilities, services, and amenities that collectively improve the speed, reliability, and identity of bus transit in a variety of environments. In many respects, BRT systems can be considered as a rubber-tired LRT (Light Rail Transit)-like systems but with greater operating flexibility and potentially lower capital and operating costs (Levinson et al. 2002).

The BRT systems started in the U.S. (United States) in the 1960s through the implementation of exclusive bus lanes. After the first truly dedicated bus way of the length of several kilometers was set up in 1972 in Lima (Peru), the step forward in developing the BRT system concept was made in 1974; the first bus-based public transport network was developed in Curitiba (Brazil) using the bus-way corridors spread as the route/line network throughout the city. Since the mid-1990s, the BRT has been intensively promoted in U.S. cities as an advanced urban transit system to alleviate the adverse effects of traffic congestion compared to the conventional urban bus transit systems at the lower investment/capital costs compared to rail-based urban transit systems such as LRT (Light Rail Transit). At the same time, it has been expected to increase the transport capacity and make the accessibility of dense urban agglomerations/regions more effective and efficient. Designed and implemented on a case-by-case basis in order to meet the specific needs and characteristics of the given urban and suburban areas, the BRT systems have been characterized by the dedicated bus corridors, terminals/stations, vehicles/buses, fare collection system, ITS technology, operational concepts (timetable), and branding elements. Consequently, they have offered more effective, efficient, faster, reliable, and punctual transport services under given conditions than conventional bus transit systems, which have approached or even exceeded the services of the rail-based systems (LRT). The main objectives behind implementation of the BRT concept have been to approach to the capacity and quality of services of LRT while at the same time benefiting from savings in infrastructure investment costs, flexibility of the bus transit system, and comparable fares for users/passengers.

Table 2.1 Some characteristics of BRT systems in the world (http://en.wikipedia.org/wiki/Bus_rapid_transit)

Region	Number of cities/urban agglomerations	Network length (km)	% of the total	Transport volume(s) (passengers/day) (million)	% of the total
Europe	42	636	17	0.937	3.8
North America	20	585	15.6	0.849	3.5
South America	50	1,250	33.4	15.694	64.1
Asia	25	882	23.6	6.439	26.3
Africa	3	62	1.7	0.238	1.0
Oceania	7	326	8.7	0.327	1.3

The BRT systems have shown flexibility in terms of feasibility of implementation in urban agglomerations with a population of between 0.2 and 10 million. As such, in many transit corridors/routes, they have represented a test bed before implementing a rail-based urban transit system such as LRT.

Depending on the layout of the city/urban agglomeration, the BRT system can operate along radial and/or star-shape corridors exclusively or as a complement/connection to the rail transit systems/lines. In addition to 'Full BRT' systems operating exclusively along dedicated bus-ways, 'BRT Lite' systems mainly operate along the mixed traffic lanes except in cases of passing through important intersections where it is given exclusive lanes.

Currently, BRT systems operate in 147 cities/metropolitan areas on all continents. The total length of the dedicated bus-ways is about 3,741 km. The total daily number of passengers using the systems is about 24.5 million. Table 2.1 gives additional characteristics of the BRT systems used around the world.

Regarding the above-mentioned characteristics, the BRT system has been developed and consequently mostly used in South America and Asia, and the least in Africa. The relative market share of the system in the total number of daily commuting users/passengers indirectly reflects such developments. In addition, the daily number of users/passengers tends to increase almost exponentially as the BRT system network is extended as shown in Fig. 2.1.

This indicates that the system is attractive for both existing users of public transit systems and those abandoning their cars for the first time.

LRT (Light Rail Transit), often considered as a strong competitor to the BRT system, can be defined as an electric railway system with a "light volume" capacity for passengers as compared to conventional (heavy) rail. Its performances are partially presented for comparative purposes. At present, 24 LRT systems operate in the U.S. In Europe, LRT systems have often been considered together with urban tramway systems. Some evidence indicates that 170 tram and LRT systems, comprising 941 lines of a total length of 8,060 km are in operation. In 21 cities, 154 existing lines have been extended by about 154 km and 21 new lines of a length of 455 km are under construction (ERRAC 2005).

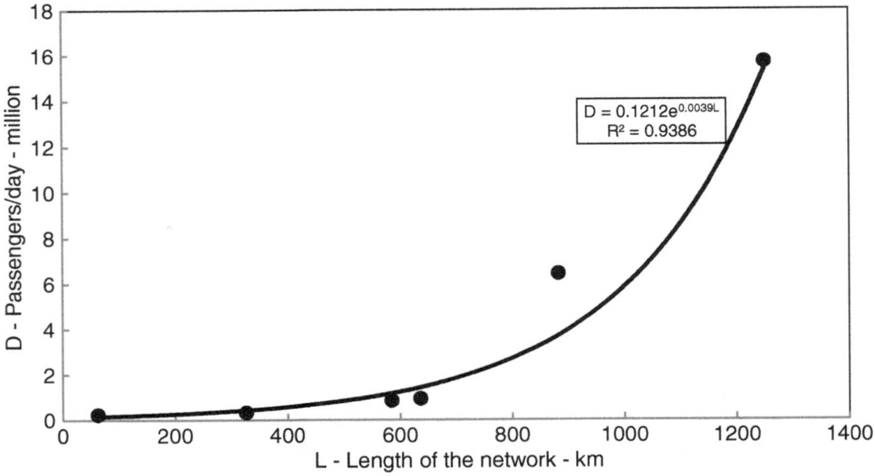

Fig. 2.1 Relationship between the daily number of passengers and the length of the BRT network in particular regions/continents (http://en.wikipedia.org/wiki/Bus_rapid_transit)

LRT systems may use shared or exclusive rights-of-way, high or low platform for users/passengers boarding/off-boarding, and multi- and/or single-car trains.

2.2.2 Analyzing and Modeling Performances

2.2.2.1 Background

The BRT systems are characterized by infrastructural, technical/technological, operational, economic, environmental, and social/policy performances. Considered together, they allow the BRT system(s) to be distinguished generally in seven features as compared to conventional/standard urban bus transit system(s) as given in Table 2.2 (GAO 2012).

The performances of the BRT system are analyzed, modeled, and evaluated using indicators and their measures. Their values are synthesized as averages from 40 BRT systems operating around the world—13 in Latin and South America, seven in Asia, three in Australia, eight in Europe, and nine in the U.S. and Canada (Wright and Hook 2007).

2.2.2.2 Infrastructural Performances

The main indicators of the infrastructural performances of BRT systems refer to the spatial layout of their networks/corridors/routes, the number/density of stations along the corridors/routes, and other characteristics.

Table 2.2 Distinguishing features of the BRT systems compared to conventional bus systems (GAO 2012; Levinson et al. 2003a, b)

Feature	Description
Running ways	Segregated and dedicated busways or bus-only roadways
Terminals/stations	Enhanced environment (information provided through real-time schedule systems and additional amenities—safety improvements, public art, landscaping, etc.)
Vehicles/buses	Standard/articulated, different engine technology (diesel, gas hybrid diesel/electric, electric), quieter, higher capacity, wider (usually low-floor) doors
Services	Faster, more frequent, punctual, and reliable
Fare collection	Prepaid or electronic passes—speedy fare collection, and boarding on/off convenience
Branding	Marketed as a distinguished service at the terminals/stations and vehicles/buses
ITS (Intelligent Transportation Systems)	Prioritization of services at intersections and traffic lights, monitoring headways between vehicles, real-time information on vehicle position and schedule

Spatial layout of the network

The BRT system networks operate under the assumption of having regular and sufficient passenger/commuter demand to be served by the relatively frequent transport (bus, trolleybus) services over a given period of time (hour, day, year) (for example, $\geq$8,000 passenger/h/direction). Consequently, the transport infrastructure network consisting of the corridors/routes with dedicated busways and terminals/ stations spread over, pass by and/or through densely populated/demand attractive areas of the given urban agglomeration—the city center(s) or CBDs (Central Business District(s)). A simplified spatial layout of the BRT network is shown in Fig. 2.2.

The BRT dedicated busways passing through the high density area continue outside it as right-of-way bus lanes. Both are connected to the freeway(s) surrounding the densely populated area(s) (CBDs). In some cases, the BRT dedicated busways or bus-only roadways are built along old rail corridors/lines. The dedicated busways are usually provided as two-way lanes in different directions in mixed traffic, as two-way lines on the same side or in the middle, or as a single line in each direction on different sides of the given corridor/route. In some cases, the bus-way is split into two one-way lanes/segments. The grade separation and elevation of BRT system routes is also provided, if needed, particularly at intersections of the routes themselves and with those of other traffic. Particular BRT busways can also be painted (red, yellow, green) in order to enhance visibility and recognition—by both the other drivers and users/passengers.

Typically, the single BRT corridor spreads between two agglomerations, one of which could be housing and the other CDB, or both CDBs. Given the length of this corridor usually defined as the distance between the initial and the end terminal/

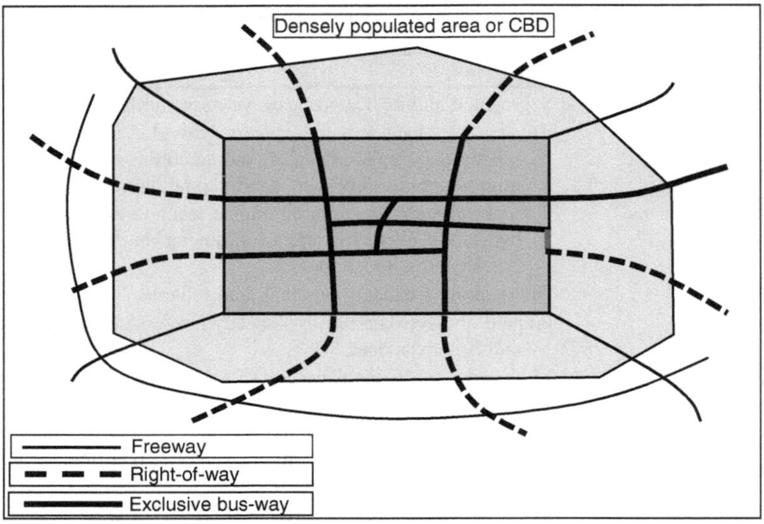

Fig. 2.2 Schematic layout of a hypothetical BRT network

station, width, and the number and area of the terminals/stations along it, the total area of land directly taken for building this infrastructure can be estimated as follows (Vuchic 2007):

$$A = L * D + n(ld) \tag{2.1}$$

where

L is the length of corridor (km);
D is the width of the corridor (m);
N is the number of stations/platforms along the corridor; and
l, d is the length and width of the plot of land occupied by the terminal/station (m), respectively.

For example, the width D of the exclusive bus-way (both directions) within the BRT corridor varies depending on the speed from 10.4–11.6 m (for moderate speeds $\leq$70 km/h) to 14.60 m (for speeds up to 90 km/h). The typical length of the bus stops varies from l = 18–26 m depending on the bus length (for a single bus). The minimum width of the bus stop at the terminal/station is about d = 3.0–3.5 m. However, the width of the area occupied by the terminal/station itself with the supporting facilities and equipment could be up to 9.0 m. For comparison, the typical (minimum) width of the corridor for building a double track LRT line respecting the vehicle's dynamic envelope is about 7.5 m. The track gauge is 1,435 mm. The typical area of the platform of the LRT station can be from 12 × 50 m (surface) to 20 × 90 m (grade separated) (Vuchic 2007; Wright and Hook 2007).

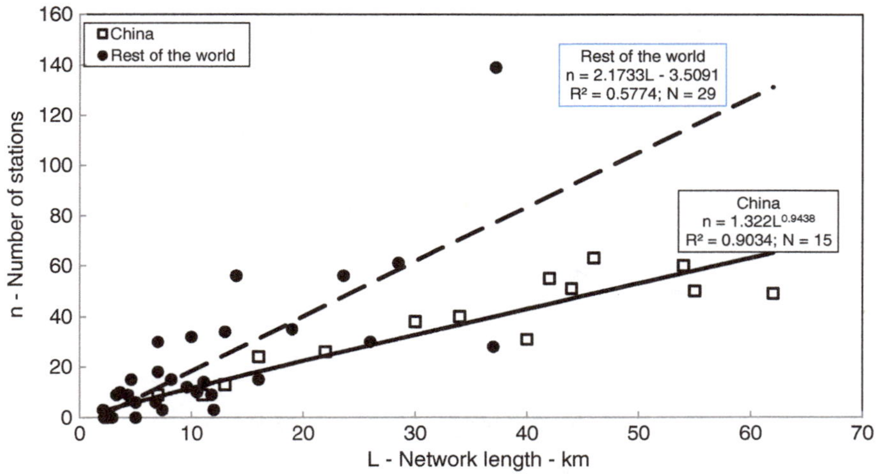

Fig. 2.3 Relationship between the number of stations and the length of the BRT system network (Levinson et al. 2003a, b; Wright and Hook 2007; http://en.wikipedia.org/wiki/Bus_rapid_transit)

Number/density of stations

The terminals/stations are important elements for the safe, efficient, and effective inter and multimodal transfers on the one hand, and for demonstrating the identity and image of the given BRT system on the other. A BRT terminal/station can be a simple stop, an enhanced stop, designated station, intermodal terminal, and/or transit center. The number and density of stations mainly depends and increases in line with the length of the BRT system network as shown in Fig. 2.3. In BRT systems around the world, except those in the People's Republic of China, this increase is of an average rate of 2.0/km. For systems in the PR of China, the average rate is 1.0/km. The network length varies from about 2 to 60 km.

BRT terminals/stations usually have passing lanes and sometimes multiple stopping/docking bays, which enable the convoying of busses in different combinations, if needed. The number of stopping/docking bays influences the required length of the given terminal/station as shown in Fig. 2.4.

Evidently, the length of the BRT terminal/station generally increases more than proportionally compared to the increase in the number of stopping/docking bays. This length and other dimensions can be larger if the BRT terminal/station is integrated with terminals/stations of other public transport modes, for example, those of the underground public transport system.

Passenger access to the BRT terminals/stations—either on foot, by bike, car/taxi, and other public transport systems—should be safe, efficient, and effective. This implies good integration including parking and short stop spaces at the rear of the stations, as well as providing convenient connections/passages to/from the bus platforms. In particular, at the BRT feeder-trunk systems, cross-platform transfers

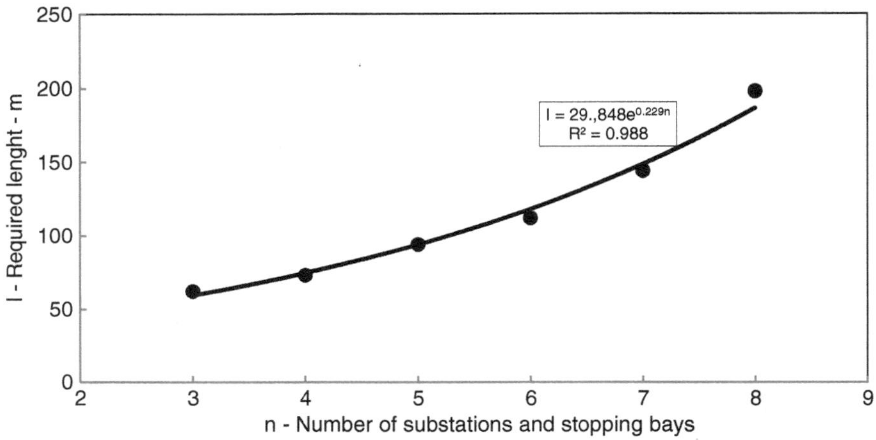

Fig. 2.4 Dependence of the required length of terminal/station on the number of substations and stopping/docking bays of the BRT system (Wright and Hook 2007)

Table 2.3 Infrastructural performances of the BRT and LRT system—infrastructure (averages) (GAO 2012; Levinson et al. 2003a, b)

Indicator/measure	System	
	BRT	LRT[c]
Number of systems	147	170
Length of the network (km/system)	25.3/15.0[c]	47.4
Number of corridors/lines per system	2	6
Number of routes per corridor	5	–
Average length of the corridor/line (km)	32[a]/12.9[b] (28)	8.6
Width/profile of the lane (m)	10.4–14.6	7.5
Number of stations (-/route)	32[a]/23[b]	–
Density of stations (-/km)	1[a]/2[b]	–
Location of the station(s) (mainly)	Side/curb/off-lane	Side
Width/length of the station(s) (m)	9.5/50–9.5/75	12.0/50–20.0/70
Type of guideways/lanes—passing lanes	Mostly yes	Yes
Platform height (at the stations)	Low	Low (or High)
Static/spatial capacity of the stations (vehicles/station)	1–3	1–2
Materials used (lanes, stations)	Concrete, asphalt	Iron/steel, concrete, asphalt
Construction time (km/year)	16–20	1–5

[a] China; [b] Rest of the world; [c] Europe

from the feeder to the trunk buses, and vice versa, should be provided (see below). Some additional indicators of the infrastructural performances of the BRT system and a comparable LRT system are given in Table 2.3.

2.2.2.3 Technical/Technological Performances

The technical/technological performances of the BRT system mainly relate to: (i) length, space (seats + stands) capacity, weight, type and power of engine(s), and riding comfort of vehicles/buses; and (ii) ITS (Intelligent Transport Systems) including the systems for managing transit services along the network/routes, providing the users/passengers with the online information, and collecting fares.

Vehicles/buses

The BRTsystems generally use standard and/or articulated transport vehicles/ buses (and trolleybuses) with a typical length of 12, 18, or 24 m, a weight of 13, 17, or 24 tons, and the corresponding capacity (seats + stands) of 75, 100, or 160 passengers, respectively. The buses have 3–4 axles. The buses of the above-mentioned 40 BRT systems are generally powered by four types of engines: diesel and diesel Euro II/III/IV (26), CNG (Compressed Natural/Propane Gas) (7), hybrid (diesel + electricity) (3), and electricity (3). The diesel/buses use diesel fuel for propulsion and electric power for auxiliary equipment. The CNG buses are powered by engines similar to diesel engines, but instead of diesel they use a methane mixture for propulsion. The hybrid diesel–electric vehicles/buses use an onboard diesel engine for producing electricity that charges their batteries. These in turn provide the electricity to run the electric propulsion motors. The electric vehicles—trolleybuses—use electricity from the overhead power supply infrastructure, i.e., from the catenary wire systems, for powering electric motors and auxiliary equipment. The typical engine power of BRT vehicles/buses is about 150–220 kW(kW–kilowatt). A summary of indicators of the technical/technological performances for typical BRT and LRT vehicles is given in Table 2.4:

An important characteristic of BRT vehicles/buses, sometimes more important than their size, is the number and width of doors. This influences utilization of vehicles/buses, consequently the route/line capacity, and otherperformances such as the average commercialspeed. Some longer buses have four doors, each about 1–1.1 m wide. Depending on the location of busways, they can be on the vehicle's right or left side.

ITS (Intelligent Transport Systems)

Systems for managing transit services

The ITS (Intelligent Transport Systems) managing the transit services of BRT systems generally include: (i) automated enforcement systems for exclusive bus lanes; (ii) an AVL (Automatic Vehicle Location) system; (iii) a CAD (Computer-Aided Dispatching) and advanced communications system; (iv) a precision docking at bus stop system; (v) a tight terminal guidancesystem; and (vi) a warning system.

Table 2.4 Technical/technological performances of the BRT and LRT system vehicles (averages) (AUMA 2007; CE 2008; STSI 2008; Vuchic 2007; Janic 2011)

Indicator/measure	System	
	BRT	LRT
Length of a vehicle (m)	12/18/24[a]	14–30[a]
Height of a vehicle (m)	3.0–3.2	4.0–6.9[b]
Width of a vehicle (m)	2.5–2.6	2.20–2.65
Cars/vehicle	1	2–4
Capacity (spaces/vehicle)	75/100/160	110–250
Seat spacing (m)	0.80	0.75–0.90
Number of axles/vehicle	3/4/4	4/6/8
Tare weight (tons)	13/17/24	25.4–38.8
Engine power (kW)	150–220	200–434
Maximum speed (km/h)	90–100	60–120
Operating speed (km/h)	27–48	40–80

[a] Vehicle can be a set consisting of few cars; [b] Including pantograph

- Automated enforcement systems for exclusive bus lanes include the transit signal priority and the queue jump system; the former changes the timing of the traffic signals in various ways in order to give priority to BRT vehicles/buses at intersections (for example, the system turns the red light to green if it "recognizes" the approachof a BRT vehicle to the intersection); the latter enables using the separate lane and receiving the green light signal upon closer approach to the intersection;
- The AVL (Automatic Vehicle Location) System is the computer-based system enabling the real-time tracking of vehicles/buses and providing them with the information for the timely schedule adjustments and equipment substitutions; at the core of this system is GPS (Global Positioning Satellite) technology and GIS (Geographic Information System) displaying the location of the vehicles/buses on the route map grids in the dispatch center;
- The CAD (Computer-Aided Dispatching) and advancedcommunications system enables adjusting dwell times at vehicle/busstops or transfer points, vehicle/bus headways, rerouting vehicles/buses, adding vehicles/buses to routes, and dispatching new vehicles/buses to replace incapacitated vehicles/buses; the drivers exchange communications with the dispatch center by radiotelephones, cellular telephones, and/or mobile display terminals;
- The precision docking system uses sensors on the vehicles/buses and on the roadside to indicate the exact place where the vehicle/bus should stop; this enables users/passengers to be in position for immediate boarding, which shortens dwell time(s) at the stops;
- The tight terminal guidance system uses sensors similar to those for precision docking to assist the vehicles/buses in maneuvering in terminals with limited space; the system can contribute to minimizing the amount of space for bus terminal operations, as well as to reducing the overall time the bus spends at the terminal/station; and

- The warning system aims at assisting/warning the vehicle/bus drivers in order to avoid collisions, pedestrian proximity, and low tire friction; this improves the-safety, efficiency, and effectiveness of the BRTsystem's operations.

User/passenger information system

The user/passenger information system at the terminals/stations and onboard the vehicles/buses provides advance information contributing to the efficiency and effectiveness of travel decisions. For the former, displays provide real-time information on forthcoming arrivals/departures, transfer times and locations, and maps of the related routes/lines. For the latter, the system automatically announces the vehicle/bus approaching its next stop, giving sufficient time for preparation, speeding up disembarking and embarking, and consequently shortening dwell time at the terminals/stations (see below).

Fare collection systems

The BRT systems generally use three mainly automatedsystems for collecting fares: (i) preboard, onboard, and free-fare collection and verification. Of the 40 of the above-mentioned systems, 16 employ preboard, 21 onboard, and 3 free-board fare collection systems. In particular, the onboard system speeds up the fare collection process and eliminates expensive cash handling operations at transit agencies using smart cards. The system uses the read-and-write technology to store the monetary value on a microprocessor chip inside a plastic card. As passengers board a vehicle/bus, the card reader determines the card's value, debits the appropriate amount for the busride, and writes the balance back onto the card, all within a fraction of a second.

2.2.2.4 Operational Performances

Operational performances of the BRT system include demand, capacity, quality of service, size of fleet, and technical productivity. They are analyzed and modeled based on the above-mentioned global BRT systems. The corresponding figures for LRT systems are also provided for comparative purposes.

Demand

In general, the volumes of demand for existing and prospective urban transport systems can be estimated under the assumption of their mutual competition. In such case, the BRT system can compete with the individual passenger carand other public transport systems such as taxi, conventional bus, tram, metro, and LRT. Under such conditions, the users/passengers are assumed to usually choose the including BRT with respect to their own characteristics (age, gender, personal income), trip purpose (work, shopping, entertainment, other), and the system's performances, both in combination reflecting the generalized travel cost. This cost, usually represented by the disutility function of using the given transit system (i) between given pair of origin and destination (k) and (l), respectively, $U_{i/kl}(T)$, can be estimated from either the aggregated trip generation data or disaggregate passenger survey data, both for a given period of time (T). The MNL (Multinomial) Logit model can then be applied to quantify the market share or the probability of choosing the system (i) as follows (TRB 2008a):

$$p_{i/kl}(T) = \frac{e^{-U_{i/kl}^2(T)}}{\sum_{i=1}^{I} e^{-U_{i/kl}^2(T)}} \tag{2.2a}$$

where

I is the number of transport systems offering transit services between the origin (k) and destination (l).

The number of users/passengers choosing the system (i) can be estimated from Eq. 2.2a as follows:

$$q_{i/kl}(T) = p_{i/\mathrm{kl}}(T) * q_{kl}(T) \tag{2.2b}$$

where

$q_{kl}(T)$ is the total number of users/passengers traveling between the origin (i) and destination (k) during the time period (T) by all available transport systems.

The user/passenger demand $q_{kl}(T)$ in Eq. 2.2b can be estimated by applying one of the causal-gravity-type models based on the trip generation/attraction socio-economic forces of the origin (i) and the destination (k), and the travel "resistance" between them (Janic 2010; Vuchic 2004).

The user/passenger demand $q_{i/kl}(T)$ in Eq. 2.2b includes the demand between the origin (k) and destination (l) as well as the demand between each pair of the vehicle/bus stops along the corridor/line (kl) as follows:

$$q_{i/kl}(T) = \overline{q}_{i/kl}(T) + \sum_{m=1}^{M} \left[q_{i/km}(T) + q_{i/ml}(t)\right] + \sum_{m=1}^{M-1} \sum_{n=m+1}^{M} q_{i/mn}(T) \tag{2.2c}$$

where

$\overline{q}_{i/kl}(T)$ is the user/passenger demand between the origin (k) and destination (l) during the time period (T);

$q_{i/kl}(T)$, $q_{i/ml}(T)$ is the user/passenger demand between the origin (k) and the station/stop (m), and the station/stop (m) and the destination (l), respectively, during the time period (T);

$q_{i/mn}(T)$ is the user/passenger demand between the stations/stops (m) and (n) during the time period (T); and

M is the number of stations/stops along the route (kl).

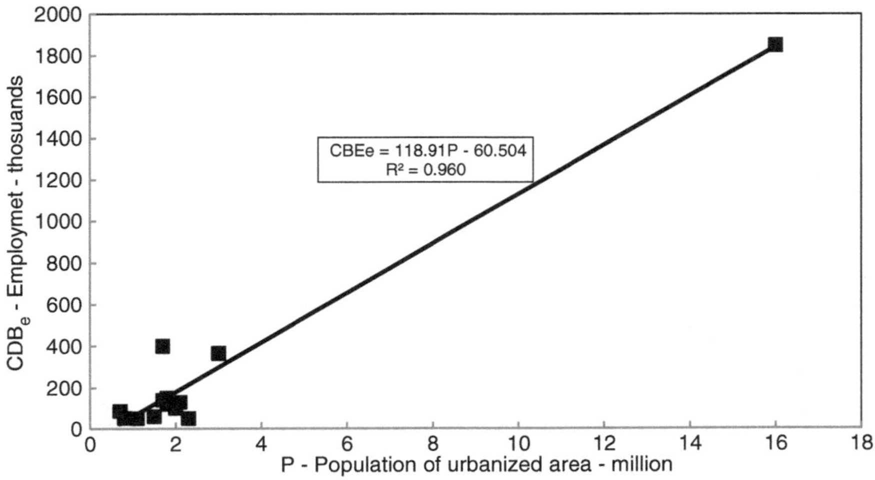

Fig. 2.5 Relationship between population of an urbanized area and employment in the corresponding *CDB* (Central Business District) (Levinson et al. 2003a, b)

Demand for BRT services mainly consists of daily users/passengers commuting from their home to the place of a given activity (work, shopping, entertainment, others), each located within or outside the given urban agglomeration (or CBD), and vice versa. The potential demand for the BRT, as well as for other urban public transport systems, can be influenced by the size of population, which in turn can influence employment and other commercial and entertainment activities in the given CBD. Figure 2.5 shows an example of the relationship between the size of urban population and employment in a CBD as the potential demand for the given BRT system.

Generally as intuitively expected, a larger urban population can generate proportionally greater employment in the CBD at an average rate of about 118 thousands employees per 1 million of population (as in the above example). Figure 2.6 shows an example of the relationship between the number of employees in CBD and the number of daily (weekday) users/passengers of BRT systems.

As expected, higher employmentin the CBD generally generates a higher daily demandfor urban transit including, in this case, the BRT.

Capacity

The transit capacity of the BRT system is one of the most important indicators of its operational performances mainly due to the requirement to transport relatively large numbers of users/passengers under given circumstances. This capacity can be considered for a single terminal/station, route/line, and the entire network providing the vehicle/bus capacity is given.

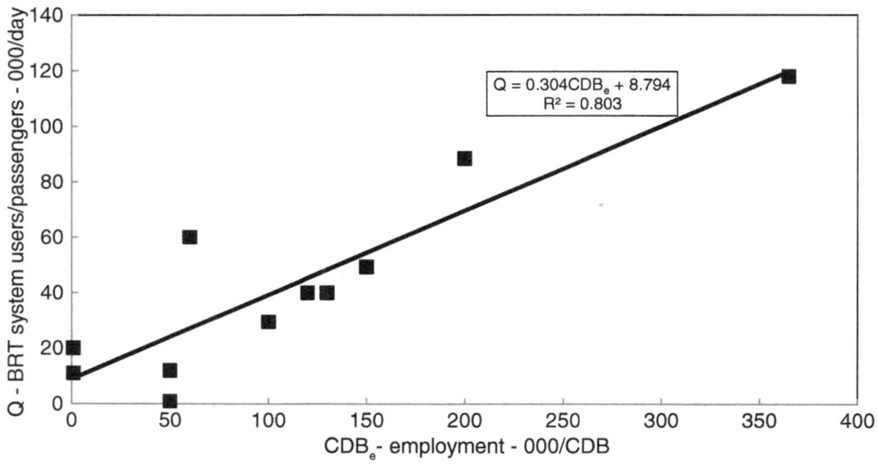

Fig. 2.6 Relationship between the daily number of users/passengers of the BRT system and employment in the CBD (GAO 2012; Levinson et al. 2003a, b; Wright and Hook 2007)

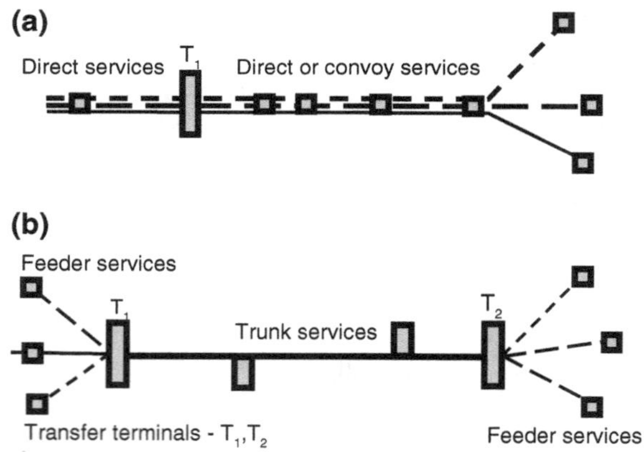

Fig. 2.7 Scheme of a BRT system service network **a** Direct or convoy network **b** Trunk-feeder network

Service network

The BRT system can generally operate as a “direct or convoy”, “trunk-feeder,” and “hybrid” service network. The layout of the former two is shown in Fig. 2.7.

- *Direct or convoy network* consists of routes and related BRT services connecting different user-passenger origins and destinations, which can be both within and outside a given urban agglomeration. In such case, many different bus services/lines connecting particular sets of these origins and destinations operate within the common/main part of the network(s) and then spread outside it toward the periphery of the given agglomeration. Consequently, a high frequency of mainly direct services (with no or few transfers) is provided to those users/passengers traveling within the parts of the common/main network. The demand served through the main part of thenetwork by the given BRT service/line during the time period (T) estimated as in Eqs. 2.2a–c is: $Q(T) \equiv q_{kl}(T)$, where (k) is the origin ($k = 1, 2\ldots, K$) and (l) is the destination of the user/passenger flow(s) ($l = 1, 2, \ldots, L$).

- *Trunk-feeder network* consists of the feeder and trunk part. The feeder part represents the local network connecting the user/passenger origins and destinations to the trunk terminal(s)/station(s) by services usually operated by lower capacity (conventional) vehicles/buses. The trunk terminals are mutually connected by services usually operated by the larger (articulated) and/or bi-articulated vehicles/buses. In any case, the size of vehicles/buses and service frequency can be easily adapted to the volumes of user/passenger demand. However, the users/passengers are forced to change at the trunk terminals/stations, which does not exclude potential necessary changes at the stations between them.
- *Hybrid network* represents a combination of a direct or convoy and trunk network set up to appropriately adjust the offered capacityto changes in user/passenger demand. This implies flexibility in adapting to the time and spatial pattern, volumes, and intensity of this demand. Consequently, this network possesses the combined features of both sub-networks it is comprised of.

The volumes of demand are generally different on the feeder and the trunk route(s) of the trunk-feeder network(s). For example, on the feeder route connecting the origin (k) and the trunk terminal/station T_1, the volume of this demand during period (T) is: $Q(T) \equiv q_{k,T1}(T) = \sum_{l=1}^{L} q_{kl}(T)$ ($k = 1, 2,.., K$). On the trunk route between the transfer terminals/stations T_1 and T_2, this volume is: $Q(T) \equiv q_{T1,T2}(T) = \sum_{k=1}^{K} \sum_{l=1}^{L} q_{kl}(T)$, for $k \neq l$. Finally, on the feeder route connecting the trunk terminal/station T_2 and the destination (l), the volume of user/passenger demand is: $Q(T) \equiv q_{T2,l}(T) = \sum_{k=1}^{K} q_{kl}(T)$ ($l = 1, 2,.., L$). The user/passenger demand $q_{k,T1}(T)$, $q_{kl}(T)$, and $q_{T2,l}(T)$ can be estimated as in Eqs. 2.2a–c. Consequently, the main differences between the above two network configurations are as follows:

- At the direct or convoy service network, the lower capacity BRTvehicles/buses directly operate between particular origin and destination terminals/stations. In this case, the volumes of user-passenger demand per origin–destination pair are usually lower, resulting in the lower service frequency and the lower loadfactor per frequency given thevehicle/bus size/space capacity; as mentioned above, the capacity of these vehicles/buses is usually 70–75 spaces (seats + stands) (12 m length).
- At the trunk-feeder service network, feeder vehicles/buses of a capacity of 70–75 spaces (12 m length) transport users/passengers between their origins and destinations and the trunk terminal(s)/station(s). At these terminals/stations, the users/passenger change for trunk vehicles/buses, usually with a capacity of

160–200 spaces (18–24 length). Thus, the volumes of users/passengers in both the feeder routes and the trunk corridor substantively increase, which generally can justify the increase in the service frequency, the vehicle/bus capacity or both simultaneously, and consequently the load factor. Such a BRT network is thus more effective and efficient.

Furthermore, all above-mentioned types of BRT networks can act as feeder systems to other mass urban (metro, tram, LRT) and inter-urban transit systems (heavyrail).

Station/terminal capacity

The capacity of a station/terminal $\mu_s(t)$ depends on the number of platforms—parking/stopping places—for the vehicles/buses and the average time they occupy them. In general, this capacity can be estimated as follows:

$$\mu_s(T) = P/\bar{t}_s(T) \tag{2.3}$$

where

P is the number of available stopping/docking bays on the given terminal/station; and

$\bar{t}_s(T)$ is the average occupancy time of a single stopping/docking bay (min) during time period T.

The average occupancy time $\bar{t}_s(T)$ of the stopping/docking bay can change over time as indicated by Eq. 2.3. Consequently, the capacity or service rate of the stopping bay can also change under conditions of having all designedspaces available during the time T (i.e., $\mu_s(t) = 1/\bar{t}_s(T)$).

Route/line capacity

The capacity of the route/line can be defined as the maximum number of vehicles/buses (sometimes also the number of passenger spaces), which can pass through its fixed point (i.e., the "reference location") during the given period of time T (usually 1 h) under conditions of constant demandfor service (Vuchic, 2007). This capacity expressed by the service frequency $f_{kl/\max}(T)$ for the route/line connecting the origin (k) and the destination (l) can be estimated as follows:

$$f_{kl/\max}(T) = \left[\frac{T}{\max\left(H_{kl/w/\min}; H_{kl/s/\min}\right)}\right] \tag{2.4a}$$

where

$H_{kl/w/\min}$ is the minimum headway between the successive vehicles/buses along the particular sections of the route/line (kl) (min); and

$H_{kl/s/\min}$ is the minimum terminal/station headway defined as the inter-arrival time of the successive vehicles/buses at the particular stations along the route/line (kl) (min).

In the case of satisfied demand with the specified average load factor per service, the frequency $f_{kl}(T)$ in Eq. 2.4a can be estimated as follows:

$$f_{kl}(T) = \left[\frac{q_{kl}(T)}{\lambda_{kl}(T)N_{kl}}\right] \quad (2.4b)$$

where

$q_{kl}(T)$ is the user/passenger demand on the route/line (kl) during the period (T) (determined according to Eqs. 2.2a–c) (passengers);

$\lambda_{kl}(T)$ is the average load factor along the route/line (kl) during time (T), and

N_{kl} is the vehicle/bus capacity operating along the route (kl) (spaces/vehicle).

In Eq. 2.4a, in most cases: $H_{kl/w/\min} > H_{kl/s/\min}$; thus, the terminal/station headway(s) determines the capacity of the given route/line. Consequently, the capacity $C_{kl}(T)$ expressed by the maximum number of vehicles/buses that can pass through a given "reference location" of the route/line (kl) during the period of time (T) can be estimated based on Eq. 2.4a as follows:

$$C_{kl}(T) = f_{kl/\max}(T) * m_{kl} \quad (2.4c)$$

where

m_{kl} is the number of vehicles/buses per each single departure/service on the route (kl).

The offered capacity of the route/line (kl), $C_{kl/0}(T)$ defined as the number of passenger spaces supplied during the given period of time T can be estimated as follows based on Eq. 2.4c:

$$C_{kl/0}(T) = C_{kl}(T) * N_{kl} \quad (2.4d)$$

where all symbols are analogous to those in the previous equations.

In Eq. 2.4d, the vehicle/busspace capacity N_{kl} depends on its constructive characteristics and is expressed by the number user/passenger spaces (seats + - stands) per vehicle/bus.

Speed

Speed as an indicator of operational performances refers to the operating and commercial speed of vehicles/buses along the routes of a given BRT system network(s). The operating speed includes acceleration, deceleration, and cruising/ operating speed, which depends on the vehicle/bus technical/design characteristics and prevailing driving/traffic conditions, while the commercial speed $v_{kl}(d_{kl})$ along the BRT route/line (d_{kl}) includes the operational speed and dwell time at stations/ terminals. It can be estimated as follows:

Table 2.5 Operational performances of BRT and LRT system—capacity (averages) (ERRAC 2004; GAO 2012; Levinson et al. 2003a, b; Vuchic 2007; Wright and Hook 2007)

Indicator/measure	System	
	BRT	LRT
Vehicle capacity (passengers seating + standing)	75–160	110–250
Route/line capacity (veh/h)	8–15	5–12
Terminal/station dynamic capacity (veh/h)	8–15	5–12
Network capacity (veh/h)	56–105	30
User-passenger capacity (pass/h/direction)	600–2400	550–3000
Commercial speed (km/h)	26.3	20–25

$$v_{kl}(d_{kl}) = d_{kl}/\tau_{kl}(d_{kl}) \tag{2.5}$$

where

d_{kl} is the length of route between the origin (k) and the destination (l) (km); and

$\tau_{kl}(d_{kl})$ is the total travel time of a vehicle/bus along the route d_{kl} (min).

Some measures of BRT and LRT system capacity have been estimated using Eqs. 2.4a–2.5 and are given in Table 2.5.

The values in Table 2.5 confirm that on average, BRT systems appear quite comparable and according to some measures even superior to LRT systems. Furthermore, the user/passenger capacityof both large systems can be much higher: 5,000–40,000 for BRT and 6,000–15,000 pass/h/direction for LRT systems, which clearly indicates the superiority of BRT systems (Wright 2003). Nevertheless, the supply of this capacity should be adapted to the volumes of demand as shown in Fig. 2.8.

In general, peak-hour capacity supply increases at a decreasing rate as the volumes of corresponding demand increase. For example, in order to serve 5,000 users/passengers/h/direction, the BRT) system needs to engage 92 vehicles/buses ($N \geq 54$). For 10,000 users/passengers, around 172 vehicles/buses are needed ($N \geq 58$). For 20,000 users/passengers, about 323 buses are needed ($N \geq 62$). This indicates that as the peak-hour demand increases, both the number and size of the vehicles/buses engaged tend to increase.

The number and size of vehicles/buses in the operator's fleet also depend on the type of service system/network. For example, the average size of the fleet operating the direct or convoy system/network is about 133 vehicles/buses and of that operating the trunk-feeder system/network 197 (80 trunk articulated and 117 feeder) vehicles/buses. Specifically, the largest BRT TransMilenio (Bogota, Columbia) operates a fleet of 1,420 (1,013 trunk articulated and 407 feeder) vehicles/buses (Wright and Hook 2007).

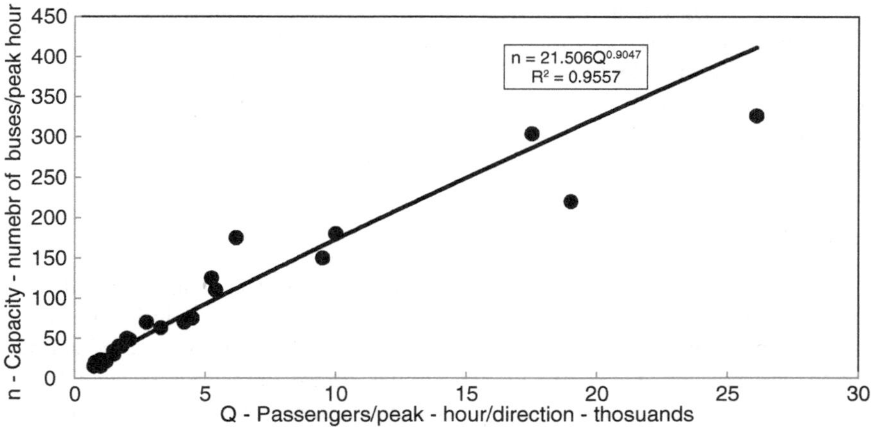

Fig. 2.8 Relationship between peak-hour capacity and demand for the selected BRT systems (Levinson et al. 2003a, b)

Quality of service

The quality of service of the BRT system is expressed by schedule delays, travel time or commercial speed, availability, reliability and punctuality, riding comfort, and the overall accessibility of a given BRT route/line and/or network.

Schedule delay

Schedule delay is defined as the time a passenger has to wait for BRT services at a given terminal/station. Assuming that the users/passengers are familiar with the timetable and arrive at the BRT station/terminal uniformly during any two successive vehicle/bus services, the average schedule delay of a user/passenger can be estimated, based on Eqs. (2.4a, b) as follows:

$$SD(T) = 1/4[T/f(T)] \tag{2.6}$$

where all symbols are as in the previous equations.

Travel time

Travel time depends on the distance of the user/passenger origins and destinations along the given route/line, the vehicle/bus operating speed, and the number and duration of intermediate stops. The duration of stops can be influenced by the number and width of doors of the vehicles/buses, and the number of user/passenger entries and exits at the particular stops (travel time can be extracted from Eq. 2.7b below). The above-mentioned commercial speed, being much higher than that of conventional bus systems (currently between 27 and 48 km/h), can be viewed as an additional measure. As such, it is close to the commercial speed of LRT. For example, the average commercial speed of a TransMilenio (Bogota, Columbia)

BRT service operating along the average route length of 13 kms is about 29.5 km/h (Levinson et al. 2003a, b; Saavedra 2011; Vuchic 2007).

Consequently, thanks to these features, BRT systems are considered "savers" of travel time. Some figures in the U.S. show that, depending on the system, these savings are about 5–35 %, which results in increases in the volumes of user/passenger demand by about 3–60 %, the latter after 1 year of operation, as compared to existing transit services (GAO 2012).

Availability, reliability, and punctuality

In general, BRT system services are considered highly available, reliable, and punctual. Availability is achieved through scheduling services during the entire day. Reliability implies operations without cancelation of the scheduled services due to any reasons. High punctuality implies minor deviations of actual from scheduled arrival times at particular locations/stations along the line(s), which is achieved thanks to operating, in the most cases, along the dedicated bus ways and applying ITS. Such high indicators are comparable to those of LRT systems.

Riding comfort

Riding comfort is usually influenced by the available space for seating and standing onboard the vehicles/buses, internal noise, and smoothness of operations depending, among other factors, on the driving regime and the quality of surface of the bus-ways. Available space per passenger is measured by the seat spacing, which is typically 0.80 m for the vehicles used in most BRT systems. The driving regime is strongly influenced by acceleration/deceleration rates due to the relatively frequent stops along the given route/line. These are about 0.8–1.6/1.1 m/s^2 for BRT vehicles/buses compared to 0.9–1.3 m/s^2 for LRT trains (Vuchic 2007).

Accessibility

BRT system services are accessible at terminals/stations located at certain distances along the routes/lines on foot, and/or by bike, car, taxi, and/or other public transport modes. In many cases, pedestrian zones lead directly to BRT terminals/stations, thus making them even more accessible. In addition, good accessibility is achieved through the convenient positioning of BRT routes/lines in the given urban context, by locating terminals/stations at easily accessible places, providing dedicated parking spaces for bikes and vehicles/cars and convenient connections/passages to BRT vehicles/bus platforms.

Table 2.6 gives some averages of the indicators and measures of the service quality of BRT and LRT systems.

This confirms that both systems are quite comparable in terms of the quality of service and as such are mutually substitutable.

Fleet size

The size of fleet of a given BRT system is expressed by the number of vehicles/buses operating during a given period of time under given conditions (service frequency and volume of user/passenger demand). This can be estimated for an

Table 2.6 Operational performances of BRT and LRT system—quality of service (averages) (GAO 2012; Janic 2011; Levinson et al. 2003a, b; Vuchic 2007; Wright and Hook 2007)

Indicator/measure	System	
	BRT	LRT
Service frequency (deep/peak-h)	8–15	5–12
Schedule delay (peak-h) (min)	1.00–1.875	1.25–3.0
Dwell time at stations (s)	24	–
Transit time (min/line)	64	–
Typical operating speed (km/h)		
Freeway/bus-way		
Nonstop	60–80	70–90
All-stop	40–55	40–60
Arterial streets	23–31	–
Acceleration/deceleration rate (m/s^2)	0.8–1.6/1.1	0.9–1.3
Reliability of services	High	High
Punctuality of services	High	High
Riding comfort	High	High

individual route/line and/or entire network. For example, for a route/line (*kl*) during the period *T*, the required number of vehicles $n_{kl}(T)$ equals, based on Eqs. 2.4a–d:

$$n_{kl}(T) = f_{kl/\max}(T) * \tau_{kl}(d_{kl}) \tag{2.7a}$$

where all symbols are equivalent to those in the previous equations.

In Eq. 2.7a, $\tau_{kl}(d_{kl})$ is the turnaround time of the vehicles/buses on the route/line d_{kl}, which can be estimated as follows:

$$\tau_{kl}(d_{kl}) = t_{kl/s1} + 2\left[\sum_{j=1}^{M_{kl}-2} t_{kl/sj} + \sum_{j=1}^{M_{kl}-1} d_{kl/j,j+1}/v_{kl/j,j+1}\left(d_{kl/j,j+1}\right)\right] + t_{kl/sM} \tag{2.7b}$$

where

$t_{skl/1}$, $t_{kl/Mj}$ is the average (scheduled) stop time of the vehicle/bus at the beginning and end station/terminal of the route/line (*kl*) (min);

$t_{skl/j}$ is the average (scheduled) stop time of the vehicle/bus at the intermediate station (*j*) along the route/line (*kl*) (min);

$d_{kl/j\ j+1}$ is the distance between the (*j*) and ($j + 1$) station along the route/line (*kl*) (km);

$v_{jkl/,\ j+1}(d_{jkl/,\ j+j})$ is the average operating speed of the vehicle/bus along the segment of the route/line (*kl*) between (*j*) and ($j + 1$) station (km/h); and

M_{kl} is the number of stations along the given route/line (*kl*).

For example, the average fleet size of BRT systems operating a direct or convoy network is 133 and of those operating a trunk-feeder network 197 vehicles/buses (the latter excludes the TransMilenio system). The average fleet size of a European LRT system is 155 vehicles.

Technical productivity

The technical productivity of an individual route and the entire BRT system network can also be determined. Based on Eqs. 2.4a–2.7b, the technical productivity of the given route/line d_{kl} can be estimated as follows:

$$TP(d_{kl}) = C_{kl/0}(T) * N_{kl} * v_{kl}(d_{kl}) = f_{kl/\max}(T) * m_{kl} * N_{kl} * v_{kl}(d_{kl}) \quad (2.8)$$

where all symbols are as in the previous equations.

For example, the average technical productivity of BRT systems varies depending on their size and scope from 15,780 to 63,120 s-km/h (excluding the TransMilenio system). The corresponding technical productivity of LRT systems in Europe varies from 11,000 to −75,000 s-km/h.

2.2.2.5 Economic Performances

The economic performances of BRT systems refer to their costs and revenues.

Costs

BRT system costs include investment costs in infrastructure, facilities, equipment and in some cases vehicles/buses, as well as operational costs.

The total costs of a given BRT system consist of investment costs and operating costs. For the period of 1 year, these costs can be estimated as follows:

$$C_T = C_I + C_o = A + 365 * V * c_v(V) \quad (2.9a)$$

where

- A is the annuity paid for investment and capital maintenance of infrastructure ($US/year);
- V is the average daily utilization of the vehicle/bus fleet (veh-km/day); and
- $c_v(V)$ is the average operating cost per unit of system output ($US/veh-km).

The average volume of vehicle kilometers carried out per day V can be determined as the product of the daily mileage of a single vehicle and the number of vehicles engaged depending on the volume of demand. Operating costs $c_v(V)$ generally decrease more than proportionally as the volume V increases. These costs include annuities on bonds for acquiring the vehicles/buses, vehicle/bus insurance costs, the wages of drivers and other support staff, the costs of vehicle/bus maintenance including wages of personnel and spare parts, energy/fuel costs, and the costs of using the infrastructure (taxes). Table 2.7 gives an example of the typical average costs for selected U.S. BRT and LRT systems.

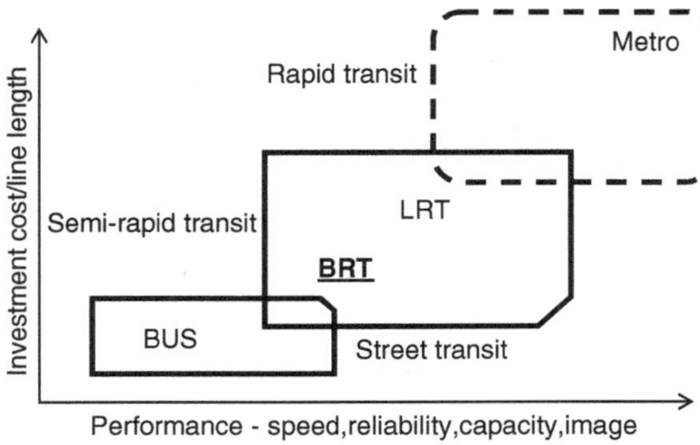

Fig. 2.9 Relationship between the investment costs and performances for selected urban and suburban passenger transport systems (Vuchic 2007)

The differences in the investment costs between BRT and LRT are mainly due to some specific components needed for LRT and not needed for the BRT system such as, for example, train signal communication, electric power systems with overhead wires to power the trains, and rails, ties, and switches. In addition, a rail maintenance facility must be built if one doesn't already exist. Furthermore, the investment costs in BRT systems differ for dedicated bus lines and for mixed traffic lines. For example, on average these amount to 1.2–6.0 $US/km for dedicated and 0.03–0.06 million $US/km for mixed traffic lane(s). The average construction time is about 16/20 km of lines per year. That said, urban and suburban transit systems with higher performances will generally require higher investment costs as shown by the linear qualitative relationship in Fig. 2.9.

In the example given in Table 2.7, the average cost per p-km and vehicle-km is lower in the case of BRT than LRT systems. However, the average cost per passenger is higher in the case of BRT than in LRT, indicating that LRT systems provide services over longer distances.

Table 2.7 Economic performances of selected BRT and LRT systems—cost (averages) (GAO 2001, 2012; Janic 2011)

Cost component	System	
	BRT	LRT
Infrastructure and vehicles		
Infrastructure (millions $US/km)[a]	8.98	18–25
Vehicle (millions/$US/unit)	0.4–1.0	1.5–3.4
Amortization period (years)	25	25
Infrastructure		
Vehicles	12–15	25
Operation		
$US/p-km[b]	0.12	0.23
$US/veh-km[c]	3.05	8.90
$US/passenger[d]	3.20	2.57

[a] In the U.S. the average investment costs for 29 LRT systems amounted to about 24 million $US/km; [b] 5 BRT and 15 LRT systems in the U.S.; [c, d] Six BRT and LRT systems in the U.S.

Revenues

Revenues from operating given BRT system are gained by collecting fares and from various subsidies. For the period of 1 year, these revenues can be estimated as follows:

$$R = 365 * q_p * p + S_u \tag{2.9b}$$

where

q_p is the daily number of users/passengers (users/passengers/day);
p is the average fare per user/passenger ($US/user/passenger); and
S_u is the annual subsidy to a given BRT system.

For example, the average fare of the above-mentioned 40 BRT systems operating around the world is 1.25$US/passenger. About 68 % of the systems (27 of 40) need subsidies at an average level of 25–30 %. Similarly, LRT systems also need subsidies at a level of 20–25 % (Tegner 2003; Wright and Hook 2007).

2.2.2.6 Environmental Performances

The environmental performances of BRT systems include energy/fuel consumption and related emissions of GHG (Greenhouse Gases), and land use/take.

Energy consumption and emissions of GHG

The energy consumption and emissions of GHG (Greenhouse Gases) by BRT system(s) can be considered as direct absolute and relative, and in terms of savings in these both thanks to the modal shift from other urban transit systems.

Direct energy/fuel consumption and related emissions of GHG are usually expressed in relative terms, i.e., as the average quantities per unit of the system's output, i.e., g/p-km or g/s-km (grams/passenger-kilometer or grams/space-kilometer). This is usually carried out for the specified vehicle size and occupancy rate (load factor) while always bearing in mind the specific conditions in which a given BRT system operates. Then, the absolute values can be easily obtained by multiplying these relative values by the corresponding volumes of output over the specified period of time, or vice versa.

In particular, relative values are convenient for comparison of BRT system(s) with other urban transport systems as given in Table 2.8.

As indicated, in both BRT systems, conventional buses (12 m long) mainly used in small and medium-sized direct or convoy systems/networks and for feeder services in larger trunk-feeder and hybrid networks consume and generate less energy/fuel and related emissions of GHG (CO_2), respectively, than their larger articulated counterparts (18 m long). In addition, in this respect, BRT systems remain inferior as compared to LRT systems on the one hand, but superior as

Table 2.8 Environmental performances of the selected urban transport systems—energy/fuel consumption and emissions of CO_2 (averages) (Vincent and Jerram 2006; VTT 2004; Wright 2003)

Impact	System				
	BRT[b]		BRT[c]	LRT[b]	Car[b]
Vehicle (length or units)	12 m	18 m	18 m	2 units	1 unit
Energy/fuel consumption (g/p-km)[a]	8.70	11.69	8.09	5.73	40.79
Emissions of GHG (CO_2) (g/p-km)	27.85	37.41	25.9	18.37	130.53

[a] Diesel fuel; [b] U.S. system(s); [c] BRT TransMilenio (Bogota, Columbia) (Based on 75 passengers per BRT and/or LRT vehicle, and 2 passengers per car)

Table 2.9 Environmental performances of the U.S. BRT and LRT systems—emissions of other than CO_2 GHG (averages) (Puchalsky 2005)

Emissions of GHG	System			
	BRT			LRT
	Diesel	Hybrid	CNG[a]	
NO_x (g/p-km//g/s-km)	0.7150	0.439/0.336	0.2300/0.1590	0.0278/0.0115
VOCs (g/p-km//g/s-km)	0.0063	0.003159/0.002418	0.0112/0.0074	0.000177/0.000073
CO (g/p-km//g/s-km)	0.0713	0.00238/0.00182	0.2570/0.1770	0.000522/0.000216

[a] *CNG* Compressed Natural Gas

compared to individual passenger cars on the other. Both BRT and LRT system are superior as compared to individual (diesel-powered) cars. Table 2.9 gives the average relative emissions of the other than CO_2 GHG—VOC (Organic Compounds), NO_x (Nitrogen Oxide), and CO (Carbon Monoxide)—generated by BRT and LRT systems in the U.S.

These values generally confirm again that BRT systems, independently on the energy/fuel used, remain inferior as compared to LRT systems in terms of relative emissions of the specified GHG. However, these emissions of GHG by LRT systems always need to be considered respecting the composition of the primary sources for obtaining electricity.

Savings in the energy/fuel and related emissions of GHG by BRT system(s) can be achieved in different ways. One can be within the system by choosing low energy/emissions vehicle/bus technologies, by designing bus ways as straight and as short as possible, and by maximizing the fuel efficiency of the vehicle/bus operations along the routes under given conditions (avoiding stops in traffic jams, minimizing the dwell time at stations, driving at fuel-optimal speeds, etc.).

The other implies keeping existing users/passengers onboard, attracting those using individual car as the mode (j) to shift to BRT system as the mode (i), and attracting new users of public transport systems. These direct savings as the average quantities per user/passenger can be estimated as follows:

$$s_{ji} = \left[(\lambda_i N_i)/\lambda_j N_j\right) * EC_j - EC_i\right] * d \quad (2.10)$$

where

N_i, N_j is the vehicle capacity of transport modes (i) and (j), respectively (spaces);

λ_i, λ_j is the occupancy rate (i.e., load factor) of vehicles of transport modes (i) and (j), respectively;

EC_i, EC_j is the average energy/fuel consumption and/or emissions of GHG of transport modes (i) and (j), respectively (g/p-km); and

d is the travel distance (km).

From Table 2.5 it follows that, for example, a BRT bus carrying 75 passengers can replace 37 individual cars, each with 2 passengers. Assuming, for example, that the average commuting distance is 19 km, the savings in the energy/fuel consumption by such car/BRT modal shift would be about 28.5 (bus—12 m long) and 28.7 (bus—18 m long) kg of diesel fuel. The corresponding savings in the emissions of GHG would be about 91.2 and 91 kgCO_2, respectively (Vincent and Jerram 2006).

Land use

Land use relates the size of land used for setting up the infrastructure for a given BRT system. This consists of land for the segregated bus-ways and stations along them, and for the buses' docking/maneuvering, short- and long-term parking (garages), repairs, and maintenance. The size of acquired land for segregated busways has already been discussed. But what are the potential savings of the parking and operating land due to the potential replacing of individual cars by BRT systems? For example, the parking space for a conventional (12 m long—70 spaces) and for an articulated BRT bus (18 m long—160 spaces) is about 36 and 54 m^2, respectively. The former can replace 18 and the latter 40 passenger cars (4 seats), each occupying 10 m^2 of space of at most. The resulting savings of parking space otherwise occupied by cars would be about 144 and 346 m^2, respectively.

2.2.2.7 Social and Policy Performances

Social/policy performances of BRT system(s) generally include noise, congestion, traffic incidents/accidents (i.e., safety and security) and contribution to social welfare.

Noise

BRT system vehicles/buses generate noise while performing transit services. As in the case of other transport systems, this noise generally depends on their constructive-technical/technological characteristics and the pass by speed. It has already been mentioned that BRT systems operate vehicles/buses powered by different engine technologies, which crucially influence levels of their noise. For example, the

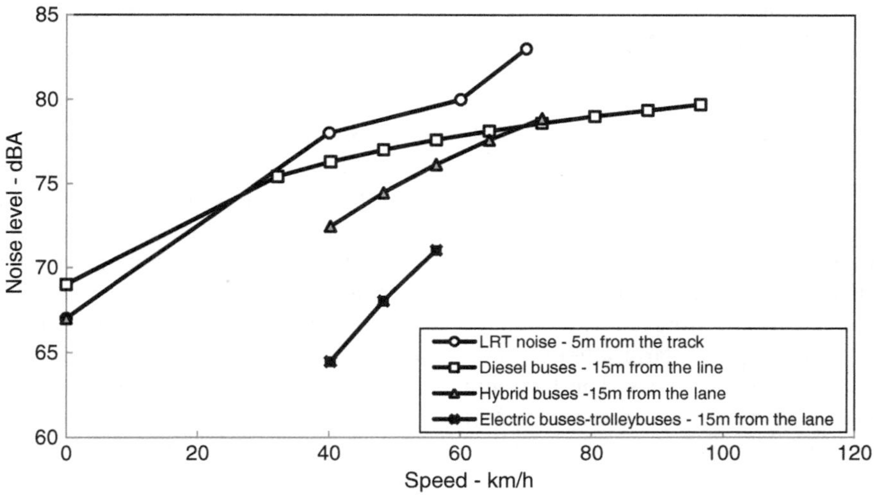

Fig. 2.10 Relationships between the noise and speed of BRT and LRT vehicles (CE 2008; Ross and Staiano 2007)

noise of BRT diesel and CNG buses comes from their exhaust system, engine block, cooling system, air intake components, and tire/pavement interaction. The noise from BRT hybrid vehicles/buses comes from both diesel and electric motors. The main noise sources of BRT trolley buses are interaction between the catenary wire and the pantograph, electric motor, auxiliary equipment, and tire-pavement interaction (Ross and Staiano 2007). In comparison, the noise by LRT vehicles primarily comes from interaction between the catenary wire and the pantograph, electric motors, auxiliary equipment, and wheel-track interaction (CE 2008). Important factors influencing received noise from both BRT and LRT system are: (i) the distance from the noise source, i.e., passing by vehicle(s), and (ii) the existence of noise barriers along the lanes. Figure 2.10 shows an example of the dependency of the noise on speed of BRT and LRT vehicles. BRT vehicles/buses are 12–18 m long, weighting 13–17 tons empty and 32 tons full (vehicle + driver + passengers) with a capacity of 75–100 spaces. LRT vehicles/trains are 20 m long weighting 37–44 tons full (vehicle-couple of cars + driver + 65–162 spaces/passengers).

In the case of BRT vehicles/buses, the noise increases in line with the operating speed at a decreasing rate. In the case of LRT systems, this rate is slightly higher. The noise level from LRT systems is higher than that of BRT buses. One of the reasons is that the distance from the source is three times shorter (5 vs. 15 m) (Urban and suburban buses operating at speeds of about 70 km/h generate noise of about 87.5–92.5 dB at a distance of about 5 m from the source (Cebrián 2008)). Nevertheless, it can be said that respecting their noise levels, BRT and LRT systems appear quite comparable. Noise barriers of a sufficient height along BRT routes built of brick or concrete contribute to decreasing noise to and below the sustainable level of about 55 dBA (Mishra et al. 2010).

Congestion

BRT system congestion can be considered from three aspects. The first implies congestion caused by interference between the BRT vehicles/buses and other traffic, and vice versa, while operating along mixed traffic bus lanes. The second implies congestion due to the clustering of the BRT buses operating along segregated bus ways—particularly those with single lanes in each direction and without passing lanes at the terminals/stations. This can happen in corridors with several BRT routes/lines operating relatively frequent services. The trunk part of the feeder-trunk network can particularly suffer from this kind of induced congestion causing delays of the affected services. The last implies the contribution of the BRT system to savings of own congestion and that of the other traffic, which both contribute to savings in the overall user/passenger travel time. For example, the savings in travel time compared to previously used transit services vary from 5 to 35 % at 16 U.S. BRT systems (GAO 2012). In addition, as compared to individual traffic, savings of 32 % at TransMilenio (Bogotá, Colombia), 35 % at Metrobús (Mexico City, Mexico), and 45 % at Metrobüs (Istanbul, Turkey) have been reported.

Safety

Traffic incidents/accidents reflect the safety and security of a given BRT system. They are caused by collisions of BRT vehicles/buses with other BRT vehicles/buses and with other vehicles, bikers, and pedestrians, all often resulting in injuries and death, as well as damages to property. The number of events per unit of the system's output—the number of passengers and/or passenger-kilometer is a convenient measure. So far, accidents in BRT systems have been relatively rare, thus indicating that the systems safe, and by all means safer than their conventional bus counterparts as shown in Fig. 2.11.

As can be seen, the rate of collisions and injuries of the conventional bus system operated in Bogota (Columbia) before the BRT system established was about 7.7/million passenger trips. Over the 2000–2005 period, thanks to the BRT TransMilenio, despite increasing the number of trips, this rate dropped to about 1–2/million passenger trips. In comparison, during the same period for the slightly higher number of passenger trips on U.S. LRT systems, this rate was about 1.5–3.0/million passenger trips (RITA 2012; http://brt.mercedes-benz.com/content/brt/mpc/Safety.html). That said, both BRT and LRT systems should always be designed and operated to be safe implying that incidents/accidents due to the already known reasons must not occur.

Social welfare

Social welfare of BRT systems relates to their urban and social effects.

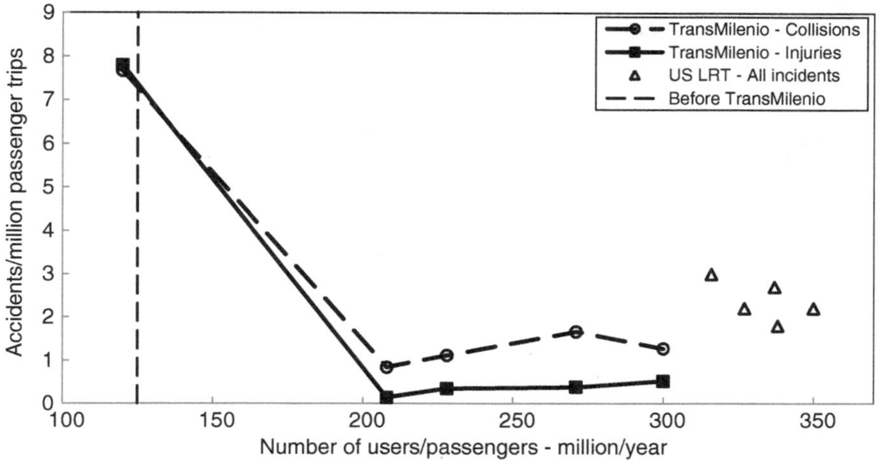

Fig. 2.11 Relationship between the accident rate and the annual volume of traffic at the selected BRT systems (RITA 2012; Saavedra 2011; http://brt.mercedes-benz.com/content/brt/mpc/Safety.html)

Urban effects

The urban effects of BRT systems include changing of the land use and the value of land and property, redistributive effects, and preceding other more efficient and effective systems.

- *Changing the land use* implies (i) taking the land for building the BRT infrastructure; and (ii) building economic and residential objects rather than parking spaces along the BRT corridors/routes and particularly around the terminals/stations.
- *Changing the value of land and property* generally implies rising their value faster due to being located closer to BRT system terminals/stations and corridors/routes. This is because the proximity of the BRT system can save time and monetary cost of commuting, thus making the properties nearby generally commercially more attractive for new developments or redevelopments than otherwise. However, in some cases, the value of land and properties can diminish due to increased noise and emissions of GHG caused by the BRT system (Levinson et al. 2003a).
- *Redistributive effects* imply the contribution of BRT systems to the potential relocation of particular businesses/firms from the suburban areas closer to the city center, and vice versa, i.e., urbanization and de-urbanization of employment.
- *Preceding other more efficient and effective systems* implies that BRT corridors are sometimes used to test the overall feasibility of LRT systems (GAO 2012).

Social effects

The main social effects of BRT systems are their contribution to direct and indirect employment, social equity, and personal meetings and interactions.

- *Direct employment* is needed for planning, designing and constructing the BRT system, and later on for its operating.
- *Indirect employment* includes institutional and supportive employment (in entertainment, other amenities-hotels, hospitals, etc.) generated purely because of the implementation of the BRT system(s).
- *Social equity* reflects the ability of the BRT system to, like other urban transport systems, facilitate accessibility and promote social equity within a city. For example, cheap BRT systems give lower income groups greater access to public services and economic opportunities (Wright and Hook 2007).
- *Personal meetings and interactions* imply that effective, efficient, safe, and cheap BRT systems can bring different groups of people in terms of age, gender, and income group to places where they can meet and interact with each other in different ways. Such interactions can diminish tensions and improve the mutual understanding between such groups.

2.2.3 Evaluation

BRT systems possess both advantages (Strengths and Opportunities) and disadvantages (Weaknesses and Threats) as compared to other potentially substitutable urban (mass) transit systems, such as conventional bus transit and LRT systems.

Comparison with conventional urban bus transit systems

Advantages

Users/passengers

- Relatively strong spatial coverage of a given urban area with corridors, routes, and lines guaranteeing relatively high quality of the spatial accessibility of the system's services;
- Relatively high service frequency, particularly during peak periods contributing to reducing schedule delay(s) and consequently the urge to shift to other systems/modes;
- Relatively fast and reliable services thanks to running along dedicated busways not affected by other traffic at higher operating speeds with minimal dwell times at the terminals/stations and stops along the routes/lines;
- Less delays due to general congestion, traffic signals, right turns, and passenger stops; and
- Higher riding comfort thanks to operating rubber-tired, low floor vehicle/buses of a suitable capacity, a sufficient number of wide doorways offering easy boarding and internal comfort, and the information system inside the vehicles/buses and at the terminals/stations/stops.

Transport operators, policy makers, and community members

- Increasing efficiency and effectiveness of operating the available vehicle/bus fleet thanks to deploying ITS;
- Offering more competitive services contributing to diminishing use of individual cars and consequently the negative impacts on the environment and society such as energy/fuel consumption and related emissions of GHG, noise, and congestion throughout the network—along the routes/lines and at parking lots;
- Affecting stronger urbanization and suburbanization along particular corridors/routes including development of new commercial and social activities;
- Contributing to increasing numbers of users/passengers switching to public urban transport systems; and
- Contributing to integrating transit and land use planning.

Disadvantages

Users/passengers

- Some confusion due to clustering too many lines/services at the same stops/stations, particularly at feeder-trunk service systems/networks; and
- Diminishing quality of service if operating along mixed traffic lanes.

Transport operators, policy makers, and community members

- Relatively substantial investments in infrastructure, vehicles/buses, and supporting facilities and equipment (for example, building tunnels or acquiring CNG (Compressed Natural Gas) or hybrid buses) including the time of commercialization;
- Affecting the space for other traffic and the urban content of the given area/city by building the BRT infrastructure, the former can temporarily contribute to increasing congestion, while the latter can deteriorate urban green areas;
- Increasing energy/fuel consumption and related emissions of GHG and noise due to more voluminous operations despite using more technologically advanced vehicles/buses (hybrid, CNG, etc.); and
- Needing subsidies for services in many cases.

Comparison with LRT (Light Rail Transit) systems

Advantages

- Flexibility and convenience of implementation in a wide range of urban and suburban areas;
- Gradual implementation and extension of the route network;
- Higher frequency and flexibility of services in terms of adjusting to variable daily demand and re-routing due to any reason, respectively;
- Comparable and flexible transport capacity; and
- Lower investment and comparable operational costs.

Disadvantages

- Generally requiring more space for setting up the infrastructure;
- Lower riding comfort due to the less sizeable and comfortable seats onboard (for passengers);
- Higher energy consumption and related emissions of GHG independently of the technology used and not absolutely free from traffic congestion;
- Rather negative image due to the general perception of LRT usually being slow, noisy, and polluting;
- Lower effects in creating greater land and property value along the corridors/ routes; and
- Lower preferences by developers to locate social-economic activities along the inherently unstable bus routes/stations rather than along more permanent LRT routes/stations.

Finally, what can be said for BRT transport systems? They are advanced public transit systems primarily characterized by the advanced organization of transport services carried out by matured but gradually improving vehicle/bus technology.

2.3 High Speed Tilting Passenger Trains

1973	The 381 series tilting trains operated by JNR (Japan National Railways) begin commercial services on the Chuo Main Line connecting Nagoya and Nagano (Japan)
1976	Fiat's ETR401 tilting trains operated by the Italian State Railways begin commercial services on the Rome-Ancona line (later extended to Rimini) (Italy)
1978	The Spanish tiling trains Talgo operated by RENFE (The Spanish National Railway) begin commercial operations (Spain)

2.3.1 Definition, Development, and Use

High-speed tilting passenger trains operate at speed of about 200 km/h on upgraded and around 250 km/h or faster on newly built tracks defined by the EU (European Union) thanks to their fully operational tilting mechanisms. This mechanism can be disabled after moving onto high-speed tracks and reaching the speed of 250 km/h. Such an advantage makes these trains highly interoperable in rail networks consisting of conventional, upgraded, and completely new (high speed) lines (http://en.wikipedia.org/wiki/Tilting_train). Some tilting trains operating on upgraded trucks at speeds of about 200 km/h include: Virgin Trains' Class 390 Pendolino operating along the West Coast Main Line in the United Kingdom (UK), the Talgo 350 train operating on the Spanish AVE high-speed lines, the Italian Pendolino 2 tilting train ETR600, the Swedish X2 tilting train, and the E5 Series Shinkansen train operating in Japan. One high-speed tilting train operating on new tracks at speeds of 250 km/h or higher is the most advanced Japanese tilting train—the N700 Series Shinkansen. This train tilts up to one degree and maintains a speed of about 270 km/h as compared to its previous speed of 255 km/h while passing through curves of a radius of 2,500 m on the Tōkaidō Shinkansen line.

2.3.2 Analyzing and Modeling Performances

2.3.2.1 Background

High-speed tilting passenger trains are characterized by their infrastructural, technical/technological, operational, economic, environmental, social, and policy performances, which all influence each other. In general, these trains are designed to primarily operate along conventional tracks at higher speeds than those of conventional passenger trains. In such cases, the tracks do not require any particular modifications. Operation at the higher speeds as the main distinguishing feature of these trains compared to their conventional counterparts is possible under given conditions thanks to the tilting capabilities.

2.3.2.2 Infrastructural Performances

The infrastructural performances of high speed tilting trains relate to the characteristics of the track design and the related standards that these trains can use.

Track design

In order to fully exploit the technical and operating speeds of high-speed tilting passenger trains, conventional tracks need to be upgraded and/or completely rebuilt anyway. The former implies partial reconstruction/redesign of the existing lines by leveling off high grades (horizontal and longitudinal slopes), increasing the radii of curved segments, and partially reconstructing tunnels, bridges, crossings, and platforms at particular stations. The latter implies building completely new lines. Both should be carried out according to the specified standards for designing rail tracks.

The main elements of rail track design relevant to understanding the operation of high-speed tilting passenger trains are as follows (Persson 2007):

- *The track gauge* as the distance between the inner faces of the rail heads of the track is measured 14 mm below the top of the rail on the inner face. The standard track gauge is approximately 1,435 mm. The track gauge has an impact on the lateral behavior of the vehicle which may lead to unstable running. In addition, it impacts the lateral behavior of the vehicle, which in turn impacts lateral ride comfort.

- *The circular horizontal curve* is a curve in the horizontal plane with a constant radius. This curve is characterized by its radius R related to the track center line and/or curvature as an inverse to the radius. The reduced radius of the circular horizontal curve increases the lateral track forces, which increases the derailment ratio. In addition, it has no impact on ride comfort.
- *The transition curve* is used to connect the straight track to the circular horizontal curve or to connect two circular horizontal curves. The transition curve is characterized by its curvature as a function of its longitudinal position. The most common transition curves have linear variation of the curvature and do not affect safety. The reduced length of the transition curve increases the rate of change of the cant deficiency and thereby also the lateral jerk perceived by passengers. It also increases the roll velocity of tilting trains, which is believed to contribute to motion sickness.
- *The track cant (or super elevation)* is the amount at which one running rail is raised above the other running rail (in the curve). The track cant is positive when the outer rail is raised above the inner rail. The UIC has proved that a cant of 180 mm is widely acceptable and safe. The track cant does not influence ride comfort.
- *The cant transitions (or super elevation ramps)* connect two different track cants. In most cases, the cant transition has the same longitudinal position as the transition curve. The cant gradient is characterized by its longitudinal distance to raise one unit (normally expressed as 1 in X, where X is the longitudinal distance in units). The ERRI (European Rail Research Institute) has showed that a cant gradient of 1/400 m/m is acceptable. The most common cant transition has a constant rate of cant change. Steep cant transitions may cause diagonal wheel unloading, which in turn may lead to derailment due to flange climbing. In addition, cant transitions do not impact ride comfort.
- *The rate of cant change* is the rate at which the cant is increased or decreased at a defined speed. The rate of cant change is characterized by the cant change per time unit. This does not impact safety.
- *Cant deficiency* arises when the installed cant is lower than the cant of equilibrium. Cant deficiency is characterized by additional cant needed to ensure equilibrium. High cant deficiency may lead to high lateral track forces. High cant deficiency also increases the risk of over-turning.
- *The rate of change of cant deficiency* is the rate at which the cant deficiency increases or decreases at a defined speed. This rate is characterized by the cant deficiency change per unit of time. The most common transition curve/cant transition has a constant rate of change of the cant deficiency, which does not affect safety. However, an increased rate of change of the cant deficiency increases the lateral jerk perceived by passengers. It also increases the roll velocity of the tilting vehicles, which is found to contribute to motion sickness while onboard the tilting vehicles.
- *The track gradient* connects tracks at different altitudes. The gradient, which affects both safety and ride comfort, is characterized as a change in altitude per unit of distance (%). In certain countries, it is represented as the longitudinal

Table 2.10 Infrastructural performances for the high-speed tilting trains in Europe—infrastructure standards (CEN 2006)

Element	Value
Cant (mm)	180
Rate of cant change (mm/s)	75
Cant deficiency (mm)	306
Rate of change of cant deficiency (mm/s)	150

distance to raise for one unit (it is expressed as 1 in X, where X is the longitudinal distance in units).

- *The vertical curve* not affecting safety and the ride comfort connects two different track gradients and is characterized by its radii.

Design standards (Europe)

The CEN (Comité Européen de Normalisation) provides guidance and standards for particular elements of geometric design of conventional rail vehicles (CEN 2002). This guidance and standards are revised latter on in order to be also convenient for high speed tilting train vehicles (CEN 2006). For high-speed tilting trains, some standards are given in Table 2.10.

In addition, CEN standards categorize the rail tracks, i.e., the rail traffic lines, based on the categories of services they accommodate, as follows:

(a) Mixed traffic lines for passenger trains traveling at speeds from 80 to 120 km/h;
(b) Mixed traffic lines for passenger trains traveling at speeds greater than 120 km/h and up to 200 km/h;
(c) Mixed traffic lines for passenger trains traveling at speeds higher than 200 km/h;
(d) Mixed traffic lines for passenger trains incorporating special technical design characteristics; and
(e) Dedicated passenger lines for passenger trains traveling at speeds greater than 250 km/h.

High-speed tilting passenger trains operate on lines (b), (c), and (d).

2.3.2.3 Technical/Technological Performances

The technical/technological performances of high-speed tilting passenger trains relate to tilting principles and tilting technology/mechanism, the vehicle/train technical specifications, signaling system, and energy consumption.

Tilting principle

When trains pass the horizontal curves along the line(s), both the vehicles themselves and passengers onboard are exposed to centrifugal force. This can be reduced by roll inwards, thus enabling trains to pass through curves at higher speeds while still maintaining passenger ride comfort. In general, this inward roll may be achieved by the track cant and/or by tilting the train. Trains composed of coaches with tilting capabilities are called tilting trains, which can generally be

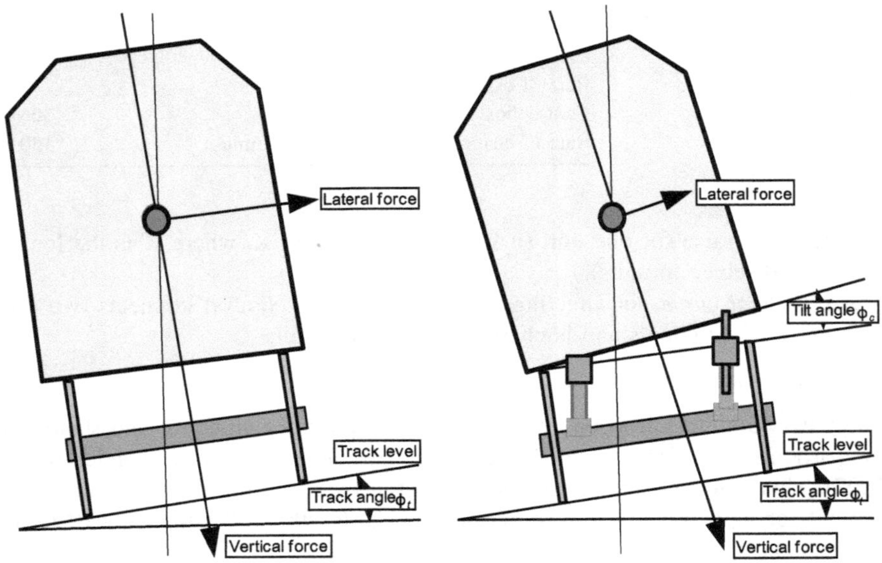

Fig. 2.12 Scheme of tilting principle of high-speed tilting passenger train(s) (Persson 2007)

categorized into two categories: (i) passively natural tilted trains (for example in Japan), and (ii) actively tilted trains (in the rest of the world including Europe). Passive tilt trains rely on the natural laws with a tilt center located above the center of gravity of the coach. Along a curve, under the influence of centrifugal force, the lower part of the coach swings outwards. Conversely, active tilt trains are based on active technology controlled by a controller and executed by an actuator. The main principle of tilting trains is rolling the coaches inwards the curve as schematically shown in Fig. 2.12 in order to reduce the lateral force affecting passengers.

Despite the higher track plane acceleration for the tilting train (right), the lateral force in the car-body is lower. When a coach/train is running along a horizontal curve, horizontal acceleration emerges. It can be expressed as follows:

$$a_h = v^2/R \tag{2.11a}$$

where

v is the operating speed (m/s); and

R is radius of the curve (m).

The acceleration in the track plane can be reduced compared to horizontal acceleration by arranging the track cant D. The angle between the horizontal plane and the track plane is a function of the track cant and the distance $2b_0$ between the two contact points of the wheel-set, as follows:

$$\phi_t = \arctan(D/2b_0) \tag{2.11b}$$

The acceleration experienced by passengers can further be reduced compared with the track plane acceleration by arranging the tilting angle of a coach ϕ_c. Acceleration in the coach is called lateral acceleration a_c, which can be determined as follows:

$$a_c = (v^2/R)\sin(\phi_t + \phi_c) - g\sin(\phi_t + \phi_c) \tag{2.11c}$$

Acceleration in the perpendicular direction is called vertical acceleration a_v, which can be determined as follows:

$$a_v = (v^2/R)\sin(\phi_t + \phi_c) - g\cos(\phi_t + \phi_c) \tag{2.11d}$$

Reducing the lateral acceleration by increasing the track cant or tilt of a coach correlates with a slightly increased vertical acceleration. Some typical values for the lateral and vertical acceleration are $a_c = 0.98$ m/s^2 and $a_v = 0.44$ m/s^2 for a train operating at speeds of about: $v = 200$ km/h, along a track curve with a radius: $R = 1{,}000$ m, track cant: $D = 150$ mm, and tilting angle: $\phi_c = 6.5°$.

Tilting technology/mechanism

As mentioned above, the tilting technologies/mechanisms are based on two principles. The first is passive or natural tilting (in Japan), while the other is active tilting (the rest of the world and Europe). Passive tilting uses natural laws with a tilt center located well above the center of gravity of the car-body. On a curve, under the influence of centrifugal force, the lower part of the car-body swings outwards. Active tilting uses tilting mechanisms based on pneumatic systems, where air is shifted from one side to the other of the air suspension. In addition, rollers and pendulums as technological innovations carry the car-body load and provide movement. Then, movement may be controlled by an actuator, which does not need to carry the car-body load, thus resulting in much lower energy consumption.

Actively tilted trains need some kind of control system. Contemporary systems include body feedback with an accelerometer placed in the car-body as a transducer. This can use different information sources. The obvious one is lateral acceleration, but also the roll and yaw velocity can be used. Most tilting trains use more than one source as the basis for their control. Consequently, they can be differentiated in light of the type of tilting mechanism/technology they use as follows: (i) trains tilting by inertial forces; (ii) trains with active tilting based on sensory information provided by an accelerometer; (iii) trains with tilting controlled by a computer; and (iv) the trains with tilting provided by active suspension (http://en.wikipedia.org/wiki/Tilting_train).

Tilting by inertial forces

Some trains tilting by inertial forces include: Talgo (Spain), UAC TurboTrain (US, Canada), and JNR 381 series (Japan). Talgo trains can achieve maximum speeds of 175–200 km/h, a tilting angle of 3–3.5°, and DC traction, while Japanese trains operate at maximum speeds of 120–160 km/h, a tilting angle of 5°, and AC and DC traction systems (DC—Direct Current; AC—Alternating Current).

Tilting actively by sensory information obtained from accelerometers

Some such trains include Light-Rapid Comfortable (LRC) trains built by Bombardier (Canada, US). Depending on the version, these trains operate at maximum speeds of 155–240 km/h, a tilting angle of 6–10°, and AC and DC traction.

Tilting actively controlled by a computer

Currently, this is the most numerous category of high speed tilting trains. Some of them are Acela Express (US), a Bombardier high-speed tilting train operating between Boston and Washington D.C.; British Rail Class 390 "Pendolino" (UK), a high-speed train run by Virgin Trains from London Euston to Liverpool/Manchester/Glasgow/Birmingham and Wolverhampton; Alfa Pendular (Portugal), ElettroTreno (Italy), ICE-T, also called ICT Technologies)(Germany), a tilting version of the German high-speed ICE; ICN (Switzerland), a new generation of tilting trains operated by Swiss Rail, a Bombardier-built high-speed tilting train operating between Zurich and Geneva; JetTrain (North America), Bombardier's experimental non-electric high-speed train; NSB Class 73 (Norway); SŽ series 310 (InterCitySlovenija), a high-speed tilting train operating among Ljubljana, Maribor, and Koper; RegioSwinger (Germany and Croatia), a diesel regional tilting train; Pendolino (Italy, Finland, the UK, and the Czech Republic), built by Alstom (formerly Fiat); Virgin Train Super Voyager, a Bombardier-built high-speed tilting train operating between London and Holyhead/Wrexham/Chester and Birmingham to Edinburgh or Glasgow; Taroko Express (Taiwan), based on the JR Kyūshū 885 Series; Tilt Train by QR, diesel and electric tilting Traveltrains (Australia) operating between Brisbane and Cairns (the Electric Tilt Train is based on the JR Shikoku 8000 series, X2 (Sweden), with tilting mechanism made by ABB; it is also used in China under the name Xīnshísù); the JR Shikoku 2000 series (Japan 1989), the first tilting DMU in the world used on many limited express services in Shikoku, including Ashizuri, Ishizuchi, Nanpū, Shimanto, Shiokaze, Uwakai, and Uzushio (the upgraded N2000 Series was introduced from 1995), the JR Hokkaido KiHa 281 series (Japan 1992), branded Heat 281 or Furico 281, and used for Super Hokuto limited express service; the JR Shikoku 8000 series (Japan 1992) is used for the limited express service on the Yosan Line, namely Ishizuchi and Shiokaze; the JE Fast E351 series (Japan 1993) is used for Super Azusa; the Chizu Express HOT7000 series (Japan 1994) is used for Super Hakuto; JR Central 383 series (Japan 1994) is used for Wide View Shinano; JR Kyushu 883 series (Japan 1994) is used for Sonic; the JR Hokkaido KiHa 283 series (Japan 1995), branded as Furico 283 is used for the Super Hokuto, Super Ōzora, and Super Tokachi limited express services. The JR West 283 series (Japan 1996) is used for Ocean Arrow; the JR Kyushu 885 series (Japan 1999) is used for Kamome and Sonic; the JR West KiHa 187 series (Japan 2001) is used for Super Inaba, Super Kunibiki, and Super Oki (these trains operate at maximum speeds of 170–250 km/h, a tilting angle of 8–10°, and AC and DC traction systems).

Tilting by active suspension

Some of these trains are: the JR Hokkaido KiHa 201 series (Japan 1996) used for rapid trains around Sapporo; the JR Hokkaido KiHa 261 series (Japan 1999), branded Tilt 261, used for Super Sōya; the Meitetsu 1600 series (Japan 1999), branded Panorama Super, mainly used for the Meitetsu Nishio Line; the Meitetsu 2000 series (Japan 2004), branded μ-Sky, used to connect Nagoya and Chūbu Centrair International Airport; the Odakyu 50000 series VSE (Japan 2005) used for Romancecar; the N700 Series Shinkansen (except N700-7000/8000 series) (Japan 2007) introduced by JR Central and JR West, used for the Tōkaidō and Sanyō Shinkansen lines; the E5 Series Shinkansen (Japan 2011) introduced by JR East, used for Tōhoku Shinkansen lines. These trains operate at maximum speeds of 120–160 km/h and a tilting angle of 5°.

Vehicles/trains

Most of high-speed tilting trains use electric energy from different voltage and current systems; such flexibility represents an important component of their interoperability. For example, in Europe, depending on the country, these are 1.5 kV and 3 kV DC, 15 kV/16.7 Hz AC and 25 kV 50 Hz AC systems. In the U.S., newly built rail lines are equipped with the 25 kV 60 Hz system. In Europe, one of the largest fleets (52 train sets) of high speed tilting trains is operated by Virgin Trains along the West Coast Main Line in the UK. The technical specifications of these trains are summarized in Table 2.11.

Signaling system

High-speed tilting passenger trains operate using the cab signaling system, which communicates the track status information to the driver's cab. In general, the system transmits information through the rails as electrical signals, which are picked up by antennas placed under the train, then processed by computers and displayed in the cab. The cab signaling system enables controlling the speed of these trains while passing through the curves along a given line in order to maintain it below or at most at the level of PS (Permissible Speed) or EPS (Enhanced Permissible Speed). The latter is slightly under the speed at which these trains can overturn. In general, the speed limits can be different for different types of tilting trains passing though the same curve but usually only one is displayed in order not to confuse the driver. In order to additionally prevent the confusion of driver, the signs for tilting train speeds must be distinctive from those of conventional trains. In addition, the total number of different speeds indicated at any given location along the line(s) must not be greater than three—one for EPS and two for conventional passenger and freight trains. Furthermore, changes in EPS on a route must be signed implying application of the continuous route signing, the positions of signs for any change in PS and EPS must be coincident, and signs for EPS must not be positioned in isolation (i.e., where provided, they must always

Table 2.11 Technical/technological performances of a selected high-speed tilting train—technical specifications (Persson 2007)

Specification	Systems
Type	Pendolino, British Class 390
Owner	Angel Trains (subsidiary of Babcock and Brown)
Operator/Franchise (until 2011)	Virgin Trains
Operations area	West coast route in UK
Number of units and configuration	Fifty two 9-car sets (468 cars) (one was lost in the accident in Grayrigg in 2007);
Delivered to use (year)	Four new 11-car train sets added over the period 2010—2012; 31 older sets received with 2 new cars, with an option for the 21 remaining units.
Configuration	Driving trailer + 7 trailers + driving trailer $1A'A1' + 1A'A1' + 2'2' + 1A'A1' + 2'2' + 1A'A1 + 2'2' + 1A'A1 + 1A'A1$; After upgrade: driving trailer + 9 trailers + driving trailer
Length of the train (m)	217.4 (9-car set); 265 (11-car set)
Maximum weight (tons)	439
Capacity (seats/passengers)	439 (145 first class; 294 standard class)
Power (MW/hp)	5.1/6840(12 × 425 kW) (in Pendolino the power is distributed throughout the train set; regenerative braking system);
Single current versions	25 kV 50 Hz overhead
Maximum technical speed (km/h)	230

have an accompanying sign for conventional trains). High-speed tilting passenger trains are permitted to operate at EPS through curves only if the speed limit information and speed supervision and control are provided by the above-mentioned cab signaling and automatic train protection system. The cab signaling system continuously displays information on the speed limit, which is consistent with EPS (where applicable), and the speed restrictions on the given line(s).

In addition to the signals along the track, the cab signaling system enables the allowable speed and information about the tracks ahead to be displayed. Furthermore, the automatic train protection system added on the top of the cab signaling system warns the driver of dangerous conditions ahead including the automatic activation of brakes able to decelerate and/or bring the train to a stop, but exclusively in cases when the driver misjudges a dangerous condition.

2.3.2.4 Operational Performances

The operational performances of high-speed tilting passenger trains include interoperability, speed, turnaround time, required fleet, technical productivity, and influence on the rail line capacity.

Interoperability

As mentioned above, the AEIF (European Association for Railway Interoperability) also provides the TSI (Technical Specifications of Interoperability) for Trans-European High-Speed Rail Infrastructure including guidance on the cant and the cant deficiency for non-tilting vehicles (AEIF 2002). However, analogous specifications and guidance for high-speed tilting passenger trains/vehicles have been left to the owners/operators/managers of the corresponding infrastructure. Under such circumstances, interoperability of high-speed tilting passenger trains can be defined as their flexibility to:

- Operate along the above-mentioned (b), (c), and (d) category of railway lines at operating speeds that are usually higher than the speeds of their conventional counterparts; and
- Use different power (electricity) supply systems (1.5, 3, and/or 25 kV; kV—kilovolt).

Both criteria have already been achieved thanks to tilting mechanism/technologies, multi-system locomotives (power units of the train sets), and particularly in Europe, thanks to the forthcoming advanced train signaling and control system developed as components of the ERTMS (European Rail Traffic Management System) (EC 2010).

Speed

As applies to the other categories, the technical speed, operational speed, and commercial speed of high-speed tilting trains can be distinguished.

Technical speed is defined as the maximum speed that a given high train can achieve under given conditions (category of rail line and power supply system). Usually, this speed is specified through the train design.

Operational speed is the maximum speed at which a given train commonly operates on the given rail line. This speed is lower than or at most equal to the technical speed.

Commercial speed is the travel speed of a given train along the given rail line including acceleration, deceleration, intermediate stops, and other maneuvers influencing the operating speed. This speed is lower than the operating speed. In addition to the length of the line, it crucially influences the turnaround time of the given train set(s).

Turnaround time

The turnaround time of a given train scheduled to operate along a given rail line is defined as the total time the train spends between its stations, i.e., from the origin to the destination station, and back. This time also includes the train's stop time at intermediate stations, which mainly depends on the pattern and volume of passenger demand to be served in both directions. In addition, the turnaround time includes the train's acceleration and deceleration time to/from the operating/cruising speed. Thus, the train line can be considered as a route consisting of several segments. If the stops are the same in both directions, the turnaround time of a given train can be estimated as follows:

$$t_{tr} = t_0 + 2^{-}\left[\sum_{i=1}^{N} t_i + \sum_{k=1}^{K} t_k\right] + t_d \qquad (2.12a)$$

where

t_o, t_d is the train's stop time at the origin and destination station (terminus), respectively (min);

t_i is the train's running time along the (i)-th segment of the given route (min);

t_k is the train's stopping time at the (k)-th intermediate station along the given route (min);

N is the number of segments of the given route; and

K is the number of stations along the given route where the train stops ($K = N-1$).

The train's running time along the (i)-th segment of the route in Eq. 2.15a can be estimated as follows:

$$t_i = v_i/a_i + s_i/v_i \qquad (2.12b)$$

where

v_i is the train's operating/cruising speed along the (i)-th segment of the route (km/h);

a_i is the train's acceleration/deceleration rate at the beginning and the end of the (i)-th segment of the route (m/s^2); and

s_i is the length of the (i)-th segment of the route (km).

In addition, the length of the route in one direction can be estimated as follows:

$$d = \sum_{i=1}^{N} s_i \qquad (2.12c)$$

Furthermore, the commercial speed of the train along a given route based on Eq. 2.15c can be estimated as follows:

$$v_c(d) = d/t_{tr} \qquad (2.12d)$$

where all symbols are as in the previous equations.

Fleet size

The fleet size of the high-speed tilting trains scheduled to operate on the route of length d during the time period T can be determined, based on Eq. 2.12a, as follows:

$$N[d, f(T,d)] = f(T,d) * t_{tr} \qquad (2.13a)$$

where

f(*T*, *d*) is the frequency of train services along the route *d* during time *T*.

The frequency *f*(*T*, *d*) in Eq. 2.13a can be determined as follows:

$$f(T, d) = Q(T, d)/[\lambda(T, D) * N(d)] \tag{2.13b}$$

where

Q(*T*, *d*) is the passenger demand on the route of length *d* during the time interval *T* (passengers);

$\lambda(T, d)$ is the average load factor of a given train service scheduled along the route of length *d* during time *T*; and

N(*d*) is the seating capacity of a given train operating along the route *d* (seats).

The other symbols are analogous to those in the previous equations.

Technical productivity

The technical productivity of a high speed tilting train can be determined as the product of its seating capacity and its technical, operational, and/or commercial speed. In particular, the latter depends on the route length *d*. Thus it follows:

$$TP[v(d), N(d)] = v(d) * C(d) \tag{2.14a}$$

In addition, the technical productivity of a given route operated by high-speed tilting trains can be estimated as the product of the total number of seats supplied during a given period of time and the average speed of train services as follows:

$$TP[f(T, d), v(d), N(d)] = f(T, d) * v(d) * C(d) \tag{2.14b}$$

Obviously, the service frequency *f*(*T*) in Eq. 2.14b is an integer as a reciprocal of the time interval(s) between scheduling the train services. Depending on the pattern of passenger demand, these intervals can typically be 1 h (*h*(*d*) = 1 h), half an hour (*h*(*d*) = 1/2 h), etc. during the day. Consequently, the technical productivity of the train services along a given line of length d can be expressed as the volume of passenger-km produced during the period *T* under given conditions.

Effects on the rail line capacity

High sped tilting trains can contribute to increasing the utilization of the available capacity of a given train line. This happens when they replace their conventional counterparts. Let's assume that high-speed tilting and conventional trains are scheduled exclusively along the given line of length d at constant intervals h_t and h_c, respectively, during the time *T*. In such case, the following conditions need to be satisfied:

$$(n_t - 1)h_t + t_{tr/t} = T \tag{2.15a}$$

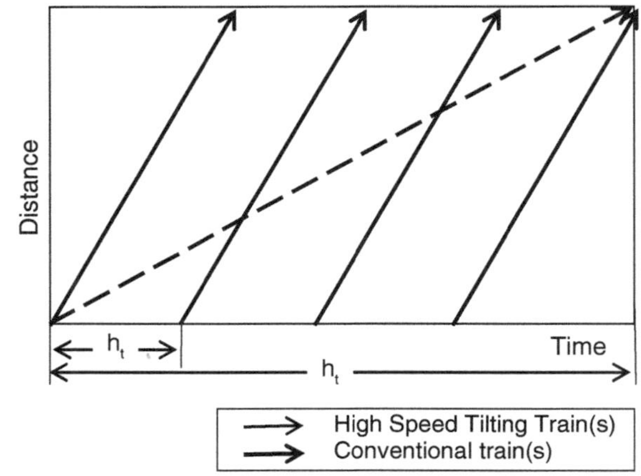

Fig. 2.13 Time-space diagram of possible replacing conventional with high-speed tilting trains on a given (rail) line

$$(n_c - 1)h_c + t_{tr/c} = T \tag{2.15b}$$

where

n_t, n_c is the number of high-speed tilting and conventional trains, respectively, scheduled on the line during the time period T.

The other symbols are as in the previous equations. From Eqs. 2.15a, b, the number of additional high-speed tilting trains as compared to the number of conventional trains that can be scheduled on the line d under given conditions can be estimated as follows:

$$n_t = (n_c - 1) * (h_t/h_c) + \left(t_{tr/c} - t_{tr/t}\right) + 1 \tag{2.15c}$$

where all symbols are as in the previous equations. A simplified scheme of the time-distance diagram is shown in Fig. 2.13.

Let the length of the line be $d = 500$ km. Both categories of trains are exclusively scheduled along the line with nine intermediate stops, each taking about 2 min. The stop time at the origin and destination station takes about 20 min. The average operating speed of a high-speed tilting train is about 190 km/h and that of a conventional train about 100 km/h. From Eq. 2.15a, the turnaround time of both categories of trains is estimated to be $t_{t\ r/t} = 5.3$ h and $t_{tr/c} = 10$ h. Both categories of trains are scheduled along the line in constant intervals of $h_t = h_c = 1$ h. From Eqs. 2.15a, b, the number of high-speed tilting trains that can be scheduled along the line is determined as: $n_t = n_c + 5$. In addition, inclusion of the seat capacity of both categories of trains can indicate the real extent of the contribution of high-speed tilting trains to increasing utilization of the capacity of a given rail line. However, we should always be aware that scheduling of any of these trains is based on the characteristics of demand along a given line, i.e., its volume(s) and time pattern(s).

2.3.2.5 Economic Performances

The economic performances of high-speed tilting passenger trains mainly imply their operational costs. In general, after the infrastructure has been upgraded according to the above-mentioned standards, these costs consist of two components: infrastructure operational costs related to its exploitation and maintenance, and train operational costs, i.e., the costs of provision of transport services using the infrastructure. In many, particularly European countries, the operation and maintenance of rail infrastructure are managed by agencies/companies independent of those providing services. Consequently, many rail service providers/operators can access and use the same infrastructure, thus competing with each other. Under such circumstances, they are charged for using the infrastructure at rates that at least enable the given infrastructure managing company to cover the maintenance costs.

Infrastructure maintenance and operating costs

Infrastructure maintenance and operating costs generally include the costs of the labor, energy and other material consumed for day-to-day maintenance and operations of the rail lines/tracks, terminals, stations, energy supplying and signaling systems, as well as the traffic management and safety systems. These costs consist of fixed and variable parts. The former depends on the volumes of operations routinely performed in accordance to technical and safety standards, while the latter depends on the intensity of traffic on the given line generating the need for different kinds of interventions. Some figures provided by UIC indicate that the labor shares the largest part of the total infrastructure maintenance costs as follows: 55 % for maintenance of electric traction installations, 45 % for maintenance of tracks, and 50 % for maintenance of equipment. For example, the average cost of maintaining new and upgraded high speed rail lines in Europe ranges from 28 to 33 thousand euros (2000) per kilometer of a single track (De Rus 2009).

Train operating costs

The operating costs of high-speed tilting passenger trains can be divided into four main categories: (i) shunting and train operations (mainly, labor costs); (ii) maintenance of the trains/rolling stock and other equipment; (iii) energy; and (iv) sales and administration. The latter vary across different rail operators depending on the expected level of traffic, since they mainly include labor costs related to ticket sales and providing information at the stations/stops. The other three components vary widely depending on the technology/type of the high-speed tilting train, and the local operating and economic conditions. For example, the average operating costs of the ETR480 high-speed tilting train operating in Italy since 1997 is about: $c = 0.1756$ €/seat-km (This does not include the costs of acquiring the train and the costs of energy consumption). The former cost appears to be rather negligible after it is spread over the annual volume of kilometers travelled (about 150–200 thousands/train) and the train's amortization period of about 30–40 years. The latter cost mainly depends on the above-mentioned factors influencing the energy consumption and the prices of electricity. The latter are often conditioned by local agreements between the suppliers and consumers (rail operator) (the ETR480 train set has a length of 296.6 m, weight of 400 tons, power 6 MW (8,000 Hp), maximum technical speed of 250 km/h, and 480 seats). In addition, the average operating costs of the ICE-T high-speed tilting train that began services in Germany in 1999 amount to about: $c = 0.1346$ €/seat-km, again excluding the costs of acquiring the train set and the costs of energy consumption (De Rus 2009) (Depending on the configuration (5 or 7 cars), the train's length is 133–185 m, its weight 270–368 tons, its engine power 3–4 MW, and seat capacity of 250–357 seats (RTR 2005)).

2.3.2.6 Environmental Performances

The environmental performances of high-speed tilting passenger trains include energy consumption and related emissions of greenhouse gases, and land use/take.

Energy consumption

High-speed tilting passenger trains consume electricity, which, in general, can be generated by electricity power plants using different primary sources. The proportion of particular sources is always region- and country-specific. For example, at present, in the EU27 (European Union) countries, these proportions are as follows: coal/lignite 28.4 %, oil 4.2 %, gas 21.0 %, nuclear 30.2 %, renewable 14.0 %, and others 2.2 % (EEA 2007; Kemp 2007).

Different factors influence the energy consumption of high-speed tilting trains. The most important are design and size of the train influencing the air and tolling resistance, the operating speed, the number of stops, and regenerative braking.

Train/vehicle design

The energy consumption of a given high-speed tilting train is directly proportional to its size. This implies that in order to lower this consumption per train seat (and/or passenger), the size of the train needs to be reduced. However, instead, particular design parameters could be adjusted during designing a new train set. The options are as follows: car body width (e.g., 5 instead of 3–4 seats per row) and bi-level cars (up to 50 % more seats per meter of the train); the mass per unit of the train length has been reduced up to about 2 tons/meter, etc. For example, the average weight across almost all existing high speed tilting trains is just above 2 tons/m. The Class 390 Pendolino is about 6 % below this average and about 30 % above the best in the class—the Shinkansen 700 high-speed tilting train (Henri et al. 1991).

Air and rolling resistance

The air resistance of high-speed tilting trains can be estimated as follows:

$$F_D = (1/2) * \rho * v^2 * A * C_d \tag{2.16a}$$

where

ρ is air density (kg/m^3);

V is operating speed (m/s);

A is reference area (m^2); and

C_d is drag coefficient.

This implies that a lower reference area at a given speed will enable lower energy consumption. However, the intention is to reduce this consumption per seat-km or passenger-km, which requires just the opposite, namely widening the train's reference area. The compromise is found in the above-mentioned design of high speed tilting trains in combination with reducing the drag coefficient from about 1.8–2.0 (ordinary trains) to 0.11 (Shinkansen 300).

Rolling resistance represents the resultant force that must be overcome by the tractive power of the locomotive to move a given train set at a constant speed along a level tangent track in still air. This force includes air resistance, train dynamic forces, bearing resistance, and rolling friction between the wheels and the track. Consequently, it appears obvious that, in order to overcome a larger resultant force, a larger quantity of energy needs to be consumed by high-speed tilting train(s).

Operating speed

The operating speed influences the energy consumption of a high speed tilting train through the energy required for acceleration and increased air resistance. In general, kinetic energy and aerodynamic resistance represent the largest part of a train's energy consumption. For example, for the selected single-deck European high-speed tilting trains, the specific energy consumption depending on the operating speed is estimated as follows:

$$E(v) = 0.00018v; \left(R^2 = 0.787; N = 11\right) \tag{2.16b}$$

where

$E(v)$ is the energy consumption (kWh/seat-km); and

V is the train operating speed (km/h).

In general, a substantial amount of energy is consumed when trains decelerate before passing through the curve segments of the line(s). However, in addition to the tilting mechanisms, the above-mentioned track cants contribute to reducing the need for substantial deceleration in the curves. Thus, the energy consumption of high-speed tilting trains remains dependent mainly on (higher) operating speed(s).

The number of stops

Stops along the route prolong the journey duration of high-speed tilting trains. Additional energy is consumed during each stop due to the train's acceleration after the stop.

Regenerative braking

High sped tilting trains are equipped with technology to convert parts of the kinetic energy back into electric energy during the deceleration and braking phase of the trip. The portion of energy that can be regained mainly depends on the braking rate and the grid. The most recent capacitor technology (ultra-caps) enables storing this energy for subsequent use, providing faster acceleration after stopping. For example, some measurements of Pendolino 390 trains operating on the West Coast Main Line (UK) have indicated that the returned energy over the period of 24 h amounts to 16–18 % of the total energy taken from the grid. This total energy includes all the electricity drawn from the overhead grid including that during train preparation and stops at intermediate stations.

Examples

An example for the energy consumption of high speed tilting trains is the above-mentioned Pendolino 390 (Table 2.11). Measurements have shown that, under different operating conditions along the West Coast Main Line (the UK), the average energy consumption has been $E = 0.040$ kWh/seat-km for a 9-car train set and $E = 0.035$ kWh/seat-km for a 11-car train set, both operated at the maximum speed of $v = 220$ km/h (The average for Pendolino trains operating in Europe is $E = 0.033$ kWh/p-km). Another example is the Swedish X2000 high-speed tilting passenger train in a 5-car configuration with 270 seats and weighting 340 tons, which consumes $E = 0.042$ kWh/seat-km. Its 6-car version, with a capacity of 310 seats and weighing 366 tons, consumes on average $E = 0.0377$ kWh/seat-km. Both trains operate at the speed of 200 km/h (Persson 2007).

Emissions of GHG

Emissions of GHG (Greenhouse Gases) depend on the energy consumption and emissions from the primary sources used to produce the electricity in question. As already mentioned, the composition of these primary sources is region/country-specific. In general, the energy consumption $E(v)$ and emissions of GHG $E_e(v)$ of the high sped tilting trains are interrelated as follows:

$$E_e(v) = e_{em} * E(v) \tag{2.17}$$

where

e_{em} is the emission rate from producing the electricity in a given region (country) ($kgCO_{2e}/kWh$).

For example, if the emission rate of GHG is $e_{em} = 0.455$ $kgCO_{2e}/kWh$ (UK), the emissions of GHG by a Pendolino 390 train operating at the speed of: $v = 200$ km/h, will be, based on Eqs. 2.16b and 2.17, equal to $E_{em}(v) = 0.455 \times 0.036 = 0.0164$ $kgCO_{2e}$/seat-km. By multiplying this amount with the number of seats per train and the running distance, the total emissions of GHG by a given train service can be obtained. Similar estimates of the energy consumption and related emissions of GHG by high-speed tilting trains operating in the other regions (countries) can be made.

Land use

High-speed tilting passenger trains can be considered as "neutral" in terms of using additional land for building new and/or upgrading existing infrastructure. This is mainly due to them operating on upgraded existing/conventional lines. In some cases, the upgrading can require additional land, but usually on a negligible scale as compared to when completely new rail lines are built.

2.3.2.7 Social/Policy Performances

The social/policy performances of high-speed tilting passenger trains relate to their noise, congestion, and traffic incidents/accidents (safety).

Noise

Noise from high-speed tilting trains is an important and sensitive issue, as the rail lines often pass close to or even through densely populated areas, as well as through areas where the ambient noise used to be very low. Recently, this noise was considered as a parameter of the trains' "interoperability" causing their speed to be limited in the TSI (Technical Specification for Interoperability) regulative document. The aim is to limit noise by all passing trains in Europe (EC 2002).

The main source of noise of high-speed tilting trains is rolling noise and aerodynamic noise. At a given speed, the former depends on the quality of wheels, and the number of axis. The latter is manly dependent on the train's aerodynamic characteristics. The train noise is usually expressed in dB(A), i.e., equivalent sound pressure *levels*—$LA_{eq,tp}$.

The lateral distance and height of the point of measurement and the source(s) of noise, i.e., passing train(s), is standardized to 25 and 3 m, respectively. Measurements throughout Europe indicate that the noise by passing conventional, high-speed tilting, and high-speed trains increases in line with their speed according to the "30 log (v)" regression rule, with dominating rolling noise, as follows (Poisson et al. 2008)

$$LA_{eq,tp}(v) = 30.465 \log(v/v_0) + 19.909; (R^2 = 0.935; N = 25) \qquad (2.18a)$$

where

v is the train operating speed (km/h); and

v_0 is the reference train speed ($v_0 = 1$)

The index (tp) in Eq. 2.18a denotes a train passing by the noise measurement location. As mentioned above, the value of 30 is commonly used in the regression equation for predicting the rolling noise of conventional trains. This confirms the fact that the contribution of the rolling noise, which is the main noise source of conventional trains, remains the same (dominating) source of noise of high-speed tilting trains, both comfortably complying with the TSI limits.

Some measurements have indicated that the ETR480 and ETR500 tilting trains generate noise of 90.5 and 88.0 dB(A) while passing at speeds of 250 km/h. This is well below the TSI limits of 92 and 94 dB(A) at speeds of 300 and 320 km/h, respectively (EC 2002; Poisson et al. 2008).

It should be mentioned that the noise from passing high-speed tilting trains experienced by the local population is always lower than that predicted by the above-mentioned regression equation. This is due to noise protecting barriers (walls) near noise sensitive areas as well as the distances of these areas from the rail line(s), which are usually greater than 25 m.

Congestion

High-speed tilting passenger trains do not cause congestion along the lines they exclusively operate. However, on lines with mixed traffic, they can cause congestion and delays of the lower prioritized conventional passenger and freight trains. Furthermore, if these trains cause a road to rail modal shift, they can contribute to mitigating local road congestion. For example, a single tilting train with a seat capacity of 440 seats and load factor of 0.5 can replace about 110 passenger cars occupied on average by two persons.

Safety

Safe operations of high-speed tilting passenger trains is provided through their scheduling and by respecting their operational constraints, particularly those concerning the maximum allowable speed along particular segments of the line(s) (along curves). Scheduling, in addition to satisfying passenger demand and the operators' perspective, also inherently implies specifying the minimum distance between any two trains moving in the same direction. This is the distance at which the running trains must stay apart in any case in order to prevent back collisions in cases of immediate and/or unpredictable braking. This safe distance mainly depends on the breaking characteristics of the trains and their operating speed as follows:

$$d_{\min} = v_{\max}^2/2b + v_{\max}/\Delta \tag{2.18b}$$

where

$v_{\max}$ is the maximum operating speed (m/s);
b is the maximum deceleration rate (m/s^2); and
Δ is the minimum time between activating and staring braking.

For example, for a high-speed tilting train operating at the speed of $v_{\max} = 200$ km/h, with a deceleration rate $b = 0.65$ m/s^2, and the braking system reaction time $\Delta = 5$ s, the minimum breaking distance will be $d_{\min} = 2{,}652$ m.

Respecting the maximum allowable speed(s) along the curves and other parts of the rail line(s) prevents derailments of the high-speed tilting trains and related damaging (incidental/accidental) events. However, except reports and descriptions of individual accidents/incidents, additional aggregate statistical figures on these events and their consequences are currently unavailable. Therefore, at this moment, judgment about the overall safety of these trains can only be made indirectly by considering the aggregate statistics for the passenger railways in the given region. For example, the number of fatalities and injuries in passenger traffic in EU27 Member States has been continuously decreasing over time (the 1997–2008 period) and reached about 3.5 fatalities per billion passenger-km in the year 2009, of which only 1 % was caused by derailments including those of high-speed tilting passenger trains (EC 2009). This suggests that high-speed tilting passenger trains in Europe have been overall very safe.

2.3.3 Evaluation

High-speed tilting passenger trains possess both advantages (Strengths and Opportunities) and disadvantages (Weaknesses and Threats).

Advantages

- Operating as the exclusive HS (High Speed) rail alternative in some regions/countries;
- Operating as an intermediate phase from conventional to the full HS trains;
- Decreasing investments and maintenance costs for building rail infrastructure as compared to that of full HS rails;
- Increasing utilization of the capacity of given rail lines;
- Enabling higher commercial speeds and thus shortening passenger journey time, which in combination with improved riding comfort makes traveling by rail more attractive for existing and prospective users/passengers;
- Stimulating internal and external modal shift; the former from conventional rail services and the latter mainly from individual passenger cars; and
- Mitigating the environmental and social impacts as compared to those of conventional trains after modal shift occurs.

Disadvantages

- Affecting interoperability at border crossings due to the diversity of standards related to the geometry of infrastructure/tracks and train sets, which are mostly country/manufacturer/rail line specific; and
- Exposing users/passengers to the inherent risk of motion sickness during tilting at high speeds through the curves.

Finally, can high-speed tilting passenger trains be considered as an advanced transport system? The answer is "yes," particularly as compared to their conventional passenger train counterparts.

2.4 Advanced Subsonic Commercial Aircraft

1935	The first flight of the piston engine-powered Douglas DC3 aircraft is carried out (U.S.)
1951	The first commercial jet engine-powered aircraft (Comet I) is launched (UK)
1957	The first flight of the first commercially successful jet engine-powered Boeing B707 aircraft is carried out (U.S.)
1969	The first commercial wide-body Boeing B747 aircraft is launched (U.S.)
2012	The Boeing B787-8 aircraft begins commercial operation

2.4.1 Definition, Development, and Use

The main priorities in aircraft design over the past 30 years for both manufacturers and operators—airlines have been improving safety while reducing operating costs. The latter indicates that technology has also been strongly driven by commercial/market driving forces. But, what about the future? Most research to date states that the same priorities will continue in the medium- to long-term future, i.e., for about 10 to 30 years ahead, without significant and revolutionary changes in technology. This implies that changes will be mostly evolutionary with the advancements likely following the current lines of development:

- In the aerodynamic design, reducing drag by about 10 % as compared to the design(s) in the 2001;
- In the operating empty weight, reduction by about 15 % thanks to the increased use of composite materials in aircraft construction (airframe, engines, and the other systems); and
- In propulsion, increasing the overall efficiency by about 8 % through improved thermal efficiency—by increasing the overall pressure ratio and turbine inlet temperature on the one hand, and improving combustion technology on the other; this should result in reducing fuel consumption by about 30 % and the related emissions of NO_x (Nitrogen Oxides) by about 8 % (ICAO CAEP/6 NO_X limits) and further by 11–19 % (the proposed EPA Tier 8 NO_X limits depending on the engine pressure ratio) as compared to aircraft/engine technologies in 2001 (EPA 2011).

Consequently, aircraft manufacturers have undertaken to design commercial aircraft that will be able to reach the above-mentioned targets. In particular, the U.S. aircraft manufacturer Boeing, using the technology previously developed for the Sonic Cruiser aircraft, announced at the end of January 2003 design of the conventional configuration, the B7E7 aircraft, which later became the B787-8. Furthermore, in July 2006, the European aircraft manufacturer Airbus began development of the advanced A350 XWB (Xtra Wide Body) aircraft family as a direct competitor to the above-mentioned Boeing B787-8 and existing B777 aircraft family. Commercial flights of these aircraft with a seating capacity of

270–350 passengers (depending on the version: A350-800, -900, -1000) are expected to begin in 2014. Because the B787-8 aircraft is already in commercial service, it is elaborated in more detail. The A350XWB aircraft is still under development and therefore it is only mentioned for comparison where reasonable.

2.4.2 Analyzing and Modeling Performances

2.4.2.1 Background

The main idea behind the design of B787 aircraft was to emphasize the convenience of smaller mid-size twin-jet compared to the large Airbus A380- and Boeing B747-400/8 aircraft, which could also drive a stronger shift from hub-and-spoke to point-to-point airline air route networks. After being postponed several times, the B787-8 aircraft began commercial operations in October 2012 (Boeing 2012).

The B787-8 aircraft is characterized by its infrastructural, technical/technological, operational, economic, environmental, social, and policy performances.

2.4.2.2 Infrastructural Performances

The infrastructural performances of B787-8 aircraft relate to its airport operations. Some of indicators and measures of these performances are given in Table 2.12. As mentioned above, those for the A350-800 aircraft are given for comparative purposes.

As far as airport operations are concerned, the Boeing 787-8 is categorized as a medium-sized twin-engine long-range aircraft. Respecting the wing span, the aircraft belongs to the E group according to the ICAO (International Civil Aviation Organization) and the V group according to the US FAA (Federal Aviation Administration) classification. Respecting the overall length, the aircraft belongs to the ICAO's RFF category 8 and FAA's ARFF Index D (Boeing 2012a). In addition, with its final approach speed of about 140 kts (kts-knots), the B787-8 belongs to the FAA category IV. The A350-800 aircraft is categorized similarly (Airbus 2012; Boeing 2012a; Horonjeff and Mckelvey 1994).

Since the B787-8 aircraft is designed to perform nonstop flights of and longer than 9,000 nm (nm—nautical mile: 1 nm = 1.852 km), the take-off runway need to be no shorter than 3,400 m (11,000 ft) (ft—feet: 1 ft = 0.305 m).

Regarding airport maneuverability, the B787-8 is more advanced than its closest counterparts, for example, the B767-300ER aircraft. Namely, the geometry of maneuvering an aircraft at the airport is characterized by its turning radii, which are a function of the nose steering angle. In principle, the larger the steering angle, the smaller the radii, and consequently the greater the maneuverability. From the standpoint of maneuvering close to buildings and other aircraft, the largest turning

Table 2.12 Infrastructural performances of the B787-8 and A350-800 aircraft—airport operations (Boeing 2012, Airbus 2012)

Specification	Aircraft type	
	B787-8	A350-800[d]
Length (m)	56.7	60.6
Wingspan (m)	60.1	64.0
Wing area (m^2)	325	443
Wing sweepback (°)	32.2	31.9
Height (m)	16.9	17.0
Fuselage constant diameter (m)	5.75	5.32
Maximum Take-Off Weight (MTOW) (t)	219.5	248
Maximum Landing Weight (MLW) (t)	168	193
Maximum payload (t)	44.5	35.7
Take-off field length (m)[a]	3,100	751
Landing field length (m)[b]	1,600–1,800	1,525
Maximum pavement width (m)	42	–
Effective steering angle (°)	65	–
Aircraft Classification Number (ACN)[c]	57–101 (flexible pavement) 57–91 (rigid pavement)	65–105 (flexible pavement) 59–93 (rigid pavement)

[a] *MTOW* Maximum Take-off Weight, Sea level pressure altitude, standard day +15 °C temperature, dry runway; [b] *MLW* Maximum Landing Weight, sea level pressure altitude, dry runway; m meter; t ton; [c] ACN is the ratio between the pavement thickness required for a given aircraft and that required for the standard aircraft single wheel load; [d] Preliminary data

radius is the most critical (Horonjeff and McKelvey 1994). The minimum turning radius corresponds to the maximum nose steering angle, which is, for example, 65° for the B787-8 aircraft. This enables it to turn on a path (runway, taxiway, apron) 42 m wide, which is about 2 m narrower than that of the B767-300ER aircraft with the maximum steering angle of 61° (44 m). Consequently, the radius of the taxiway centerline for B787-8 is 32.9 m as compared to that of B767-300ER of 33.8 m (AT 20).

In addition, the size of parking stands at the apron gate complex depends mainly on the aircraft's overall size, the required buffer space between the aircraft and the permanent fixtures (buildings), temporarily static objects (other aircraft and traffic at the airport), and the type of parking scheme (nose-in, parallel, angled) (Horonjeff and McKelvey 1994). Respecting its dimensions in Table 2.12, the size of parking stand of a nose in parked B787-8 aircraft is approximately 5,385 m^2 (this includes the aircraft footprint and a buffer space of 7.5 m around it). For comparison, the size of the parking stand of nose in a parked B767-300ER aircraft is approximately 4,376 m^2, which is about 23 % lower than that of the B787-8 aircraft whose scheme of airport/ground servicing is shown in Fig. 2.14.

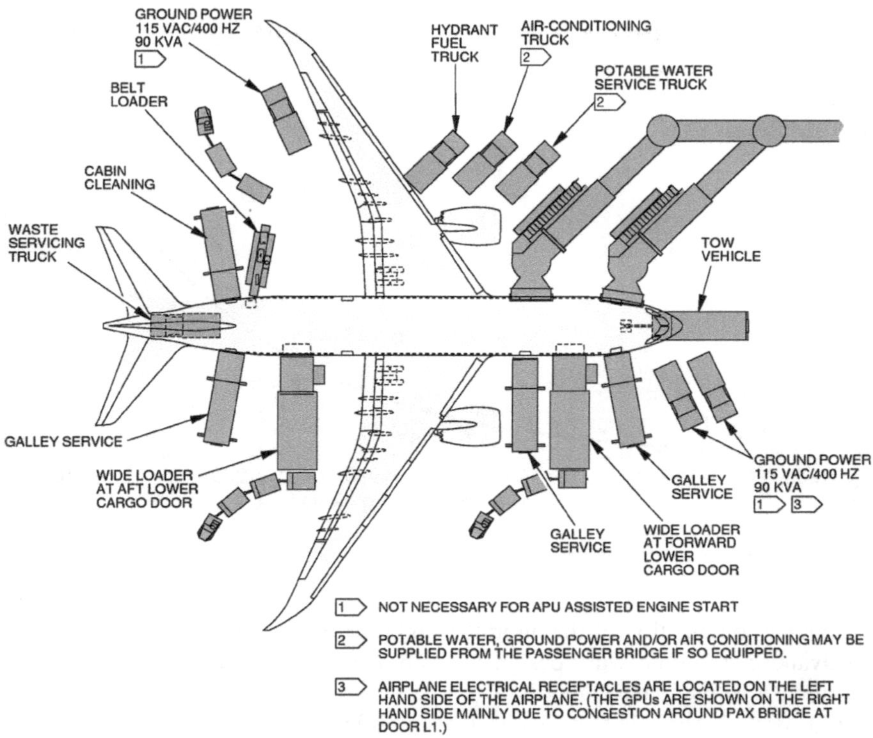

Fig. 2.14 Scheme of ground servicing of the B787-8 aircraft—regular conditions when *APU* Auxiliary Power Unit is used (Boeing 2012a)

A can be seen, the facilities and equipment for the airport/ground servicing of B787-8 are similar to those of B777 aircraft. The maximum turnaround time of the aircraft at the apron gate stand is specified to be 41 min and the through time 28 min (Boeing 2012a).

2.4.2.3 Technical/Technological Performances

The technical/technological performances of B787-8 aircraft relate to its aerodynamic design and materials used for its construction, engines, and aircraft systems.

Aerodynamics and materials

Aerodynamic

The aerodynamic design of the Boeing 787-8 aircraft is "advanced" due to the following reasons:

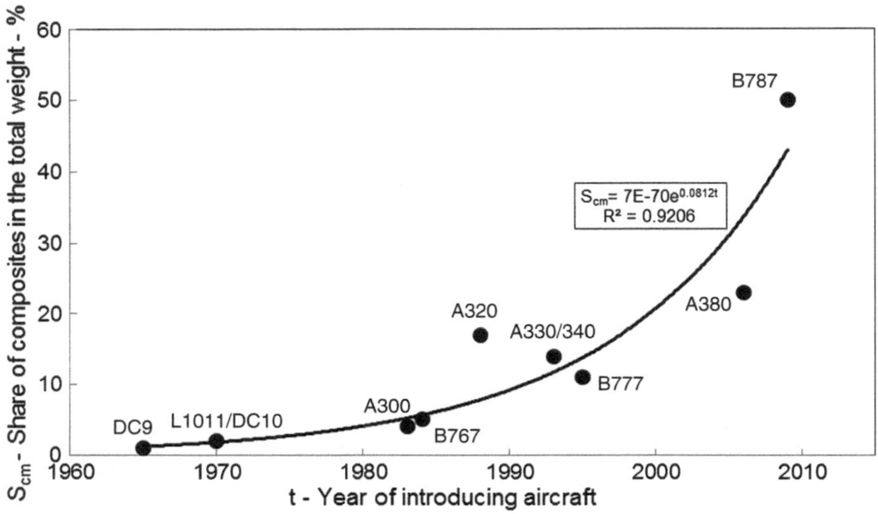

Fig. 2.15 Share of composite materials in the commercial aircraft over time (GAO 2011)

- Optimal combination of the aircraft weight, drag, and engine performance;
- Advanced transonic wing design for improved speed and lift (raked wingtips optimal for long-haul flights);
- High performance, but mechanically simplified high lift system of high reliability and reduced maintenance costs;
- Tightly integrated packaging of systems to reduce the size of aerodynamic fairings for reduced weight and drag; and
- Final nose configuration (four windows, fewer posts, pilot vision similar as in a B777 aircraft, non-opening windows, crew escape door, vertically stowed wipers, etc.).

Materials

The structures of commercial aircraft have been continuously upgraded with an increased use of composite materials as shown in Fig. 2.15.

As can be seen, the share of composites in the total weight is the greatest in the latest B787 aircraft. In addition, Table 2.13 shows shares of different materials in the weight of Boing 787-8 and its current and forthcoming counterparts.

The share of composite materials in the B787-8 is the second greatest—after that of the forthcoming A350-800 aircraft. In terms of volume, composite materials account for about 80 % of the B787/8 aircraft total. One of the main reasons to use substantial amounts of these materials is to reduce the aircraft weight while retaining the required strength of the construction. In turn, this improves the aircraft efficiency primarily in terms of reduced fuel consumption and simpler maintenance, thus reducing both corresponding costs. Specifically, in the B787-8 aircraft, composites (carbon laminate, carbon sandwich, and fiberglass) are mainly

Table 2.13 Share of particular materials in the weight of selected aircraft (%) (Boeing 2012; Airbus 2012)

Material	Aircraft type			
	B777	A330-200	B787-8	A350-800[c]
Composites[a]	11	10	50	53
Steel	11	19	10	10
Titanium	7	8	15	14
Aluminum/Al-Li[b]	70	58	20	19
Others	1	2	5	8

[a] Carbon laminate, Carbon sandwich, fiberglass (Fiberglass = Carbon fiber); [b] Alloys of Aluminum and Lithium: [c] Preliminary data

used for the largest part of the fuselage, wings, horizontal, and vertical part of the tail. Aluminum is used on the wing and tail leading edges. Titanium is used mainly on the engines and fasteners, while steel is used in various places.

In order to illustrate the possible effects of substituting different materials in aircraft construction, let p_i be the current share of the material (*i*) with a specific gravity of g_i (g/cm^3) to be partially substituted by material (*j*) with the current share p_j and a specific gravity of g_j (g/cm^3). The proportion of material (*i*) to be substituted by material (*j*) is assumed to be q_{ji}. Consequently, the relative change of aircraft weight due to changing the shares of these two materials on account of each other can be estimated as follows:

$$\Delta w_{ji} = 1 - \left[(p_i - q_{ji})g_i + (p_j + q_{ji})g_j\right] / (p_i g_i + p_j g_j) \tag{2.19a}$$

The value Δw_{ji} in Eq. 2.19a can take positive and negative values. The former implies a decrease and the latter an increase in the aircraft weight in relative terms by the given substitution of materials. For example, if the current share of composite materials of $p_j = 50$ % with a specific gravity of $g_j = 2.1$ g/cm^3 was further increased in the construction of B787 by about $q_{ji} = 5$ % on the account of aluminum with a current share of $p_i = 20$ % and a specific gravity of $g_i = 2.7$ g/cm^3, the aircraft weight would be reduced by about 4.5 %. If steel (specific gravity $g_i = 7.83$ g/cm^3), with a current share of 10 % was reduced by about 5 % on account of composites whose share was increased to 55 %, the aircraft weight would be further reduced by about 16 %.

Engines

General

The most important performances of contemporary turbofan jet engines are thrust, fuel efficiency, and SFC (Specific Fuel Consumption).

- *Thrust (T)* is generally derived from the change in momentum of the air through the engine and the thrust that occurs due to the static pressure ratio across the final (exhaust) nozzle. Analytically, it can be expressed as (Jenkinson et al. 1999):

$$T = m(v_1 - v_0)/g + (p - p_0)/A \tag{2.19b}$$

where

m is the air flow through the engine (kg/s);
v_1 is the velocity of exhaust jet (m/s);
v_0 is the velocity of air entering the engine (m/s);
g is gravitational acceleration (m/s^2);
p, p_0 is the pressure at the intake and the exhaust station, respectively, (N/m); and
A is the nozzle cross sectional area (m^2).

The thrust T in Eq. 2.19b is usually expressed in kN (kiloNewton) (SI units) or Libras (lb) (British units).

- *Efficiency (η_e)* directly expresses the rationale of the engine fuel consumption, i.e, higher efficiency implies lower fuel consumption per unit of the engine thrust, and vice versa.
- *SFC (Specific Fuel Consumption)* expresses the ratio of fuel burned per hour per ton of net thrust (Janic 2007). It is expressed in kg of fuel per kg of thrust/h. The SPC and engine efficiency η_e are interrelated as follows:

$$SFC = M/4 * \eta_e \tag{2.19c}$$

where

M is the Mach number.

Engines with higher bypass ratios usually have lower SPC. For most contemporary aircraft turbofan jet engines, this amounts about 0.25–0.30 kg of fuel/kg of thrust/h. In addition, SFC relates to the jet engine bypass ratio (BR). The nature of this relationship is illustrated using the data for the cruising phase of the flight of 20 different engines produced by different airspace manufacturers. The regression relationship in which the bypass ratio BR is considered as the independent and SFC as the dependent variable is as follows (Janic 2007):

$$SFC = 0.3435BR^{-0.1624}; (R^2 = 0.425; N = 20) \tag{2.19d}$$

The B787-8 aircraft is powered either by RR (Rolls Royce) TRENT 1000 or by General Electric GEnx1B engines.

RR (Rolls Royce) TRENT 1000 engine

The RR (Rolls Royce) TRENT 1000 engine is considered an advanced turbofan jet engine as compared to its counterparts due to the following features: no-engine-bleed systems, higher bypass ratio and higher pressure ratio compressor, high-flow slower-speed fan, advanced materials and coatings, architecture, low-noise nacelles with chevrons, and interchangeability at wing/pylon interface. Table 2.14 gives some important performances of the RR TRENT 1000 engine.

Table 2.14 Characteristics of the RR (Rolls-Royce) Trent 1000 aircraft engine (RR 2011)

Parameter	Value
Type	Three-shaft high bypass ratio turbofan engine
Dimension/weight	
Length (m)	4.769
Fan diameter (m)	2.85
Dry weight (tons	6.018
Performances	
Maximum thrust (kN)	307–330
OPR (Overall Pressure Ratio)	33
SFC (Specific Fuel Consumption)[a] (kg-fuel/kg-thrust/h)	0.224
BR (Bypass Ratio)	10.0–11.0
Thrust-to-weight ratio (kN/ton)	51.01–54.84

[a] Based on the performance of the RR Trent 800 engine (reduction for about 15 %); t ton

The no-engine-bleed systems of the RR Trent 1000 engine and its tiled engine combustor suit the need for more electricity in Boeing 787-8 s, and thus enable reduction in the overall aircraft weight, and consequently fuel consumption. In addition, the RR Trent 1000 engine has a bypass ratio of about 10.0–11.0, which gives an average efficiency rate of about $\eta_e = 0.908$ during the cruising phase of flight at the speed of: $M = 0.85$. For example, for the BR (Bypass Ratio) = 11, the SFC (Specific Fuel Consumption) will be SPC = 0. 224 kg-fuel/kg-thrust/h (cruise).

Aircraft systems

The systems of the B787-8 aircraft include: (i) Efficiency Systems; (ii) Highly Integrated Avionics, and (iii) e-Enabled Airplane Systems (Nelson 2005; Boeing 2012a).

Efficiency Systems

Efficiency systems aim at generating, distributing, and consuming energy efficiently and effectively. They include the following subcomponents: Advanced Energy Management (The More Electric Aircraft), and Flight Controls (Variable Camber Trailing Edge and Drooped Spoilers).

Highly Integrated Avionics

Highly integrated avionics systems enable the efficient and effective navigation during flight. They include the following subcomponents: Common Core Systems open architecture, Integrated Flight Controls Electronics, Integrated Communication/Navigation/Surveillance equipment, and Integrated Aircraft Systems Control.

Table 2.15 Operational performances of the B787-8/9 and A350-800 aircraft (Airbus 2012; Boeing 2012a)

Indicator/measure	Aircraft type		
	787-8	787-9	A350-800[a]
Cockpit crew	Two	Two	
Seating capacity (seats)	242 (3-class) 264 (2-class)	250–290 (3-class) 280 (2-class)	
MTOW (tons)	219.5	251	248
MLW (tons)	168	193	190
MZFW (tons)	156	181	181
OEW (tons)	110	115	105
Cruising speed (Mach/kts)	0.85/490 at 3,5000 ft/10.700 m		0.85/490 40,000/12,190
Maximum speed (Mach/kts)	0.89/515 knots at 35,000 ft/10,700 m)		0.89/515 40,000/12,190
Service ceiling (ft/m)	43,000 /13,100		–
Range, fully loaded (nm/km)	5,550/10,280	6,500/12,036	5,000/9,375
Maximum fuel capacity (000 l)	126	139	129
Engines (×2) Type	RR Trent 1000	RR Trent 1000	Trent XWB
Thrust (kN)	2 × 307−330	2 × 330	2 × 351

MTOW Maximum Take of Weight; *MLW* Maximum Landing Weight; *MZFW* Maximum Zero Fuel Weight; *OEW* Operating Empty Weight; *l* liter; *ft* foot; *kts* knots; *kN* Kilo Newton; *RR* Rolls Royce

[a] Preliminary data

e-Enabled Airplane Systems

e-enabled aircraft systems provide the flight crew with wireless communication both inside and outside the aircraft. They include broadband connectivity within the aircraft and with the ground (Flight Deck, Crew Information System, Onboard Health Maintenance, and Cabin systems).

2.4.2.4 Operational Performances

The main operational performances of the B787-8 aircraft include its payload-range characteristics and technical productivity. They are influenced by the aircraft relevant parameters/indicators given in Table 2.15.

Payload-range characteristics

The payload-range characteristics of the Boeing 787-8 aircraft can be analytically expressed as follows:

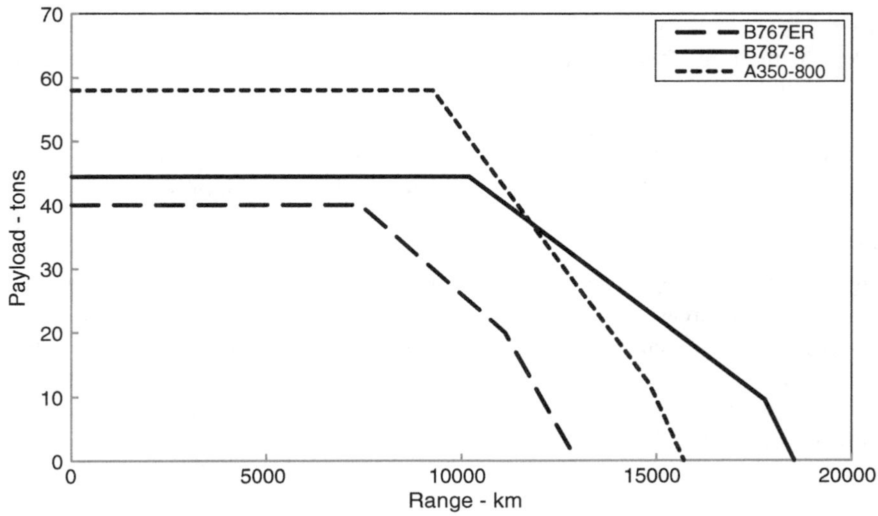

Fig. 2.16 Payload-range characteristics of the selected commercial aircraft (Airbus 2012; Boeing 2012a; http://nl.wikipedia.org/wiki/Airbus_A350)

$$PL(R) = \begin{bmatrix} 44.5, & 0 < R < = 5550 \\ 44.5 - 0.00864 * (R - 5550), & 5550 < R < = 9600 \\ 9.5 - 0.023 * (R - 9600), & 9600 < R < = 10000 \end{bmatrix} \quad (2.20a)$$

where

R is the range (nm); and

$PL(R)$ is the payload (tons).

In addition, the payload-range characteristics for the B787-8, B767-300ER, and A350-800 aircraft are shown in Fig. 2.16.

Evidently, the B787-8 appears superior to its B767-300ER counterpart as it is able to operate along longer nonstop distances with greater payload. The forthcoming A350-800 will at least be comparable to the B787-8 aircraft by carrying seemingly higher payloads on shorter nonstop distances.

Technical productivity

The technical productivity of commercial aircraft including the B787-8 can be estimated as the product of their operational/cruising speed and payload. As mentioned above, the maximum payload of each aircraft changes and depends on the range. Consequently, technical productivity can be calculated as follows:

$$TP[v(R), PL(R)] = v(R) * PL(R) \quad (2.20b)$$

where

$v(R)$ is the aircraft operating/cruising speed depending on the range R (km/h or kts (nm/h)).

The other symbols are analogous to those in previous equations. Thus, for example, if the B787-8 aircraft performs a flight of the length of $R = 9{,}280$ km at an average cruising speed of $v(R) = 900$ km/h, its technical productivity will be $TP\ [v(R), PL(R)] = 44.5$ tons $\times$ 900 km/h $=$ 40,050 ton-km/h. In case of longer flights carried out approximately at the same speed, this productivity will decrease in line with decreasing of the payload carried. It would not be reasonable to investigate the influence of the operating/cruising speed on much shorter flights with the maximum payload since this aircraft is just designed to operate long-haul flights within the airline point-to-point network. The similar seems to apply to the forthcoming A350-800 aircraft and the rest of its family.

2.4.2.5 Economic Performances

The main economic performances of the B787-8 aircraft are its costs and revenues, which can also be said for its conventional counterparts.

Costs

Aircraft costs are roughly divided into operating and non-operating cost. In particular, operating cost can be divided into DOC (Direct Operating Cost) and IOC (Indirect Operating Cost) (Janic 2007). The former consist of the costs of aircraft depreciation, insurance, maintenance and overhauling (airframe, engines and avionics), and the cost of flight operations (crew, fuel/oil, airport, and navigation charges). The latter roughly include the costs of aircraft and traffic servicing, promotion and sale, passenger services, general and administrative overheads, and maintenance and depreciation of the ground property and equipment. In general, both DOC and IOC have shown to increase in line with the aircraft size (i.e., seating capacity) and stage length.

The aircraft DOC are usually expressed in average monetary units per flight or per unit of flight output (US$ or € per ASK (Available Seat-Kilometer) or PKM (Passenger Kilometer) passenger-km). The ATA (Air Transport Association) of American method with the necessary modifications of the values of inputs is still relevant for estimating and comparing aircraft DOC, particularly of aircraft that are just at the beginning of their full commercialization such as the B787-8. An example of application of this method is shown in Fig. 2.17 (AC 2005; Janic 2007).

As can be seen, the average cost per ASK (Available Seat-Kilometer) decreases more than proportionally as the stage length increases, thus indicating economies of distance in the case of the selected aircraft. At the same time, the average costs of the B787 s aircraft are by about 3–11 and 8–11.5 % lower than those of the

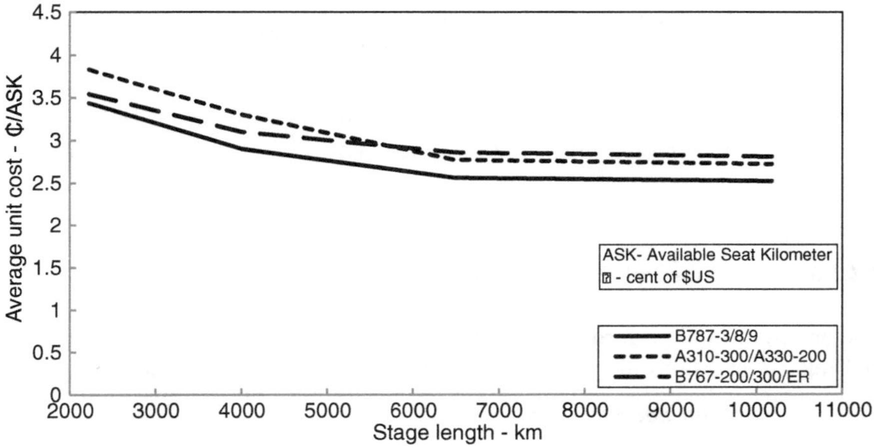

Fig. 2.17 Relationships between the average operating cost and stage length of the selected commercial aircraft (AC 2005)

B767-200/300/ER and A310/330-200 aircraft, respectively (The cost components include aircraft price and interests on bonds, i.e., finance charges, navigational and landing charges, flight attendant and crew costs, and maintenance and fuel cost). The beginning of commercial operations of the B787-8 aircraft has confirmed the above-mentioned expectations mainly thanks to the lower maintenance airframe and engine costs and lower fuel consumption of increasingly expensive fuel. These last two cost components will likely continue to influence the short-term variations of the DOC. For example, let's assume that the share of fuel costs in the total operating cost of a given long-haul flight is about 30 %. Since the B787-8 aircraft is supposed to consume about 20 % less fuel than its counterparts (B767-300ER, A310-300/A330-200/300), the share of its fuel costs in DOC will decrease from 30 % to about 24 %. If the other cost components and their influencing factors remain unchanged, this will decrease the total cost per flight by about 6 %. The savings in DOC from operating the B787-8 aircraft will increase, for example, by about 10 % if the share of fuel costs mainly caused by increasing of the fuel prices rises (for example, from 30 to 50 %).

Revenues

As with other aircraft, revenues from operating B787 aircraft are obtained by charging passengers on the given routes. For example, the average price/airfare covering the cost of a flight in time T can be estimated as follows:

$$p_{ij}(T) \geq \frac{c_{ij}(N_{ij}, d_{ij})}{\lambda_{ij} N_{ij}} \tag{2.21}$$

where

$c_{ij}(N_{ij}, d_{ij})$ is the cost per flight carried out by an aircraft of the seat capacity N_{ij} on the route of length d_{ij} in time T; and

λ_{ij} is the average load factor of a flight on the route d_{ij} carried out by an aircraft of the seat capacity N_{ij} in time T ($0 < \lambda_{ij} \leq 1$).

As Eq. 2.21 indicates, the average price/airfare depends on the cost and the average load factor of a given flight carried out by any aircraft, including the B787-8 aircraft.

2.4.2.6 Environmental Performances

The environmental performances of the B787-8 aircraft include fuel consumption and related emissions of GHG (Greenhouse Gases), and land use/take.

Fuel consumption and emissions of GHG

Multiplying the SFC with the thrust per engine and the number of engines per aircraft can give an estimation of the aircraft total fuel consumption per unit of time (tons/h). In addition, fuel consumption per seat-kilometer can be relevant for comparison of different aircraft with respect to fuel consumption/efficiency. Some estimates given in Fig. 2.18 indicate that the B787-8 aircraft could be more fuel efficient by about 8, 9, and 18 % than its B767-200ER, B777-200ER, and 777-300ER counterparts, respectively.

The most recent figures obtained by ANA (Air Nippon Airlines, Japan) show that the fuel savings by operating B787-8 powered by RR Trent 1000 engines on short-haul routes are 15–20 % and up to 21 % on long-haul (international) routes as compared to the B767-200/300ER aircraft. Some additional savings of up to about 3 % have been reported by JAL (Japan Airlines) using the B787-8 aircraft powered by GEnx 1B engines (AW 2012). In addition, Airbus expects the fuel consumption of the A350-800 to be about 6 % lower than that of B787-8 aircraft as shown in Fig. 2.18.

The related emissions of GHG (Greenhouse Gases) from burning JP1 or JP-A fuel are CO_2 (Carbon), H_2O (Water vapor), NO (Nitric Oxide) and NO_2 (Nitrogen Dioxide) (together called NO_x (Nitric Oxides)), SO_x (Sulfur Oxides), and smoke. The emission rates of CO_2, H_2O, and SO_2 are relatively constant—3.18 and 1.23 kg/kg, and up to 0.84 g/kg of fuel, respectively (Janic 2007). The emission rate of NO_x (Nitrogen Oxides) changes (increases) in line with increasing of the OPR (Overall Pressure Ratio[1]) of the given turbofan jet engine. The above-mentioned higher fuel efficiency thanks to the higher engine BR (Bypass Ratio) (11.0) and better combustion makes the related emissions of GHG (particularly CO_2 and NO_x) by the B787-8 aircraft proportionally lower than those of its closest counterparts.

[1] The OPR (Overall Pressure Ratio) is defined as the ratio of the total pressure at the compressor discharge and the pressure at the compressor entry (Hunecke 1997; Janic 2007).

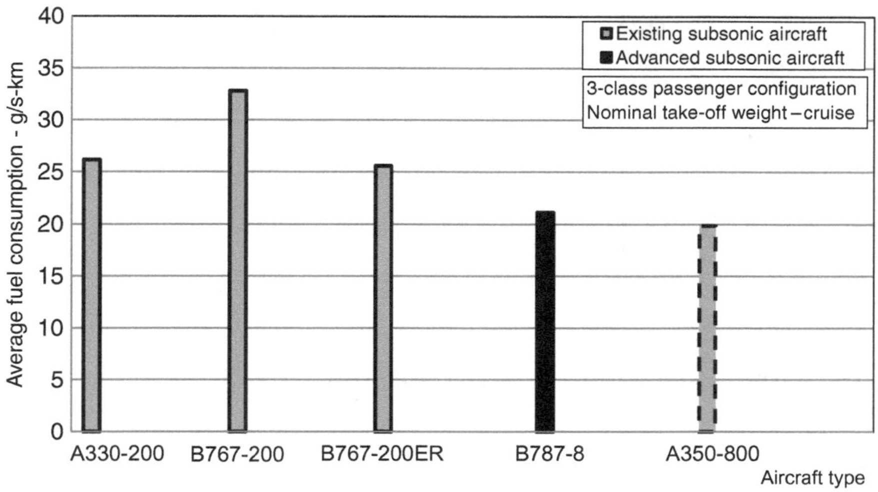

Fig. 2.18 Fuel efficiency of the selected commercial aircraft (http://www.lissys.demon.co.uk/boeing787.html)

Land use

Thanks to its advanced maneuverability, accommodating the B787-8 aircraft at certified airports does not require use of additional land. Eventual modifications of taxiways and apron-gate parking stands are to be carried out within the airport area, i.e., over the already taken land. Thus, the B787-8 aircraft can be considered as a land use/take "neutral" aircraft.

2.4.2.7 Social/Policy Performances

The social/policy performances of the B787-8 aircraft relate to noise, congestion, and traffic incidents/accidents (safety).

Noise

Noise primarily comes from the aircraft engines while flying near the ground, i.e., during approach and landing, flyover, and take-off. The noise spreads in front of and behind the aircraft engine(s). The front noise-spreading generators are the engine(s) compressor and fan. The back noise-spreading generators are the turbine, fan, and jet efflux. The aircraft are certified for noise at various noise certification locations around the airport. The maximum noise at these locations must not exceed 108 EPNLdB (this noise is equivalent to about 96 dB (A) measured by A-noise weighted scale) (Huenecke 1997; Horonjeff and McKelvey 1994;

Table 2.16 Social performances of the selected commercial Aircraft—noise (Cohen-Nir 2010; EASA 2011, 2012)

Aircraft type		Noise level (EPNLdB)[c]		
	TOW[a]	Lateral	Flyover	Approach
B767-200	144	95.7	91.5	102.1
B767-200ER	168	97.8	91.1	98.6
B767-300	158	96.0	91.3	98.5
B767-30ER	180	95.7	91.5	99.7
A330-200	230	97.0	94.4	96.8
A330-300	217	97.6	91.6	98.9
B787-8	220	90.5	83.0	96.2
A350-800[b]	259	89.0	83.0	95.0

[a] Typical *TOW* Take-Off-Weight; [b] Preliminary data; [c] *EPNLdB* Effective Perceived Noise Level in decibels (typical engines)

Janic 2007). Table 2.16 gives the noise characteristics of selected aircraft at the noise certification locations.

Evidently, the B787-8 aircraft generates about 5–7, 7 and 0.6–6 dB lower certificated noise than its counterparts while taking off, flying over, and approaching, respectively (EASA 2011, 2012). In addition, making a broader judgment concerning mitigation of noise by B787-8 can be made by assuming it replaces the B767-200/200ER aircraft. This implies the gradual increase in the number of replacing aircraft (B787-8) on the account of gradually replaced aircraft (B767-200/200ER). The total number of B767-200/200ER aircraft to be replaced is assumed to be 800 (based on the current orders of B787-8). Regarding operating long-haul flights, each aircraft of both fleets is assumed to perform the same number of flights (2/day). The noise level is considered to be the certified EPNLdB (the case in Europe) (EASA 2011). The total noise exposure by take-off or approach/landing operating fleet of both aircraft during the day can be estimated as follows (Smith 2004):

$$\overline{EPNLdB} = \frac{1}{n}\sum_{i=1}^{2} n_i[EPNLdB_i + 10\log_{10}(n_i * \theta_i)] \tag{2.22}$$

where

$EPNLdB_i$ is the Effective Perceived Noise Level of the aircraft in decibels (i);
n_i is the number of aircraft (i) in operation (per day);
θ_i is the average number of flights per day of aircraft (i); and
n is the total number of aircraft in operation (per day).

The example is shown in Fig. 2.19.

As can be seen, the total potential noise exposure by the entire replacing and replaced fleet decreases more than proportionally with increasing of the proportion of advanced B787-8 aircraft. By full replacement, this exposure could be reduced by about 3 dB and 6 dB during approach and take-off, respectively. Consequently,

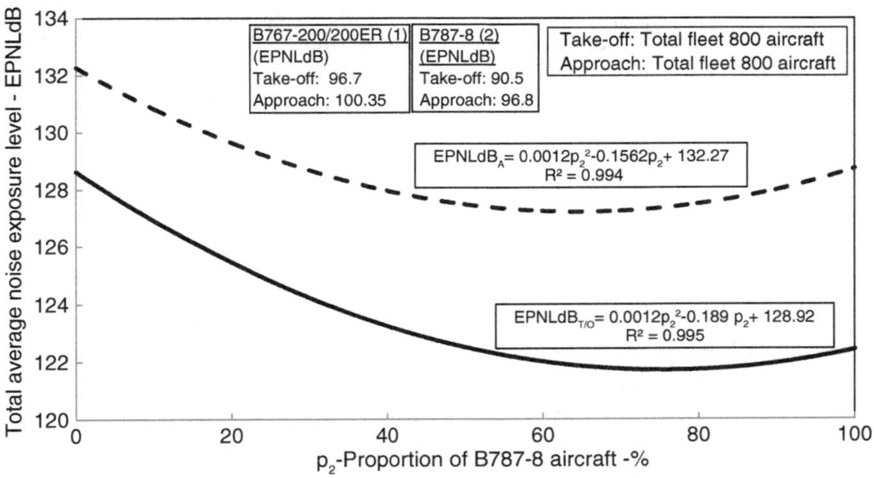

Fig. 2.19 Relationship between the average noise exposure and the proportion of B787-8 aircraft in the fleet

the "noise contour" or "noise footprint" as the area of constant noise generated by B787-8 aircraft around an airport can be by about 60 % smaller than those of its counterparts, thus ensuring that noise above the level of 85 dB certainly does not spread outside the airport boundaries. This is achieved mainly thanks to the improved aerodynamics design on the one hand, and the lower fan speed and low jet velocity of the RR Trent 1000 engines, on the other. The forthcoming A350-800 aircraft is expected to be even quieter (Cohen-Nir 2010).

Congestion

The B787-8 aircraft is categorized as a heavy aircraft. This implies that as being the leading aircraft in the landing sequence, it needs to be separated from A380, other Heavy, Upper and Lower Medium, Small, and Light aircraft by 4, 4, 5, 6, and 7 nm, respectively. Currently, the separation amounts to 10 nm (nautical mile; 1 nm = 1.875 km). As the leader in the take-off sequence, the B787-8 aircraft needs to be separated from all other aircraft by 3 min (FAA 2011). Consequently, like the increased proportion of other heavy aircraft in the airport arrival and departing streams, an increased proportion of B787-8 aircraft can contribute to increased overall separation interval(s), which reduces the runway landing and take-off capacity almost proportionally. Consequently, at saturated capacity-constrained airports, the overall congestion and aircraft delays may increase.

Safety

The B787-8 is an advanced aircraft designed to operate absolutely safely implying its immunity to incidents and accidents due to all previously known reasons. The forthcoming more intensive operations will certainly confirm such expectations since the aircraft is also designed in light of the overall objectives to make the present and future air transport system safer despite its continuous growth. For example, in the case of an engine failure during the cruising phase of flight, the ETOPS (Extended Range Twin-Engine Operational Performances) capabilities enable B787-8 aircraft to stay in the air for up to 180 min, which is, like its counterparts, sufficient to reach the closest airport and land safely.

2.4.3 Evaluation

The B787-8 aircraft possesses advantages (Strengths and Opportunities) and disadvantages (Weaknesses and Threats) as compared to its closest counterparts—the Boeing B767-300ER and the Airbus A330-200/300.

Advantages

- Advanced aerodynamic design;
- Substantial use of lighter but very strong composites (about 50 and 80 % of the weight and volume, respectively);
- Superior technical productivity over the entire range;
- Advanced navigational systems onboard being a part of the forthcoming NextGen and SESAR research and development initiatives in the USA and Europe, respectively;
- Increased efficiency in terms of fuel consumption and related emissions of GHG (Greenhouse Gases), but still not sufficiently convincible to be in line with the widely advertised 20 % decrease; and
- Seemingly convincible reduction of noise at source—aircraft engines—but again slightly lower than advertised.

Disadvantages

- Relatively high price;
- Rather modest reduction of direct operating costs due to improvements in the fuel efficiency, i.e., lower fuel consumption and share of fuel costs in the total operating costs;
- Contribution to increasing congestion and delays at capacity constrained airports because of reducing the airport runway capacity affected by increased overall separation within the arrival and departure stream(s); and
- An inherent uncertainty in the technical and operational reliability of the innovative technologies and particularly electrical systems and composites during the aircraft life-cycle of about 25–30 years.

Finally, the B-787-8 aircraft is certainly an advanced subsonic commercial aircraft. At present, most of its infrastructural and technical/technological performances are known. However, its remaining operational, economic, environmental, and social/policy performances and consequently further advantages and disadvantages will only be able to be analyzed in more detail after more intensive aircraft use.

References

AC. (2005). Can the 787 & A350 transform the economics of long-haul services?. *Aircraft Commerce*, No. 39 (February/March), pp. 23–30.

AEIF. (2002). *Trans-european high-speed rail system, technical specification for interoperability (TSI), infrastructure*. European association for Railway interoperability, international union of Railways. Paris, France.

Airbus. (2012). *A350XWB sharping efficiency*. Blagnac Cedex: Airbus Company.

AUMA. (2007). *Towards a light rail transit program: Analysis and recommendations*. Alberta, California: Alberta Urban Municipalities Association.

AW. (2012). ANA, JAL boeing 787 fuel burn performance beating expectations. *Aviation week & space technology*, AW&ST.com, (June), pp. 20–22.

Boeing. (2012a). *787 Airplane characteristics for airport planning*. Seattle, Washington: Boeing Commercial Airplanes.

Boeing. (2012). *Boeing 787 dreamliner being designed for environmental performance*. Seattle, Washington: Boeing Commercial Airplanes.

CE. (2008). *Vehicle technology review: City of Edmonton, LRT Projects*. Retrieved from www.edmonton.ca/LRTProjects

Cebrián, P. (2008). CNG vehicles for urban transport: An available solution for cleaner air in cities. *BESTUFS (Best Urban Freight Solution) II-8th Workshop*, IVECO trucks and commercial vehicles, March, Madrid, Spain.

CEN. (2002). Track alignment design parameters—track gauges 1435 and wider—part 1: Plain line. *Railway Applications*. EN13803-1, Comité Européen de Normalisation, Brussels, Belgium.

CEN. (2006). Track alignment design parameters—track gauges 1435 and wider—part 1: Plain line. *Railway Applications*. Enquire Version EN13803-1, Comité Européen de Normalisation, Brussels, Belgium.

Cohen-Nir, D. (2010). *Airbus Program Update, Presentation*. Airbus America Inc., Airbus SAS. Herndon, Virginia, USA.

De Rus, C. J. G. (Ed.). (2009). *Economic analysis of high speed rail in Europe*. Bilbao: BBVA Foundation.

EASA. (2011). *Type certificate data-sheet for noise no. EASA. A004 for Airbus A330, EASA IM.A. 115 for boeing 787-8, B767*. Koln, Germany: European Aviation Safety Agency.

EASA. (2012). *Type certificate data sheet data sheet for noise no. EASA.IM.A.115 for boeing 787-8*. Koln, Germany: European Aviation Safety Agency.

EC. (2002). Technical specification for interoperability relating to the rolling stock subsystem of the trans european high-speed rail system, 30/05/2002. *Official Journal of the European Commissions*, No. 12 (September).

EC. (2009). *EU energy and transport in figures*. Statistical pocket book, directorate general for energy and transport, Brussels, Belgium.

EC. (2010). *Signal*. The newsletter of the european rail traffic management system ERTMS, No. 21. European Commission, Brussels, Belgium.

EEA. (2007). *The emission factors: Technical annex*. Copenhagen, Denmark: European Environmental Agency.

EPA. (2011). *Control of air pollution from aircraft and aircraft engines; proposed emission standards and test procedures*, Part II, 40 CER Part 80 and 1068. Environmental Protection Agency, Federal Register (Vol. 16, No. 144), Washington D.C., USA.

ERRAC. (2004). *Light rail and metro systems in Europe: Current market perspectives and research implications*. Brussels, Belgium: The European Rail Research Advisory Council.

FAA. (2011). *Interim procedures for boeing 747-800 and boeing 787 flights*. Notice N JO 7110.542, Federal Aviation Administration, US Department of Transportation, Washington, D.C., USA.

GAO. (2001). *Mass transit: Bus rapid transit shows promise*. GAO-01-984, Report to Congressional Requesters, United States Government Accountancy Office, Washington, D.C., USA.

GAO. (2011). *Aviation safety: Status of FAA's actions to oversee the safety of composite airplanes*. Report to Congressional Requesters, Government Accountability Office, GAO-11-849, Washington, D.C., USA.

GAO. (2012). *Bus rapid transit: Projects improve transit service and can contribute to economic development*. GAO-12-811, Report to the Committee on Banking, Housing, and Urban Affairs, United States Government Accountancy Office, Washington, D.C., USA.

Henri, L., Guy, M., & Jürg Z. (1991). Les Structure des Véhicules. *Revue Générale des Chemins de Fer*. No. 7–8 (July/August).

Horonjeff, R., & McKelvey, F. R. (1994). *Planning and design of airports* (3rd ed.). New York: McGraw-Hill Book Company.

Hunecke, K. (1997). *Jet engines: Fundamentals of theory, design and operation*. UK: Airlife Publications Ltd.

Janic, M. (2007). *The sustainability of air transportation: Quantitative analysis and assessment*. UK: Ashgate Publishing Company.

Janic, M. (2011). Light rail rapid transit (LRRT) system for more sustainable ground accessibility of airports. *Transportation Planning and Technology, 34*(6), 569–592.

Jenkinson, L. R., Simpkin, P., & Rhodes, D. (1999). *Civil jet aircraft design*. London: Arnold Publisher.

Kemp, R. (2007). *T618: Tracton energy metrics*, No. 2, Rail Safety and Standards Board. London, UK.

Levinson, S. H., Zimmerman, S., Butherford, G. S., & Glinger, J. (2002). Bus rapid transit: An overview. *Journal of Public Transportation, 5*(2), 1–30.

Levinson, S. H., Zimmerman, S., Clinger, J., Gast, J., Rutherford, S., Smith, L. R., Cracknell, J., Soberman, R. (2003a). *Bus rapid transit, Volume 1: Case studies in bus rapid transit*, Report, TCRP (Transit Cooperative Research Program), TRB (Transportation Research Board), Washington, D.C., USA.

Levinson, S. H., Zimmerman, S., Clinger, J., Gast, J., Rutherford, S., Bruhn, E. (2003b). *Bus rapid transit, Volume 2: Implementation Guidelines*, Report, TCRP (Transit Cooperative Research Program), TRB (Transportation Research Board), Washington, D.C., USA.

Mishra, K. R., Parida, M., & Rangnecar, S. (2010). Evaluation and analysis of traffic noise along bus rapid transit corridor. *International Journal of Environmental Science and Technology, 7*(4), 737–750.

Nelson, T. (2005). *787 Systems and Performance*, Flight operations engineering. Boeing Commercial Airplanes, Seattle, USA.

Persson, R. (2007). *Tilting trains: Description and analysis of the present situation, Literature study ISBN 978-91-7178-608-1*. Stockholm: Royal Institute of Technology.

Poisson, F., Gautier, P.E., Letourneaux, F. (2008). Noise sources for high speed trains: A review of results in the TGV case. In B. Schulte-Werning et al. (Eds.), *Noise and vibration mitigation.* NNFM 99, springerlink.com, Springer, Heidelberg,Germany, pp. 71–77.

Puchalsky, C. M. (2005). Comparison of emissions from light rail transit and bus rapid transit. *Transportation Research Record 1927: Public Transportation*, pp. 31–37.

RITA. (2012). *Safety and security time series data (2002–2011)*. Research and Innovative Technology Administration, U.S. Department of Transportation, National Transit Database Washington, D.C., USA.

Ross, C. J., Staiano, A. M. (2007). A comparison of green and conventional diesel bus noise levels. Paper presented at *Noise-CON Conference*, Reno, Nevada, USA, p. 8.

RR. (2011). *TRENT100 Factsheet: Optimized for B787 Dreamliner Family*. Rolls Royce, COM13797, Issue 9, November 2011, London, England.

RTR. (2005). High speed and network extension. *Rail technical review*, No. 2. pp. 11–18.

Saavedra, S. N. (2011). *The evolution of transportation planning in bogota, Report*. Traffic and Transport Research Program, Universidad Nacional de Colombia, Bogota, Columbia.

Smith, J. T. M. (2004). *Aircraft noise*. Cambridge: Cambridge University Press.
STSI. (2008). *Light rail vehicles*. Siemens Transportation Systems Inc., Rolling Stock Division, Sacramento, California, USA.
Tegner, G. (2003). Comparison of costs between Bus, PRT, LRT, and Metro/Rail. In EC project *European Demonstration of Innovative City Transport (EDICT)*. European Commission 5th Framework Program, DG VII, Brussels, Belgium.
TRB. (2008). *Airport ground access system mode choice models: A synthesis of airport practice*. ACRP Synthesis 5. Washington, D.C., USA: TRB.
Vincent, W., Jerram, C. L. (2006). The potential for bus rapid transit to reduce related CO_2 emissions. *Journal of Public Transportation,* Special Edition, pp. 219–237.
VTT. (2004). *Transit bus emissions study: Comparison of emissions from diesel and natural gas buses*. Research Report, Technical Research Center of Finland, Tampere, Finland.
Vuchic, R. V. (2007). *Urban public transportation systems and technology*. New York: Willey.
Wright, L. (2003). *Bus rapid transit-mode 3b, Deutsche Gesellshaft fur Technische Zusammenarbeit (GTZ) GmbH*. Germany: Eschbom.
Wright, L., & Hook, W. (2007). *Bus rapid transit planning guide* (3rd ed.). New York: Institute for Transportation & Development Policy.

Chapter 3
Advanced Transport Systems: Operations and Economics

3.1 Introduction

This chapter describes advanced freight collection/distribution networks, road mega trucks, LIFTs (Long Intermodal Freight Train(s)) as components of the advanced freight/goods system in Europe, and large commercial freight aircraft. Advanced freight collection/distribution networks can be operated by a single or few different (integrated) transport modes moving freight/goods consolidated into loading units such as pallets, containers, swap-bodies, and/or semi-trailers between the doors of particular shippers and receivers. When these networks are operated exclusively by road, standard or mega trucks are exclusively used. The road mega truck is the largest, i.e., the longest and heaviest, commercial freight vehicle proposed to carry fright/good shipments throughout the EU (European Union) countries. As such, it is longer and heavier that its current largest counterpart—the standard truck with a length of 18.75 m and a weight of 40–44 tons.

In rail-operated networks, CIFTs (Commercial Intermodal Freight Train(s) or LIFTs can be used along the main route(s). Specifically, LIFTs (Long Intermodal Freight Train(s)) are proposed to carry out large freight/goods shipments on long distances in Europe. Due to being longer and heavier than their conventional counterparts, they are expected to be powered by two instead of one locomotive. Operating as the main mode of the rail/road intermodal transport system, they are expected to be competitive in terms of operational (internal) costs to the road transport system operating either standard or mega trucks.

In intermodal networks, CIFTs, LIFTs, or barges can be used along the main routes, and road trucks for the collection and distribution of loading units at the beginning and from the end intermodal terminal(s), respectively. The intermodal rail/road freight collection/distribution networks operated have been initiated by the ERRAC (European Research Advisory Council) and the rail freight operators in France, Belgium, Germany, and the Netherlands. The aim has been to improve the market position of freight nonroad transport modes, and consequently mitigate the impacts of the road freight transport mode on the environment and society.

M. Janić, *Advanced Transport Systems*,
DOI: 10.1007/978-1-4471-6287-2_3, © Springer-Verlag London 2014

The Boeing B747-8F and Airbus A380-800F are examples of large commercial freight aircraft. They are advanced mainly due to their size, i.e., payload capacity, and their related operational/economic advantages, both as compared to their closest smaller counterpart such as the B-747-400F.

3.2 Advanced Freight Collection/Distribution Networks

1954	The ICC (Interstate Commerce Commission) (U.S.) concludes that the transportation of piggyback trailers on flatcars is not a motor-carrier service and stopped requiring railroads to have a motor-carrier certificate in order to operate Trailer-On-Flat Car (TOFC) services (U.S.)
1956	Intermodal TOFC (Trailer-On-Flat Car) services are launched by the Trailer Train Company set up by the Pennsylvania Railroad and the Norfolk and Western Railways (U.S.)
1961	The ISO (International Organization for Standardization) set standards for containers (Switzerland)
1972	The "landbridge" intermodal service, where containers are transferred from a ship to a train on one coast, then transported across the country to the other coast and transferred back from train to another ship, is introduced by the Southern Pacific railroad and Sea-Land (U.S.)
1994/ 1997	The ten Pan-European transport corridors are defined at the second Pan-European Transport Conference in Crete; additions are made at the third conference in Helsinki (Europe)
1996	Guidelines for the TEN-T (Trans-European Networks-Transport) are initially adopted by the European Parliament (Europe)
2001	Guidelines for the TEN-T with respect to the seaports, inland ports, and intermodal terminals are amended (Europe)
2004	More fundamental changes of the TEN-T policies are intended to accommodate the enlargement of EU (European Union) and consequent changes in the pattern of freight/goods flows are made (Europe)

3.2.1 *Definition, Development, and Use*

Non-road freight transport modes as the main components of the intermodal or multimodal freight transport networks in Europe have steadily grown stimulated by both market forces and EC (European Commission) transport policy. For example, during the 1990–1999 period, intermodal or multimodal transport within the EU (European Union) 15 Member States grew from an annual volume of about 119 to about 250 billion t-km[1] (tons-kilometers), which resulted in an increase of

[1] Of this total about 91 % was international and 9 % domestic traffic. Rail carried about 20 %, inland waterways 2 %, and short-sea shipping 78 % of international traffic. About 97 % of domestic traffic was carried by rail and 3 % by inland waterways (EC 1999, 2002).

its market share from about 5 to 9 %. This mainly happened after enhancement of operations in the Trans European transport corridors along distances of 900–1,000 km where about 10 %t of the total volumes of freight/goods (tons) were transported. On distances up to about 900 km where about 90 % of these freight/goods volumes were transported, the market share of intermodal or multimodal transport was negligible—only about 2 % in the total volumes of t-km, and 2–3 % in the total volumes of tons (EC 1999, 2002). Since 1999, the above-mentioned market shares have not substantially changed mainly due to the following reasons: (i) the rather low overall unitarization rate[2] of freight/goods of only about 10–15 %; (ii) the frequent deterioration of the quality of services of intermodal or multimodal transport main mode(s)—primarily rail—as a result of insufficient frequency and rather low reliability of service; (iii) the low door-to-door freight/goods delivery speed of only about 18–20 km/h, and consequently long delivery time; and (iv) further improvements in efficiency and effectiveness of the road freight transport sector including initiatives for introducing road mega trucks on a wider European scale (CNT 2006; EC 1999, 2000a, 2001a, 2002, 2007).

The above-mentioned developments reflect a part of the visionary policy for development of the European freight transport sector by 2020 and beyond implying further consolidation and more intensive use of rail/barge/short-sea on the account of the road transport mode aiming at making freight and consequently the entire transport sector more sustainable (EIRAC 2007). Specifically, in the scope of the above-mentioned policy, the transport and logistics operators have considered the advanced freight collection/distribution networks as options to provide users/freight shippers/receivers advanced door-to-door transport services as compared to other transport (freight) service networks. These networks can be operated by single or few transport modes. In both cases, the freight shipments are consolidated into compact units such as pallets and loading units—containers, swap-bodies and semi-trailers. They can be exchanged between vehicles of different size/capacity operated by the same or different transport modes at dedicated terminals. In case when the networks are operated by a single mode, the shipments from smaller incoming vehicles can be consolidated into those to be carried conveniently by larger outgoing vehicles such as road standard or mega trucks (with electronic coupling), CIFTs (Conventional Intermodal Freight Train(s)), LIFTs (Long Intermodal Freight Train(s)), large containers ship(s), or large commercial freight aircraft, and vice versa at the dedicated uni-modal freight terminals. When the network is operated by different transport modes, the vehicles of particular modes meet at the multimodal terminals where seamless transshipment of freight shipments between them takes place. The combination of modes can be different, but most frequently the inland systems include road and one of the nonroad transport modes (rail, and/or barge).

[2] The unitarization rate refers to the proportion of goods carried in loading units—containers, swap-bodies, and semi-trailers.

3.2.2 Analyzing and Modeling Performances

3.2.2.1 Background

Advanced freight collection/distribution networks are based on the advanced organization of transport and transshipment services, in many cases supported by advanced technologies (EC 1997a, b; Janic et al. 1999; Janic 2005). The advancements in organization of transport services imply operating demand-driven direct-fixed or "shuttle" services connecting shippers and receivers, directly and/or via single- or multimodal terminals. Advancements in transport technologies include those of transport means/vehicles and the transshipment of loading units in terminals. The former implies more intensive use of advanced transport means/vehicles such as standard and mega trucks, multisystem locomotives, special rail cars, barges of different capacity, etc., and loading units such as containers, swap-bodies, and semi-trailers. The latter includes technologies based on increased automation enabling both horizontal and vertical transshipment of loading units. These technologies provide higher capacity, efficiency, and effectiveness of terminal operations.[3]

Advanced freight collection/distribution networks can be classified into five categories as follows: P–P (Point-to-Point) networks, TCD (Trunk line with Collecting/Distribution forks) networks, HS (Hub-and-Spoke(s)) networksnet-works, L/R (Line or Ring) networks, and M (Mixed) networks (EC 1997a, 1998a; Janic et al. 1999; Janic 2005).

- A P–P (Point-to-Point) network serves relatively substantial and regular freight flows. Loading units from different shippers are collected by road and concentrated at the origin (begin-) terminal. After being transshipped from the trucks to the direct-fixed or shuttle train(s) (or barge(s)), they are transported to the destination (end-) terminal. There, they are transshipped from the train/barge to the trucks and transported to the final receiver;
- A TCD (Trunk line with Collecting/Distribution forks) network is actually an expanded P–P (Point-to-Point) network covering a wider area. This network consists of two types of terminals: the origin/destination trunk terminals and CD (Collection/Distribution) terminals. Feeder rail or barge services are operated between the origin/destination and CD terminal, and between the CD terminals themselves. At the CD terminal(s), these feeder trains/barges can be partly loaded/unloaded by loading units from/to the other origin/destination terminals, respectively. Then, they proceed toward the other CD/destination terminals, respectively. The loading units to/from each terminal of the network are usually collected/distributed from/to the doors of shippers/receivers, respectively, by road (EC 1997a, 1998a);

[3] For example, horizontal transshipment can be eased by the automatic locking on the container castings or on the tray castings, in combination with the automatic positioning of the train along the loading/unloading floor.

- A HS (Hub-and-Spoke(s)) collection/distribution network is particularly operated by the rail and/or barge freight transport mode. The network consists of several begin-/end-terminals (spokes) and one centrally located (intermediate) terminal (hub). The loading units sent by shippers arrive at the spoke terminal(s) usually by road. There, they are transshipped from the trucks to the rail wagons or barges. Then, they proceed to the hub, where they are transshipped between the incoming and the outgoing trains/barge shuttle services. The transshipment takes from several minutes to several hours depending on the inbound and outbound timetable and the applied type of transshipment operations. After the outgoing trains/barges from the hub arrive at the spoke(s), the loading units are again transshipped to the trucks and delivered to the door(s) of receivers (EC 1997a, 1998a);
- A L/R (Line or Ring) collection/distribution network consists of a begin- and end-terminal located at both ends of the line/ring and several intermediate (line/ring) terminals located in between. Line train and/or barge services are scheduled between the begin- and end-terminal. They may stop at one and/or few intermediate terminals where transshipment of loading units between the same and/or different transport modes can take place. This implies that the road can also be used between the terminals of the network and the doors of shippers/receivers of loading units (EC 1997a, 1998a); and
- A M (Mixed) network usually consists of a combination of elements of the above-mentioned four basic networks.

Advanced freight collection/distribution networks possess network-generic and the network-specific performances. The former include inventories, delays, and the costs of processing loading units in the network(s), while the latter include infrastructural/spatial, technical/technological, operational, economic, environmental, and social/policy performances. Both types of performances can be considered from the aspects of particular actors involved such as: users of network services (freight/goods shippers/receivers), transport and terminal operators, manufacturers of transport and terminal infrastructure facilities and equipment, and DM (Decision-Making) authorities and communities at different institutional levels (local, national, international). Modeling the network-generic performances implies developing analytical equations for estimating the level of inventories and related delays of freight/goods shipments, and the total operating and inventory costs. Modeling the network-specific performances implies quantifying particular indicators and measures and then using them as attributes/criteria in the multicriteria ranking of particular network configurations (EC 1997a, 1998a; Hay 1977; Manheim 1979; Tarski 1987).

3.2.2.2 Generic Performances

Network generic-performances such as inventories, delays, and the costs of processing loading units in the network(s) imply analyzing the network structure and operations, and then developing the corresponding analytical models based on the following assumptions:

- Demand is expressed by the number of loading units requesting service in the network during a given period (a day, week);
- The available transport and terminal capacity are always sufficient to satisfy this demand;
- Terminals represent the network nodes where loading units are exchanged between the same and/or different transport modes; in addition, the loading units can enter and/or leave the network through these nodes, i.e., terminals;
- Loading units enter and leave the network in batches, which can form inventories at particular locations such as the network nodes, i.e., terminals and the network links, i.e., routes;
- Formation of inventories of loading units causes their delays; delays can be repetitive at regular time intervals called network cycles (day or a week) depending on the pattern of demand and supply;
- The level of inventories of loading units during a cycle represents the network state;
- The total costs of particular networks consist of inventory costs, transport costs, and terminal costs; inventory costs relate to handling inventories of loading units while in a given network; transport cost consists of the operators' cost to transport loading units between their origin(s) and destination(s); terminal cost includes handling cost of loading units while in and through the terminal(s). (Hall 1987, 1993; Janic et al. 1999; Janic 2005)

P–P (Point-to-Point) network

Inventories and delays

A P–P (Point-to-Point) collection/distribution network operates as follows: the loading units are delivered from their local origins (i.e., doors of the shippers) to the origin intermodal terminal by road haulage. Then, they are loaded onto a direct or shuttle train or barge (vessel) and transported to the destination terminal. From there, they are distributed (again by road) to their final destinations (i.e., doors of the receivers). Figure 3.1 shows a simplified layout of the network and inventories of loading units in it during a cycle.

The flow of loading units q_{AB} is transported between terminal A and B. The length of the route connecting the two terminals is l_{AB} and the average speed of transport units is $v_{AB}(l_{AB})$. The intensity of collection of loading units at the terminal $(A \equiv i)$ is λ_i. The loading units are assumed to arrive there by road some time before departure of the train, barge, or vessel. This time may vary from the time of preceding departure to the time of closing "new" departure for loading. The period between the arrival of the first and the last loading unit of the batch q_{AB} is τ_{0i}.

The total loading time is dependent on the size of batch $q_{AB} \equiv q_{ij}$ and loading rate μ_i. After the batch q_{ij} is loaded, the transport unit is inspected and dispatched after the period τ_{1i}. The transport unit arrives at terminal $(B \equiv j)$ after time $l_{AB}/v_{AB}(l_{AB})$. Then, the batch q_{ij} is unloaded. The intensity of unloading at terminal

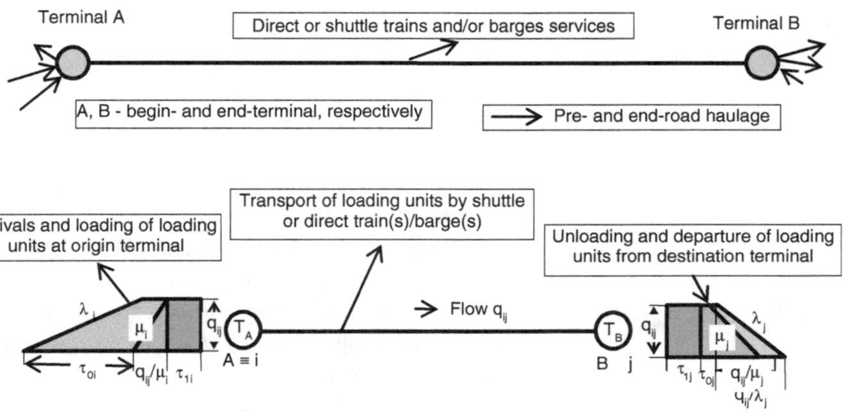

Fig. 3.1 Scheme of a P–P (Point-to-Point) network (Janic et al. 1999; Janic 2005)

($B \equiv j$) is μ_j. The inspection and preparation of transport unit for unloading is carried out in time τ_{1j} and τ_{0j}, respectively. Similarly as at the origin terminal, the loading units that arrive at the terminal (j) can be either placed in the terminal's "storage" area waiting for being picked-up latter or directly transshipped to the road trucks, and vice versa. The intensity of their leaving the terminal (j) is λ_j.

The time for processing batch q_{ij} through the network defines the network's cycle, which can be repetitive in both directions. If it is assumed that the successive cycles do not overlap and thus do not affect each other, the total delay of loading units accommodated by the network during a single cycle can be estimated as follows:

$$\begin{aligned} D_{ij} &= D_{bi} + D_{Tij} + D_{bj} \\ &= \left[\tau_{oi} + \tau_{1i} + q_{ij}\left(\frac{1}{\mu_i} - \frac{1}{2\lambda_i}\right)\right] q_{ij} + \left(\frac{l_{ij}}{v_{ij}}\right) q_{ij} + \left[\tau_{1j} + \tau_{0j} + \frac{1}{2}\frac{q_{ij}}{\min(\lambda_j; \mu_j)}\right] q_{ij} \end{aligned} \tag{3.1a}$$

The symbols D_{bi}, D_{Tij}, and D_{bj} denote the "inventory" delay at the origin terminal (i), delay along route (ij), and the "inventory" delay at the destination terminal (j), respectively.

Costs

The total costs of a P–P (Point-to-Point) network can be estimated based on Eq. 3.1a as follows:

$$\begin{aligned} C_{ij} &= C_I + C_{TR} + C_{TE} = \left(D_{bi} + D_{Tij} + D_{bj}\right) * p_{ij} * r_{ij} + c(d_{0i}, W_{0i}) * f_{0i} \\ &\quad + c\left(l_{ij}, W_{ij}\right) * f_{ij} + c\left(d_{j0}, W_{j0}\right) * f_{j0} + \left(\frac{b_{0i} + b_{1i}}{u_i Q_i} + \frac{b_{0j} + b_{1j}}{u_j Q_j}\right) * q_{ij} \end{aligned} \tag{3.1b}$$

where

D_{bi}, D_{Tij}, D_{bj} is the delay of loading units while passing through the terminal (i), route (ij), and terminal (j), respectively (LU-h) (LU—Loading Unit; h—hour);

p_{ij} is the average value of a shipment belonging to the batch q_{ij} (€/LU);

r_{ij} is the average "inventory" charge of a shipment belonging to the batch q_{ij} (€/h);

$c(d_{i0}, W_{i0})$, $c(l_{ij}, W_{ij})$, $c(d_{j0}, W_{j0})$ is the cost of transport unit of capacity W_{i0}, W_{ij}, and W_{j0} along the route d_{i0}, l_{ij}, and d_{j0}, respectively (€/dispatch);

f_{i0}, f_{ij}, f_{j0} is the transport service frequency on the routes d_{i0}, l_{ij}, and d_{j0}, respectively (departures/h);

b_{0i}, b_{1i}, b_{0j}, b_{1j} is the total handling and operation cost of the terminals (i) and (j), respectively, (€/year);

Q_i, Q_j is the volume of demand planned to be handled in terminal (i) and (j), respectively, (LU/year); and

u_i, u_j is the average utilization of the terminals (i) and (j), respectively.

The transport cost function can be expressed either in linear or log-linear form as follows:

$$c(d, W) = a_0 + a_1 * d + a_2 * W \quad \text{or} \quad c(d, W) = a^0 * d^{a_1} * W^{a_2} \tag{3.1c}$$

where

a_0 is the fixed transport cost (€/dispatch);
a_1 is the average cost per unit distance (€/km);
a_2 is the average cost per unit of capacity of a transport unit (€/tone or €/LU);
d is the length of the route (km), and
W is the capacity of transport unit (tons/TU or LU/TU; TU—Transport Unit).

For any transport mode, the service frequency on the routes connecting the particular spokes can be determined based on the assumption that the demand is always satisfied as follows:

$$f = q/\lambda * W \tag{3.1d}$$

where

q is the volume of loading units on a given route (LU/period), and
λ is the average utilization of transport units running on the route.

The other symbols are analogous to those in previous equations.

TCD (Trunk line with Collecting/Distribution forks) network

Inventories and delays

A TCD (Trunk line with Collecting/Distribution forks) network consists of origin and destination "local" terminals, which are assigned to the corresponding trunk terminals. The loading units move between the "local" terminals via the CD terminals, which can all be single- or multimodal. Figure 3.2 shows a simplified spatial layout and inventories of loading units in the network during a cycle.

The origin "local" terminals and the destination "local" terminals are: T_{di} ($i = 1, 2, \ldots, N$) and T_{aj} ($j = 1, 2, \ldots, M$), respectively. The flow of loading units between the origin and destination "local" terminals (i) and (j), respectively, is q_{ij}. Handling of these loading units at terminals is carried out as in a P–P (Point-to-Point) network. The loading units are transported by direct or shuttle "feeder" trains (or barges) from the "local" to the trunk terminals, and vice versa. For example, at the trunk terminal T_A, the loading units are regrouped into batches convenient for longer direct or shuttle transport units (trains or barges), which are then dispatched along the trunk route l_{AB} to the trunk terminal T_B.

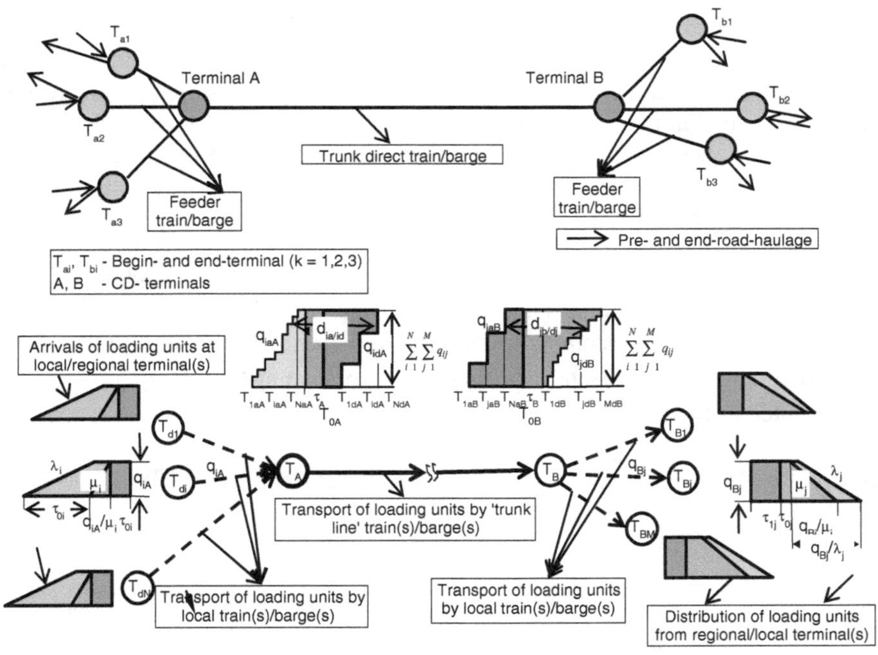

Fig. 3.2 Scheme of a TCD (Trunk line with Collecting/Distribution forks) network (Janic et al. 1999; Janic 2005)

In rail-based networks, both "local" and "trunk" terminals may operate like "local" and "regional" shunting yards, respectively. In particular, the delays of loading units during their passing through the "trunk" terminals are dependent on the number, capacity, and timetable of the inbound "local" trains and the outbound "trunk" trains. In addition, these delays can be significantly influenced by the strategy and "speed" of shunting, both depending on the size and type of shunting yard.

In Fig. 3.2, the arrival time of *i*-th "local" train at the terminal T_A ($i = 1, 2, \ldots, N$) is T_{iaA}, the departure time of *i*-th "trunk" train from the terminal T_A is T_{idA}, the arrival time of *i*-th "trunk" train at terminal T_B is T_{iaB}, and the departure time of *i*-th "local" train from the terminal T_B is T_{idB}. In case when all outbound trunk trains have to wait for all inbound trains in order to exchange wagons and/or loading units, and vice versa, their accumulation and related delays may become rather long. The maximum number of loading units that can be accumulated in the terminals T_A and T_B can be determined as: $\sum_{i=1}^{N} q_{iA} = \sum_{i=1}^{N}\sum_{j=1}^{M} q_{ij} = \sum_{j=1}^{M} q_{Bj}$; q_{iA} is the batch of loading units moving between the fork terminal (*i*) and the trunk terminal T_A ("*i*" is assigned to "T_A") ($i = 1, 2, \ldots, N$), i.e., $q_{iA} = \sum_{j=1}^{N} q_{ij} (i = 1, 2, \ldots, N)$; q_{iB} is the batch of loading units moving from the trunk terminal T_B to the fork terminal (*j*), ("*j*" is assigned to "T_B"), ($j = 1, 2, \ldots, M$), i.e., $q_{Bj} = \sum_{i=1}^{M} q_{ij} (j = 1, 2, \ldots, M)$; *N* and *M* are the number of origin and destination (local) terminals assigned to the trunk terminals T_A and T_B, respectively. The loading units arrive at the origin "local" terminal (*i*) at the rate of λ_i ($i = 1, 2, \ldots, N$). They may be either directly transshipped to the feeder train(s) or placed in the terminal storage area. The intensity of loading the feeder trains at the origin terminal (*i*) is μ_I, while the intensity of unloading the feeder trains at the destination terminal (*j*) is μ_j. The loading units leave the destination (local) terminal (*j*) at the rate of λ_j ($j = 1, 2, \ldots, M$). The times T_{0A} and T_{0B} are the moments at which the trunk terminals T_A and T_B, respectively, change the operating regime. At the time T_{0A} the terminal T_A is closed for further arrivals of incoming feeder trains and opened for the departure of the outgoing trunk trains. At the time T_{0B}, the terminal T_B is closed for arrivals of incoming trunk trains and opened for the departures of the feeder outgoing trains. The total delay of the batch of loading units $\sum_{i=1}^{N}\sum_{j=1}^{M} q_{ij}$ can be computed as follows:

$$
\begin{aligned}
D = D_B + D_T = {} & \sum_{i=1}^{N}\left[\left(\tau_{0i} + \tau_{1i} + q_{iA} * \left(\frac{1}{\mu_i} - \frac{1}{2\lambda_i}\right)\right) + (T_{0A} - T_{iaA}) + (T_{idA} - T_{0A} + \tau_A)\right] * q_{iA} \\
& + \alpha_A * \frac{\left(\sum_{i=1}^{N}\sum_{j=1}^{M} q_{ij}\right)^2}{\mu_A} + \alpha_B * \frac{\left(\sum_{i=1}^{N}\sum_{j=1}^{M} q_{ij}\right)^2}{\mu_B} + \sum_{j=1}^{M}\left[\begin{array}{c}(T_{0B} - T_{jaB}) + (T_{jdB} - T_{0B} + \tau_B) + \\ \left(\tau_{1j} + \tau_{0j} + q_{Bj} * \left(\frac{1}{\mu_j} - \frac{1}{2(\min(\lambda_j; \mu_j))}\right)\right)\end{array}\right] * q_{Bj} \\
& + \left(\sum_{i=1}^{N} q_{iA} * \frac{l_{iA}}{v_{iA}} + \sum_{i=1}^{N}\sum_{j=1}^{M} q_{ij} * \frac{l_{AB}}{v_{AB}} + \sum_{j=1}^{M} q_{Bj} * \frac{l_{Bj}}{v_{Bj}}\right)
\end{aligned}
\tag{3.2a}
$$

where

τ_{0i}, τ_{0j} is the time between the arrival and loading of the batch of loading units q_{iA} at the origin (local) terminal (i), and the time between the arrival and departure of the batch q_{Bj} from the destination (local) terminal (j), respectively (minutes, h);

τ_{1i}, τ_{2j} is the time of preparing the transport unit to depart from the origin (local) terminal (i), and the time of preparing the transport unit for unloading after arrival at the destination (local) terminal (j), respectively, (minutes, h);

α_A. α_B is a binary variable taking the value "1" if the service of loading units at the trunk terminals T_A and T_B is realized just after the arrival of the whole batches q_{iA} and q_{iB}, respectively, and the value "0," otherwise;

l_{iA}, l_{AB}, l_{Bj} is the length of route connecting the origin (local) terminal (i) and the trunk terminal T_A, the trunk terminals T_A and T_B, and the trunk terminal T_B and the destination (local) terminal (j), respectively, (km); and

v_{iA}, v_{AB}, v_{Bj} is the average speed of transport unit (train/barge/vessel) on the routes l_{iA}, l_{AB}, and l_{Bj}, respectively (km/h).

Costs

The total costs of a TCD (Trunk line with Collecting/Distribution forks) network can be estimated based on Eq. 3.2a as follows:

$$C = C_I + C_{TR} + C_{TE}$$
$$= \sum_{i=1}^{N}\sum_{j=1}^{M}\left[\begin{array}{l} \left(D_{biA} + D_{TiA} + D_{bA} + D_{TAB} + D_{bB} + D_{TBj} + D_{bBj}\right) * p_{ij} * r_{ij} \\ + c(d_{i0}, W_{i0}) * f_{i0} + c(d_{iA}, W_{iA}) * f_{iA} + c(d_{AB}, W_{AB}) * f_{AB} \\ + c_{Bj}\left(d_{Bj}, W_{Bj}\right) * f_{Bj} + c\left(d_{j0}, W_{j0}\right) * f_{j0} + \dfrac{b_{0i} + b_{1i}}{u_i Q_i} * q_{iA} \\ + \left(\dfrac{b_{0A} + b_{1A}}{u_A Q_A} + \dfrac{b_{0B} + b_{1B}}{u_B Q_B}\right) * \left(q_{iA} + q_{Bj})\right) + \dfrac{b_{0j} + b_{1j}}{u_j Q_j} * q_{Bj} \end{array}\right] \quad (3.2b)$$

where all symbols are analogous to those in previous equations.

HSs (Hub-and-Spoke(s)) network

Inventories and delays

A HSs (Hub-and-Spoke(s)) network usually consists of a hub (the central node—terminal) and several spokes (peripheral nodes—terminals). A simplified layout and inventories of loading units in this network during a cycle are shown in Fig. 3.3.

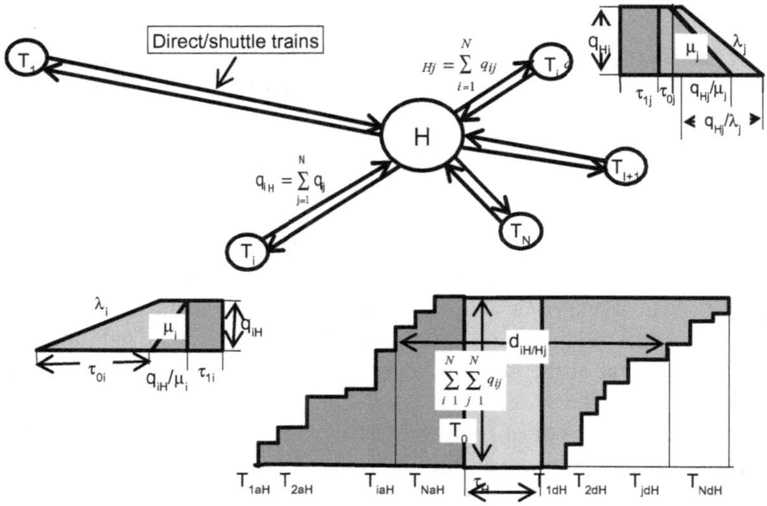

Fig. 3.3 Scheme of a HS (Hub-and-Spoke) network (Janic et al. 1999; Janic 2005)

The spoke terminals T_i ($i = 1, 2, \ldots, N$) can be connected directly to the hub H and indirectly through the hub H between themselves by direct and/or shuttle trains/barges transporting loading units between their origins and destinations (hub and spoke nodes). The loading units enter and/or leave the spoke terminals (i.e., the network) usually by road. There, they are transshipped from the trucks to direct or shuttle train(s)/barges which are then dispatched to the hub H, or vice versa. Direct trains or barges operate between particular spokes by passing through the hub, while shuttle trains operate exclusively between spokes and the hub. Loading units can pass through the hub H either by staying on the same transport units (wagons/barges) all the time (direct trains/barges) or by changing them (shuttle trains/barges). In the former case of rail, wagons are exchanged between trains, thus implying that the hub terminal actually operates as a conventional shunting yard. In the latter case of rail, loading units are exchanged between different transport units (wagons) while being in the hub, thus implying that the hub operates as an advanced automated rail–rail terminal with vertical transshipments of loading units. The "inventories" of loading units have a similar pattern as those at a TCD network, i.e., the transshipment of loading units at spoke terminals T_i and shunting of trains at the hub terminal H are analogous to the corresponding operations carried out at the local terminals T_{ai} and T_{aj}, and the trunk terminals T_A and T_B, respectively.

If the flow of loading units between spokes (i) and (j) is q_{ij}, the resulting flow of loading units on the routes (iH) and (Hj) will be q_{iH} and q_{Hj}, respectively. The total number of loading units passing through the hub H is equal to the sum of all individual flows q_{ij} ($i, j \in N; I \neq j$). The total delay of loading units served by the network operated by rail during a cycle can be estimated as follows:

$$D = D_B + D_T = \sum_{i=1}^{N}\left[\tau_{0i} + \tau_{1i} + q_{iH} * \left(\frac{1}{\mu_i} - \frac{1}{2\lambda_i}\right) + (T_0 - T_{iaH})\right] * q_{iH} + \alpha_H * \left(\sum_{i=1}^{N}\sum_{j=1}^{N} q_{ij}\right)^2 / \mu_H$$
$$+ \sum_{j=1}^{N}\left[(T_{jdH} - T_0 + \tau_H) + \tau_{0j} + \tau_{1j} + q_{Hj} * \left(\frac{1}{\mu_j} - \frac{1}{2[\min(\lambda_j; \mu_j)]}\right)\right] * q_{Hj} + \sum_{i=1}^{N}\sum_{j=1}^{N} q_{ij} * \left(\frac{l_{iH}}{v_{iH}} + \frac{l_{Hj}}{v_{Hj}}\right) \quad (3.3a)$$

where

λ_i, λ_j is the intensity of arrival and departure of loading units at/from the spoke terminals (*i*) and (*j*), respectively (LU/h);

μ_i, μ_j is the intensity of loading/unloading of loading units at the spoke terminals (*i*) and (*j*), respectively (LU/h);

T_{iaH} is the arrival time of a train operating between the spoke (*i*) and hub (*H*);

T_0 is the time when all trains arrived from *N* spokes are ready to be shunted at the hub *H*;

α_H is a binary variable taking the value "1" if shunting starts after the arrival of the last among *N* trains at the hub (*H*), and the value "0," otherwise;

μ_H is the average rate of shunting assumed to be approximately constant and not dependent on the train and wagon characteristics (trains/h);

T_{jdH} is the departure time of a train operating between the hub *H* and the spoke (*j*);

τ_H is the total time needed for shunting *N* trains; shunting is assumed to start just after the arrival of the last train from the batch of *N* trains (hours);

l_{iH}, l_{Hjj} is the length of route connecting the spoke (*i*) and (*j*), via the hub (*H*), respectively (km); and

v_{iH}, v_{Hj} is the travel speed of a train along the routes l_{iH} and l_{HJ}, respectively (km/h).

Costs

The total costs of a HSs (Hub-and-Spoke(s)) collection/distribution network can be determined based on Eq. 3.3a as follows:

$$C = C_I + C_{TR} + C_{TE}$$
$$= \sum_{i=1}^{N}\sum_{j=1}^{N}\left[\begin{array}{l}(D_{bi} + D_{TiH} + D_{bH} + D_{THj} + D_{bj}) * p_{ij} * r_{ij} + c(d_{i0}, W_{i0}) * f_{i0} \\ + c(d_{ih}, W_{ih}) * f_{iH} + c_{Hj}(d_{Hj}, W_{Hj}) * f_{Hj} + c(d_{j0}, W_{j0}) * f_{j0} \\ + \left(\frac{b_{0i} + b_{1i}}{u_i Q_i} * q_{iH} + \frac{b_{H0} + b_{H1}}{u_H Q_H}\right) * (q_{iH} + q_{Hj}) + \frac{b_{0j} + b_{1j}}{u_j Q_j} * q_{Hj}\end{array}\right] \quad (3.3b)$$

where all symbols are analogous to those in the previous equations.

L/R (Line or Ring) network

Inventories and delays

A L/R (Line or Ring) network is a line or ring configuration where single- and/or intermodal terminals are located in a line or ring in relation to the direction in which the loading units are moving. Rail and inland navigation (barges/vessels) direct and/or shuttle services usually connect these terminals where the exchange of loading units can takes place between different transport modes: rail/rail, truck/rail, truck/barge, and truck/short and deep-sea vessels. The loading units can enter and leave the network at any terminal, which can also be their origins and destinations. Figure 3.4 shows a simplified layout and inventories of loading units in this network.

As in the networks presented above, the "inventories" of loading units emerge in the terminals where they are collected and/or distributed. For example, at the terminal T_i, the loading units intended to be sent to some other location, are collected with the intensity λ_{ai} (loading units/unit of time). After some time, the batch of loading units q_{ai} is formed. Two possibilities for proceeding this batch q_{ai} to its final destination can be applied. First, if a transport unit is immediately available, the batch can be directly loaded (transshipped) to it. Otherwise, the batch will be stored in the terminal and wait for a free transport unit to come and pick-up it. In that case, the waiting time of the batch q_{ai} will be τ_{ai} (min/h/days). The intensity of loading depending on the available facilities and equipment in the terminal and/or transport unit itself is μ_{ai} (LU/h). After being transported to the terminal T_i, the batch q_{di} is unloaded with the intensity μ_{di} (LU/h). Loading and unloading can be carried out in two ways, sequentially and simultaneously. Sequential loading/unloading implies that the whole batch q_{di} is unloaded and then the whole batch q_{ai} is loaded. Simultaneous transshipment is carried out

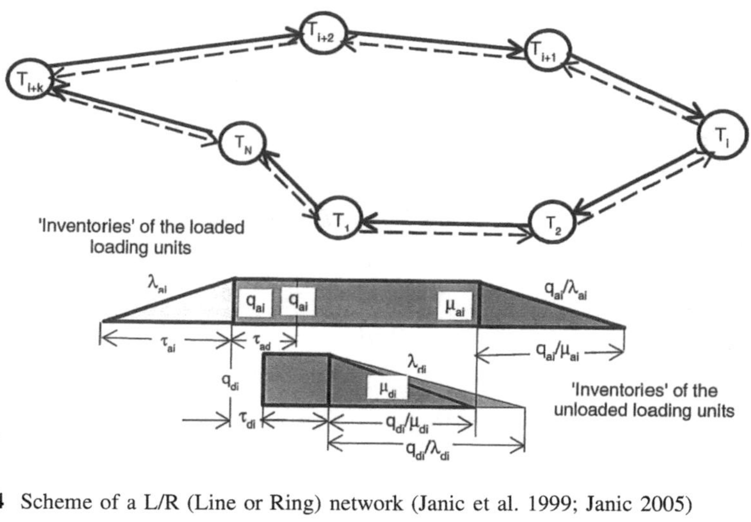

Fig. 3.4 Scheme of a L/R (Line or Ring) network (Janic et al. 1999; Janic 2005)

"in parts,"e.g., one loading unit from the batch q_{di} is unloaded first and then one loading unit from the batch q_{ai} is loaded. After the batch q_{ai} is loaded, it is transported to the final destination. The unloaded batch q_{di} is either directly transshipped to the other transport units already being in the terminal (trucks, trains, barges, or vessels) or stored in the terminal. The waiting time of the batch to be picked-up is τ_{di}. The rate of emptying the terminal T_i is λ_{di} (LU/h). The batch q_{ai} consists of loading units having their origin at the terminal T_i and destination at other terminals T_j $(i \leq j \leq N)$, e.g., $q_{ai} = \sum_{j=1}^{N} q_{ij}$. The batch q_{di} consists of loading units having their origin at the terminal T_j and destination at terminals T_i $(1 \leq j \leq i)$, e.g., $q_{di} = \sum_{j=1}^{i} q_{ij}$. Thus, the number of loading units transported between any two terminals T_i and T_{i+1} is equal to $q_{i,i+1} = \sum_{k=1}^{i} \sum_{j=i+1}^{N} q_{kj}$. The delay of loading units while passing through the network can be estimated as follows:

$$
\begin{aligned}
D &= D_B + D_T \\
&= \left[\sum_{i=1}^{N} \left(\begin{array}{l} \alpha_{ai} * \frac{a_{ai}^2}{2\lambda_{ai}} + (\tau_{ad} + \tau_{di} + \beta_i * q_{di}/\mu_{di}) * q_{ai} \\ + \left(\tau_{di} + \frac{1}{2} \frac{q_{di}}{\min(\lambda_{di};\mu_{di})} \right) * q_{di} \\ \qquad - \gamma_i * \left(\frac{q_{ai}^2}{2\mu_{ai}} * \left(\frac{\mu_{di}}{\mu_{ai} + \mu_{di}} \right) \right) \end{array} \right) \right] + \sum_{i=1}^{N} \frac{l_{i,i+1}}{v_{i,i+1}} * q_{i,i+1}
\end{aligned}
\tag{3.4a}
$$

where

$$q_{ai} = \sum_{j=1}^{N} q_{ij}^{0};\ q_{di} = \sum_{j=1}^{i} q_{ji}^{0};\ \text{and}\ q_{i,i+1} = \sum_{k=1}^{i} \sum_{j=i+1}^{N} q_{kj}^{0} \quad \text{for} \quad i = 1, 2, \ldots, N-1$$

τ_{ad} is the inter-arrival time of batches q_{ai} and q_{di} (minutes/h);

α_{i} is a binary variable, which takes the value "1" if the batch q_{ai} instantly emerges at the terminal (i), and the value "0," otherwise;

β_{i} is a binary variable, which takes the value "1" if the batch q_{ai} is loaded after unloading of a complete batch q_{di}, and the value "0," otherwise;

γ_{i} is a binary variable, which takes the value "1" if loading and unloading of the batches q_{ai} and q_{di} at the terminal (i) are simultaneously performed, and the value "0," otherwise;

q_{ij}^{0} is the local flow of loading units originating from the terminal (i) and sinking in the terminal (j) $(i \neq j;\ i, j \in N)$ (LU);

$l_{i,i+1}$ is the length of the route connecting terminals (i) and $(i + 1)$ (km); and

$v_{i,i+1}$ is the average speed of transit unit (train, barge, vessel) operating between terminals (i) and $(i + 1)$ (km/h).

Costs

The total costs of a L/R (Line or Ring) network can be estimated based on Eq. 3.4a, as follows:

$$C = C_I + C_{TR} + C_{TE} = \sum_{i=1}^{N} [D_{bi} * p_i * r_i + c(d_{i0}, W_{i0}) * f_{i0} + c(d_{0i}, W_{0i}) * f_{0i}] + \sum_{i=2}^{N} c(d_{i,i-1}, W_{i,i-1}) * f_{i,i-1} + \sum_{i=1}^{N} \left(\frac{b_{0i} + b_{1i}}{u_i Q_i}\right) * (q_{ai} + q_{di}) \tag{3.4b}$$

where all symbols are analogous to those in the previous equations.

3.2.2.3 Specific Performances

Network-specific performances are expressed by indicators and measures such as the network size, average route length, capacity of transport vehicles, service frequency, schedule delay, average transport speed and time, terminal time, average delivery distance, speed and time, punctuality, reliability, coefficient of terminal time, transport work, intensity of network services, technical productivity, costs, and externalities. As analyzed, modeled, and quantified, they can be used as the attributes/criteria for ranking different configurations of advanced freight collection/distribution networks.

Indicators and measures

Size of the Network (SN)

The network size reflects its physical features such as the number of inter-modal terminals (nodes), routes (links), as well as the number of single- and/or multimodal subnetworks. In particular, the number of terminals reflects the spatial accessibility of network services; their geographical (spatial) distribution defines the network's area coverage. The number of routes reflects the spatial extension of network services. The number of single- and/or multimodal subnetworks indicates the level of network complexity. This indicator is always preferred to be as high as possible.

Average Route Length (ARL)

The average route length reflects the average travel distance of transport means/ vehicles throughout a given network. This can be determined either for the entire network, its subnetworks, and/or for particular routes operated by the same and/or different transport modes. This indicator is preferred to be as low as possible and as high as possible at the same time. This implies that the network should be able to simultaneously offer services along very short (previously exclusively reserved for road) to extremely long freight delivery distances (where rail and inland waterways/barge transport modes are supposed to dominate).

Capacity of Transport Vehicle (CTV)

The capacity of a transport vehicle reflects the quantity of freight/goods shipments in terms of weight and/or volume that can be loaded onto it. This is of relevance for both transport operators and users-goods shippers/receivers. Generally, the capacity is preferred to be as high as possible. Transport operators can benefit from using transport units of a greater capacity since they can reduce the service frequency and consequently operating costs of serving the given volume of demand. Users may benefit and suffer losses at the same time. The benefits may include the increased possibility to place their shipments on larger transit units, thus enabling them to depart as soon as possible (i.e., at the desired time), while losses can include increased schedule delays of goods due to the reduced service frequency.

Service Frequency (SF)

Service frequency is the number of departures scheduled on a given route of the network during a given period of time (hour, day, and week). It is usually setup to fit the time pattern of the expected demand and provide feasible operations for transport operators. The service frequency is proportional to the volumes of freight demand and inversely proportional to the capacity of transport unit(s) and rate of their utilization, i.e., load factor. Users prefer the service frequency to be as high as possible.

Schedule Delay (SD)

Schedule delays reflect the way of interaction between the demand represented by the number of freight/goods shipments (containers, swap-bodies and semi-trailers) and the service frequency on particular routes of a given network. It is defined as the time between the desired (most convenient) and the actual departure of loading units from a given origin to a desired destination. This is actually the waiting time of loading units at the origin terminal(s) for the nearest departure of transport service. This time is mostly dependent on the time pattern of demand, service frequency (i.e., headway between the successive departures), and punctuality of services (delays). Users and terminal operators prefer the schedule delay do be as short as possible: the former because shorter schedule delays can make their networks more attractive and competitive thanks to more frequent services, while the latter because longer schedule delays may require larger special capacities for handling inventories of loading and transport vehicles/units during the network/route cycle.

Average Transport Speed (ATS)

The average transport speed is the operating speed of transport means/vehicles (i.e., units) in a given network. Shippers/receivers of loading units prefer this speed to be as high as possible simply due to its contribution to shortening the door-to-door delivery time of shipments Transport and terminal/logistics operators prefer this speed to be as high as possible due to its contribution to generally shortening the turnaround time of both transport and loading units. Such shorter turnaround time requires engagement of smaller fleets of transport and loading units under given circumstances, which generally contributes to decreasing operational costs.

Average Transport Time (ATT)

The average transport time is the time the transport units spend on the routes between given origin and destination terminals. Generally, this time may have two effects on transport operators: first, a shorter time of transport units can contribute to shortening the total delivery time of loading units; second, the vehicle fleet needed to serve given volumes of demand can be smaller than otherwise. As in the case of operating speed, the networks providing the shortest transport times are always preferable.

Terminal Time (TT)

The terminal time is the time that both transport and loading units spend in the terminal(s) on-route between their origins and destinations. This time is usually spent transshipping loading units between transport units operated by the same or different transport modes. The terminal time influences the delivery time of loading units and consequently the total delivery cycle of both transport and loading units. Transport and terminal operators prefer this time to be as short as possible since transport operators can offer faster and more reliable services and thus increase their attractiveness, and at the same time increase utilization of their fleet and reduce operating costs. Terminal operators can improve efficiency and effectiveness of their services and consequently make them more attractive for transport operators. In addition, more efficient and effective terminals require less space (land) and are considered more environmentally friendly by the relevant decision makers, authorities, and public.

Average Delivery Distance (ADD)

The average delivery distance is the distance between the origin and destination of loading units measured along the transport line that connects them. It is mostly dependent on the size of a given network and the spatial pattern of demand. The average delivery distance is preferred to be as short as possible in order to increase the competitiveness of advanced nonroad freight collection/distribution networks compared to dominating road transport services on short distances.

Average Delivery Speed (ADS)

The average delivery speed is the speed of movement of loading units between their origins and destinations. This is mainly influenced by the average transport speed and terminal time. For a given delivery distance, on the one hand, it increases with increasing of the transport speed, which decreases the average transport time; on the other, it decreases with increasing of the terminal time. This speed is always preferable to be as high as possible similarly as the average transport speed.

Average Delivery Time (ADT)

The average delivery time is the door-to-door time, i.e., the time loading units spend between their origins and destinations in the given network. By definition, this time will be shorter if loading units are delivered on shorter distances at a higher delivery speed, and vice versa. It is always preferred to be as short as possible. Generally, shorter average delivery times can yield several macro- and micro-economic benefits. The main macro-economic benefit implies savings of time, which can be used either for additional production, i.e., accelerating the production and turnover processes, or for other nontransport activities. Transport operators and users of their services can generate analogous micro-economic benefits. In addition, the principal social benefit from shortening the average delivery time is reducing the "frozen" working capital during its time in the network, which may contribute to improved production and consumption processes. Consequently, the advanced freight collection/distribution networks offering the shortest delivery times are always preferable.

Punctuality (P)

Punctuality reflects delivery of loading units from shippers to receivers on-time, i.e., according to the schedule. This can be expressed by delays expressed as differences between the actual and the scheduled (planned) time of departures and arrivals of transport and loading units at the particular "reference locations" in a given network. Usually, these include terminals, temporary storage facilities, and the doors of receivers. In any case, these delays are preferred to be as short as possible.

Reliability (R)

Reliability is the ratio between the realized and planned (scheduled) transport services in the given network. In this context, the transport services can be realized either in accordance to the timetable or at the time most convenient for users/shippers/receivers of loading units. The reliability is considered to be zero if the scheduled departures do not take place and the loading units do not reach their destinations. This can be caused due to the various impacts such as bad weather, technical defects/failures, traffic incidents/accidents, and other reasons (e.g., the staff industrial actions). The transport and terminal operators, and users prefer this indicator to be as high as possible.

Coefficient of Terminal Time (CTT)

The coefficient of terminal time expresses the share of the terminal time in the total door-to-door delivery time of loading units. On the one hand, it is relevant for network and terminal operators, while on the other, it is important for terminal manufacturers. The users/shippers/receivers of loading units consider terminal operations inherently unproductive. However, they are necessary for exchanging loading units within the same or between different transport modes. As mentioned above, this time is not exclusively influenced by the terminal's capacity but also by the timetable(s) of incoming and outgoing transport services. This time and the corresponding coefficient are preferred to be as short and as low as possible, respectively. In order to achieve this, terminal operators should speed-up terminal operations by better using existing or introducing NG (New Generation) increasingly automated transshipment technologies. In addition, transport operators need to properly match incoming and outgoing services at these terminals.

Transport Work (TW)

Transport work is the product of the number of loading units and their transport distance. It increases in line with the number of the above-mentioned components individually and/or both simultaneously. For network operators, this represents one of the simplest measures of output and is preferred to be as great as possible. As such, this output can reflect networks with (i) smaller volumes of loading units transported over long distances; (ii) greater volumes of loading units transported over shorter distances; and (iii) larger volumes of loading units transported over longer distances. In some freight transport markets, higher output may provide a greater market power and thus make the operator's position more competitive.

Intensity of Network Services (INS)

Intensity of network services is defined as the transport work carried out over the unit of the network or its route length per unit of time. It is preferred to be as high as possible and can be particularly relevant for the local and central authorities aiming to ensure balanced development and proper utilization of transport infrastructure and co-ordination of land-use planning.

Technical Productivity (TP)

Technical productivity reflects the transport work carried out by one or a fleet of transport units per unit of time. It is estimated as the product of the capacity of transport unit (or fleet) and transport (operating, commercial) speed. By definition, this indicator increases as either or both factors increase. From the standpoint of network (transport) operators, it is preferred to be as high as possible.

Costs (C)

The costs reflect the total expenditures to operate a given network. These costs directly influence the overall profitability and indirectly competitiveness of the network (transport) and terminal operators. In many cases, both local and central authorities can partially subsidize the networks with lower total cost. Consequently, this indicator is generally preferred to be as low as possible.

Externalities (E)

Externalities represent the costs of impacts of a given network on the environment and society. These include energy consumption and related emissions of GHG (Green House Gases), land use, noise, congestion, and traffic safety (incidents/accidents) (EC 1996). Communities and authorities at different (local, national, and international) levels are particularly interested in as low as possible above-mentioned impacts. Consequently, as in case of other transport concepts, advanced collection/distribution networks with lower impacts will be preferable.

Table 3.1 Indicators and measures of performances as ranking attributes/criteria of advanced freight collection/distribution networks (Janic et al. 1999; Janic 2005)

Indicator	Unit[a]	Character	
		'Benefit'	'Cost'
1. Size of the network (SN)	–	+	
2. Average route length (ARL)	(km)	+	
3. Capacity of transit vehicle (CTV)	(TEU/TU)	+	
4. Frequency (F)	(TU/h)	+	
5. Schedule Delay (SD)	(h)		–
6. Average transport speed (ATS)	(km/h)	+	
7. Average transport time (ATT)	(h)		–
8. Terminal time (TT)	(h)		–
9. Average delivery distance (ADD)	(km)		–
10. Average delivery speed (ADS)	(km/h)	+	
11. Average delivery time (ADT)	(h)		–
12. Punctuality[b] (P)	(min, h)	+	
13. Reliability[b] (R)	(%)	+	
14. Coefficient of terminal time (CTT)	–		–
15. Technical productivity (TP)	(TEU-km/h)	+	
16. Transport work (TW)	(TEU-km)	+	
17. Intensity of network services (INS)	(TEU-km/km)	+	
18. Cost (C)	(€/TEU-km)		–
19. Externalities[b] (E)			–

[a] *TEU/TU* Twenty Foot

[b] Indicators identified as relevant but not applied in the ranking

Equivalent Unit/Transit Unit; (km/h) kilometers/hour, (h) hour, (TU/h) Transit Units/hour, (TEU-km/h) TEU-kilometers/hour, TEU-km TEU-kilometers, (TEU-km/km) TEU-kilometers/kilometer of network length; (€/TEU-km) €//TEU-kilometer; *TEU* is an equivalent for 20′ (foot) container

3.2.2.4 A Methodology for the Multicriteria Ranking of Networks

The methodology applied to the multicriteria ranking of advanced freight collection/distribution networks enables assessing their overall feasibility as alternatives respecting the preferences of particular actors/stakeholders involved. As such, this methodology includes definitions of the particular attributes/criteria of network performances, the ranking method, and its application to the selected network cases.

Attributes/criteria

The above-mentioned network-specific indicators of performances are used as the attributes/criteria for ranking different configurations of advanced freight collection/distribution networks. For such a purpose, they are classified as "benefit" and "cost" attributes/criteria given in Table 3.1.

The (+) sign denotes a "positive" ("beneficial") preference, i.e., as the indicator's value is higher, the corresponding network concept will be more preferable, and vice versa. The (−) sign denotes a "negative" ("cost") preference of the indicator, i.e., if the value of this indicator is greater, the corresponding network will be less preferable, and vice versa.

Ranking method

The ranking method consists of its basic structure, techniques for normalizing the values of attributes/criteria in order to make them convenient for comparison across different network configurations, and the methods/models for assigning weights to attributes/criteria reflecting their preferences for DMs (Decision Maker(s)).

Structure

The simple additive weighting (SAW) method is used as one of the best-known and widely used multiple-criteria methods for ranking different alternatives, in this case advanced freight collection/distribution networks. Its basic structure is as follows (Hwang and Yoon 1981):

Let m represents the number of alternatives to be ranked and n the number of criteria per alternative. Suppose that DM has assigned a set of importance weights $w = (w_1, w_2, \ldots, w_n)$ to particular criteria. Then the most preferable alternative A^* is selected such that:

$$A^* = \max_i\{A_i\} = \max_i \sum_{j=1}^{n} w_j r_{ij} / \sum_{j=1}^{n} w_j \Bigg\} \quad \text{for} \quad i = 1, 2, \ldots, m. \tag{3.5}$$

where $\sum_{j=1}^{n} w_j = 1$, and r_{ij} is the normalized outcome of the alternative A_i with respect to j-th "benefit" criterion (the "cost" criterion is converted to the "benefit" criterion by taking the reciprocal of the former before normalization).

Normalizing the values of attributes/criteria

Normalizing the values of attributes/criteria can be carried out by using different techniques such as, for example, the "vector normalization" and the "linear scale transformation" technique. In the given case, the latter technique is applied by dividing the value of a given attribute/criterion by its maximum value, provided that all attributes/criteria are defined as "benefit' criteria (i.e., the larger the value of the attribute/criterion, the greater its preference, and vice versa). The 'cost' attributes/criteria can be treated as the "benefit" attributes/criteria by taking the inverse of their outcomes, and vice versa. Let x_{ij} be the outcome of the "benefit" attribute/criterion (j) of the alternative (i). Then, the transformed outcome of x_{ij} will be equal to:

$$r_{ij} = \frac{x_{ij}}{\max_i x_{ij}} \tag{3.6a}$$

It is clear that: $0 \le r_{ij} \le 1$ and that the outcome is more favorable as r_{ij} approaches 1.0. The advantage of this scale-transformation is that all outcomes are transformed in a linear (proportional) way. For the "cost" attribute/criterion x_{ij}, the transformed outcome will be as follows:

$$r_{ij} = \frac{1/x_{ij}}{\max_i(1/x_{ij})} \tag{3.6b}$$

Thus, $r_{ij} = 0$ implies the worst outcome and $r_{ij} = 1$ the best outcome of a given attribute/criterion (Hwang and Yoon 1981).

Assigning weights to attributes/criteria

Assigning weights to particular attributes/criteria to reflect the DM's preference for a given alternative(s) for can be carried out by different methods. In general, weights can be assigned by experts and/or DM (Decision Maker) according its own preferences. If expert weights are lacking, analytical methods can be applied, one of which is the entropy method used in this chapter but described in detail in Chap. 5, (Hwang and Yoon 1981).

3.2.2.5 Application of the Methodology

Input

Sixteen of the 19 indicators of performances of the above-mentioned collection/distribution networks have been estimated in Table 3.1 for the given cases of the above-mentioned advanced collection/distribution networks. This has uncovered a high diversity of particular indicators and measures across the routes and sub-networks of the same and/or different networks. Therefore, the average (i.e., typical) values of particular indicators and measures have been used. Consequently, with the exception of the indicators "Size of the network' (1), "Transport work" (17), and "Intensity of network services" (16), which relate to the entire network, all other indicators relate to an average (typical) route of the given network. This guarantees consistency of the approach, which, at the same time, possesses advantages and disadvantages. The main advantage seems to be the high level of simplification and relative easy use of the selected ranking method/model. The main disadvantage seems to be the risk of losing some important details on network performances due to the high level of aggregation of values of particular indicators. The values of indicators and measures of performances of the selected cases of collection/distribution networks used as the ranking attributes/criteria are given in Table 3.2.

Results

The SAW method applied to the above-mentioned 15 advanced collection/distribution networks as alternatives has provided estimates of their performances and ranks given in Table 3.3.

Among the three P–P collection/distribution networks, the RoadRailer network appears the one with the best performances followed by the RailRoads and Jämsänkoski network. The TCD network concepts have not been identified in the considered sample. Among the six evaluated HS collection/distribution networks,

Table 3.2 Indicators and measures of performances of the selected advanced freight collection/distribution networks (according to the SAW method) (Janic et al. 1999; Janic 2005)

Network	Indicator/measure of performances[a]															
	SN	ARL	CTV	F	SD	ATS	ATT	TT	ADD	ADS	ADT	CTT	TW	TP	INS	C
Point-to-Point																
Jämsänkoski	5/4	2009	26	1	84.0	26.6	75.5	48.0	2009	15.9	126.1	0.381	104437	728	259	–
RailRoads	8/5	1121	42	7	12.0	38.0	29,5	12.0	1121	20.9	41.5	0.289	1647870	1597	294	–
RoadRailer	2/2	607	45	4	22.6	58.0	10.5	3.7	607	42.7	14.2	0.261	399526	2594	355	–
Trunk line-TCD	–	–	–	–	–	–	–	–	–	–	–	–	–	–	–	–
Hub-and-Spoke																
Wembley EFOC	16/15	300–1670	40	7	12.0	40.3	24.1	1.0	766	31.8	25.1	0.382	3564760	1733	232	–
Hub of Metz and the Quality Net	37/37	60–2400	44	28	3.0	35.5	22.7	4.3	1619	31.4	51.7	0.083	9576292	1564	1186	–
"Drehscheiben'	–	–	–	–	–	–	–	–		–	–	–	–	–	–	–
The NEN	8/7	100–750	65	4	21.0	32.3	5.4	28.8	370	9.2	40.3	0.715	4667612	2522	603	
Bahntrans	11/10	100–750	40	7	12.0	20.6	5.8	6.0	350	15.2	23.0	0.261	9100000	1200	4550	–
GT Hupac[b]	13/12	320–1250	43	9	9.3	45.2	19.7	24.0	862	19.7	43.7	0.549	5414264	1932	638	–
Line (or) Ring																
Piggy-back	–	–	–	–	–	–	–	–	–	–	–	–	–	–	–	–
RingZug	10/1	167	82	35	2.4	23.7	7.0	1.6	167	19.4	8.6	0.186	623200	1946	5438	–
The "Linienzug'	6/7	259	34	24	3.5	95.7	2.7	3.5	259	41.8	6.2	0.565	280376	3378	537	–
Mixed																
Voltri	9/8	140–1500	27	7	18.9	60.0	4.2	33.0	251	6.8	37.2	0.877	373329	1620	164	–
Sogemar	26/19	100–500	27	3	28.0	45.7	20.3	28.0	647	14.3	45.3	0.552	1052973	1231	168	–
FlexNode	–	–	–	–	–	–	–	–	–	–	–	–	–	–	–	–

[a] See Sect. 3.2.2.3 and Table 3.1 for an explanation of the abbreviations
[b] *GT* Gateway Terminal

Table 3.3 Performances and ranks of the selected collection/distribution networks in the given example (Janic et al. 1999; Janic 2005)

Network	Performance	Rank
P–P (Point-to-Point)		
Jämsänkoski	0.198	3
RailRoads	0.686	2
RoadRailer	0.745	1
TCD (Trunk line-CD forks)	–	–
HSs (Hub-and-Spoke(s))		
Wembley EFOC	0.430	4
Hub of metz and the quality net	0.569	1
Drehscheiben	–	–
The NEN	0.314	5
Bahntrans	0.550	2
GT Hupac	0.472	3
L/R (Line or Ring)		
Piggy-back	–	–
Ring zug	0.695	1
The "Linienzug'	0.529	2
M (Mixed)		
Voltri	0.585	2
Sogemar	0.744	1
Flex-node	–	–

the rail-based Hub of Metz and the Quality Net emerges as the preferable network. The Ring-Zug Rhein-Rhur network appears the preferable L/R collection/distribution network, while the Sogemar network is identified as preferable among the three M collection/distribution networks.

3.2.2.6 Preferable Network(s)

The multicriteria ranking of the above-mentioned advanced freight collection/distribution networks enables us to synthesize the generic performances, which should be similar to those of the RoadRailer (P–P), Hub of Metz and the Quality-Net (HSs), Ring Zug Rhein-Rhur (L/R), and Sogemar (M) networks.

The RoadRailer network as the most preferable P–P network suggests the following generic performances of similar networks:

- The network consists of a begin- and end-terminal where the pure sequential exchange of loading units between road and rail (and vice versa) takes place;
- The network serves relatively large and predictable freight flows in both directions, on transport distances of between approximately 400 and 600 km;
- The large volumes of predictable and regular freight flows justify operating shuttle (or direct) trains at a regular frequency; this frequency should be matched with

demand (typically, one service per day or one service per 2 days seems to be quite acceptable);
- Flexibility in adapting the transport capacity to demand through changing the service frequencies and capacity of transport units (trains) makes this network both beneficial for its operators and users (shippers/receivers);
- The relationships between transport speed, delivery distance, and total terminal time provide the service quality A (delivery in 24 h) and/or B (delivery in 48 h); and
- Operating trains of a relatively large capacity at relatively high transport speeds produces the concept's high technical productivity, intensity of network services, and transport work; these make such networks competitive to road-haulage in equivalent freight transport markets (characterized by similar volumes of freight and delivery distances) with respect to the total delivery time, utilization of the available infrastructure, and reduction of the negative impacts on the environment and society.

The Hub of Metz and Quality Net network as the most preferable HSs network suggests the following generic performances of similar networks:

- The network should be as large as possible in terms of the number of terminals (nodes) and routes (links). The hub terminal should be located at the intersection of important rail-lines ("axis"), i.e., in a central location with respect to the location of the other terminals (spokes); it needs to provide a sequential exchange of rail wagons and/or wagon groups (and loading units) between the incoming and outgoing trains; the exchange time and the total terminal time of loading and transport units can be shortened by ensuring proper balance of the inbound and outbound train timetable(s);
- The spoke terminals should be located over a large area (such as half of Europe) in order to provide high spatial accessibility of the network services and delivery of loading units over a wide spectrum of distances (routes) (from several tenths—let's say 60 km to a few thousand—2,500 km); the spatial location of terminals and high diversity of route lengths allows the establishment of different subnetworks (and thus segmentation of markets and related services), which begin and end at the central/common hub;
- Regular services of direct trains should be provided on each route connecting the spokes and the hub; the frequencies should be sufficient, at least one per day, but adjusted to demand, capacity, and utilization of trains;
- Regarding the frequency (i.e., schedule delay), delivery distance, average transport speed (20–40 km/h), and the average terminal time (about four hours), different types of quality of service can be offered: A/B (delivery up to 24 h on distances up to 800 km), A/C (delivery up to 48 h on distances from 800 to 1,600 km), and A/D (delivery up to 72 h on distances over 1,600 km);
- With respect to transport and delivery distances and distribution of the route lengths in the network (long routes prevail), the terminal time is expected to have a rather negligible influence on the total delivery time of loading units; therefore, introduction of NG terminal transshipment technologies to speed-up

the exchange of loading units between incoming and outgoing trains seems to be of rather low added value; and
- A network of such a configuration and frequency of services is capable to replace the high volumes of transport work on the particular routes, which otherwise would be carried out by road; thus, it can offer significant reduction of impacts on the environment and society over a wide area as compared to road haulage.

The Ring Zug Rhein-Rhur network as the preferable L/R network suggests the following generic performances of similar networks:

- The network should consist of a reasonable number of terminals (say about 10), located at relatively short distances (about 30–40 km) along the line or ring. The begin-, end-, and intermediate terminals should provide sequential exchange of loading units between the same and/or different transport modes. The network should cover a relatively small area (region) and should be able to generate and attract big, regular, predictable, and dense freight flows; this justifies frequent shuttle train services (a few per day) delivering loading units to very close destinations, up to 30–40 km (regional) and up to 200 km international transport;
- The high frequency, i.e., very short schedule delay, short delivery distance, acceptable transport speed, and the short terminal time guarantee A-quality services (delivery in up to 24 h); the terminal time may be further reduced by introducing NG terminal transshipment technologies enabling simultaneous (pure and batch) exchange of loading units; this may additionally rise the competitiveness of the network(s) in short distance markets compared to the currently dominating road haulage; and
- The high frequency and capacity of trains operating along short distances produce large quantities of transport work, which would otherwise be realized by road; this could significantly mitigate the environmental and social impacts of freight transport in the given region while improving utilization of the rail and terminal infrastructure.

The Sogemar network as the preferable M network suggest the following generic performances of similar networks:

- The network should consist of a great number of terminals where the sequential exchange of loading units between the same and/or different transport modes takes place; this makes the layout and corresponding through time of these terminals more dependent on the inbound and outbound timetable(s) than on the characteristics and efficiency of their operations; different subnetworks operating independently from each other and capable to cover a relatively wide area(s) can be established between particular terminals; each subnetwork may include different transport modes such as short-sea, road, and rail; the rail (inland) portion of the specific subnetwork can cover a wide range of distances, from about 100 to 1,500 km;
- Regular and frequent services by shuttle trains/barges/vessels matching the offered capacity to demand as much as possible should be provided on the particular inland routes; and

- the length of delivery distances (routes), delivery speed(s), and the terminal time(s) guarantee A-quality services (delivery in up to 24 h); introducing NG terminal transshipment technologies in order to improve the overall terminal and network efficiency and effectiveness can be reasonable; all these make the particular rail-based inland subnetworks competitive to road haulage.

3.2.3 Evaluation

Advanced freight collection/distribution networks possess both advantages (Strengths and Opportunities) and disadvantages (Weaknesses and Threats).

Advantages

- Larger batches of loading units transported at higher speeds require a smaller number of larger but highly utilized transport units (means/vehicles), lower service frequency, and less labor on the transport side of logistics chain(s); this increases the efficiency of operations, thus leading to reduction of transport-related operational cost in both absolute and relative terms;
- The said reduction in transport operational costs partially compensates the additional costs of loading units due to their inventorying at terminals; and
- Deploying larger trains and barges, or their more frequent services on main routes-markets/corridors brings: (i) increased utilization of capacity of the rail and inland navigation infrastructure; and (ii) specialization of transport operators for the specific services and consequent improvement of the competitiveness in the corresponding markets.

Disadvantages

- Formation of inventories of loading units in the network nodes-terminals causing their delays as a generic feature of these networks;
- Requiring as great as possible inventories of loading units in order to fill reasonably frequent and higher capacity direct and shuttle train or barge services on the main routes of these networks;
- Requiring more storage space, handling facilities, and equipment, energy and workforce in terminals for handling larger inventories of loading units, thus increasing the related costs;
- Diminishing the commercial speed and related door-to-door delivery time of freight/goods shipments due to their inventorying at terminals;
- Increasing time of spending larger higher-valued (time-sensitive) freight/goods shipments in the network due to inventorying of loading units, which proportionally depreciates them; and
- Inability to contribute to mitigating impacts on the environment and society even more particularly at the local level due to the inevitable use of trucks for delivering loading units to/from the begin- and end- terminal(s).

Finally, advanced freight collection/distribution networks can be quite rightly considered advanced transport systems mainly thanks to their organizational, operational, and economic advancements as compared to their conventional counterparts.

3.3 Road Mega Trucks

1950s	Direct-injection turbo-charged diesel engines become standard; standard-sized steel truck containers (that can be fixed to a trailer chassis for use on the road) are developed by Malcom McLean (U.S.)
1996	Common standards/limits/measures on the maximum length and weight of road trucks operating throughout EU (European Union) Member States are set up by the EC (European Commission) (Directive 96/53/EC) (Europe)
2008	25.25 m long road mega trucks (as compared to the standard length of 18.75 m) are allowed to operate in Finland, Sweden, Norway, Denmark, the Netherlands, and Germany for the period of 3 years (Europe)
2009	The EC (European Commission) begins considering raising the maximum weight of road trucks to 60 tons and their length to 25.25 m (as compared to the standards of 40 tons and 18.75 m, respectively) (Europe)

3.3.1 Definition, Development, and Use

A road mega truck is defined as the largest truck currently or planned to be used for transporting freight/goods by road in a given region (country, continent). The attribute "largest" refers to the truck size, i.e., its gross weight and length. In this case, the size mainly contributes to considering these trucks as "advanced". The analysis below refers to these largest and consequently most advanced road trucks in Europe. Their implementation is currently a very pressing issue for the EU (European Union) transport policy makers as they face the dilemma whether to allow their wide implementation throughout the Member States, similarly as in the USA. Beginning in 1992, the gross weight, number of axles, and length of road trucks have been limited (harmonized and standardized) to 40–44 tons, 5–6 axles, and 18.75 m, respectively. This gives a maximum axle load of these vehicles of 40–44/5, i.e., between 8 and 8.8 tons/axis. As such, these vehicles have always been referred as the "standard" trucks and used as "term(s) of reference" for other trucks. Thus, trucks heavier or longer than the above-mentioned standard ones can be can be classified as mega trucks, "road trains," or "gigaliners." As far as the scale of using these trucks is concerned, from the beginning of 2009, some categories of mega trucks have been in operations in the Netherlands (weight of 50 tons and length of 18.75 m), Sweden (weight of 60 tons and length of 25.25 m), and Finland (weight of 44/48 tons and length of 25.25 m), with some pilot trials also being carried out in Germany. Consequently, the EC's (European Commission's) Commissioner for Transport has left each Member State set up its own limits on the gross weight and length of road mega trucks, but exclusively with respect to operations in their own countries. In parallel, the EC has been considering a proposal based on scientific research to allow raising the maximum gross weight of trucks to 60 tons, 7–8 axes, and length to 25.25 m (as it is the case in Sweden). This would represent an increase of the truck's gross weight by 50 % and length by almost 40 %; the average axle load would remain almost the same or even decrease in the case of vehicles with 8 axles to 7.5 tons/axle as compared to standard trucks that typically carry 8 tons/axle (EC 1996; UIC 2008).

3.3.2 Analyzing and Modeling Performances

3.3.2.1 Background

The road mega tracks are characterized by infrastructural, technical/technological, operational, economic, environmental, and social/policy performances.

The infrastructural performances relate to suitability for using the existing road network in Europe. The technical/technological performances include the truck design and carrying capacity, engine, use of ITS/ICTTechnologies) (Intelligent Transport System/Information Communication Technologies), and technical productivity. The economic performances related mainly to the operational cost. The environmental performance relate to the fuel consumption and emissions of GHG (Green House gases), and land use. The social/policy performances of mega trucks include noise, congestion, and safety (traffic incidents/accidents).

3.3.2.2 Infrastructural Performances

In general, the European road network is designed to meet the existing standards on the weight and length of the heaviest standard trucks. However, there is evidence that this network is not completely suitable to efficiently, effectively, and safely accommodate trucks heavier and longer than the standard ones of a weight of 40–44 tons and a length of 18.75 m (European Directive 96/53/EC (EC 1996)). Consequently, the more widespread use of heavier and longer trucks (mega trucks), with a maximum gross weight of 60 tons and length of 25.25 m, will likely require modification of the existing road infrastructure as follows (ASECAP 2010; UIC 2008):

- *Bridge-bearing structures* will need to be significantly reinforced in order to accommodate higher loads and maintain the current safety standards (crashes of mega trucks inherently create higher dynamic stresses, which need to be absorbed by crash barriers of greater dimensions; in addition, since these dynamic forces are also absorbed by the bearing structure, this also needs to be reinforced);
- *Tunnels* will need to be changed structurally including re-sizing the profiles of the parking niches/breakdown bays; since the increased load of trucks inherently

increases the risk of fire almost proportionally, this will be need to be carried out in several European countries (Alpine and Pyrenees regions, etc.);

- *Access to and the parking areas* used for the mandatory breaks of drivers during their journeys will need to be reconstructed;
- *Interconnections between primary and secondary roads* such as junctions, roundabouts, etc., designed and built according to the regulations related to the above-mentioned "standard" trucks, will need also to be partially reconstructed and modified; and
- *Experience and evidence including data on the true impact of mega trucks on the wearing and tearing of the road infrastructure* will need to be obtained through experimental trials and simulations under different conditions.

Facilitating these shortcomings requires rather substantial investment, despite uncertainty about exactly how much investment is really needed. For example, do the roads and turning areas need to be widened? The turning cycle of road mega trucks is also the subject of EU regulation, which states that any vehicle independently of its configuration must be able to make a turning circle within a corridor bounded by two circles, the inner with the radius of 5.3 m and the outer with the radius of 12.5 m (EC 1996; Larsson 2009).

Nevertheless, the more widespread use of road mega trucks could bring some benefits through increased utilization of the capacity of existing road infrastructure. This capacity is expressed by the maximum quantity of goods (tons) transported over a given road link/line or passing through its "reference location" during a given period of time under given conditions. For a single mega truck replacing its standard counterpart, the increase could be about 50 %. This could mitigate the need for expanding the road infrastructure in order to facilitate the forthcoming expected growth of road freight/goods transport demand efficiently, effectively, and safely. Consequently, the pressure for additional land use/take for building new roads could be mitigated.

3.3.2.3 Technical/Technological and Operational Performances

The technical/technological and operational performances of road mega trucks relate to their design influencing capacity, i.e., payload, engine influencing fuel consumption, ITS/ICTTechnologies) (Intelligent Transport System/Information Communication Technologies), and technical productivity.

Design and capacity

The design of mega trucks appears to be relatively simple. In addition to a stronger engine, a larger trailer is usually attached to a standard truck weighting 40 tons. Figure 3.5 shows the typical configuration of a standard and mega truck. Contrary to the length, the frontal area of a mega truck is almost the same as that of a standard truck, i.e., about 7.7 m^2, which generates an aerodynamic drag coefficient of about 0.6.

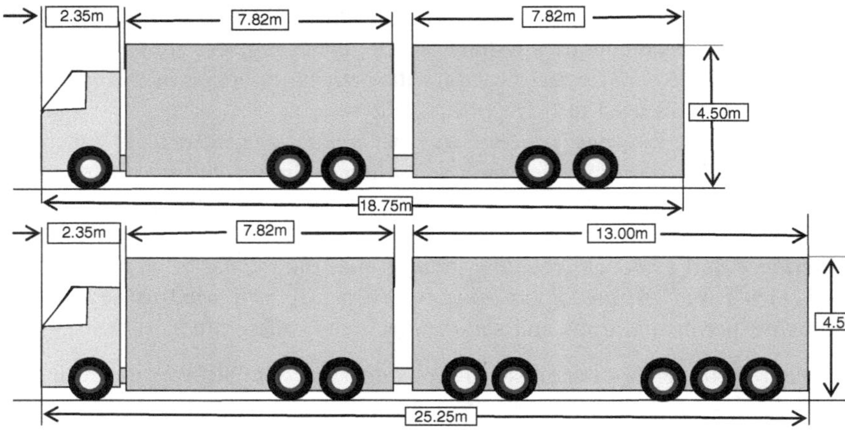

Fig. 3.5 Scheme of typical configuration of a standard and mega truck (Fraunhofer 2009)

The rolling resistance coefficient is about 0.0068 (AEA 2011). Increase in the gross weight of 50 % increases the payload capacity of a mega truck as compared to its standard counterpart by about 60 % in terms of weight (40 vs. 25 tons), 50 % in terms of the number of pallets (52 vs. 34), and 42 % in terms of the volume of freight/goods (160 vs. 105 m^3). In addition, the ratio between the payload capacity and the gross weight of a mega truck is: 40/60 = 0.667, which is about 6.7 % greater than that of a standard truck: 25/40 = 0.625, which contributes to its higher overall efficiency. Consequently, the transport capacity of the operator's fleet will incrementally increase by about 60 % as soon as the fleet of mega trucks is deployed instead of the fleet of their standard counterparts.

Engine

The engines of both standard and mega trucks is designed respecting the Euro V emissions legislation[4] with different volumes of 10, 11, 12, 13, or 16 l in line or 6, 12 l V6 or 16 or 18 l V8 configuration. Following the emissions legislation, these diesel engines require the use of both EGR (Exhaust Gas Recirculation) and SCR (Selective Catalytic Reduction) after treatment to meet the prescribed emission limits.

---

[4] In Europe, two solutions to meet emissions legislation are widely adopted at Euro V: one using high levels of EGR along with DPF (Diesel Particulate Filter); and the other using SCR (Selective Catalytic Reduction). At Euro VI, it is anticipated that OMs (Overall Emission(s)) will follow the current technologies in Japan and USA using both EGR and SCR to meet the new prescribed emission limits. Both will have an impact on the fuel consumption, which could be reduced through engine power transmission technologies available. In Europe, AMT (Automatic Manual Transmission) is favored while in the U.S. and Japan the standard fit transmissions are both manual. In particular, AMT can reduce fuel consumption by up to about 7 % depending on the driver's skills.

Road mega trucks are expected to have generally stronger engines than standard trucks, with the power of about: 290–450 kW (1 kW = 1.36 hp; kW—kilowatt; hp—horse power) at 1,700 rpm (rotations per minute). This power can be adjusted during operation by either ATMS (Automated Manual System) offering 12 to 16 or MS (Manual System) with 12 speed changes.

ITS/ITC (Intelligent Transport System/Information Communication Technologies)

ITS/ICT (Intelligent Transport System/Information Communication Technologies) are deployed to enable efficient, effective, and safe trips of mega trucks. They include adaptive cruise control, navigation system, fleet board telematics system, and forward collision warning system. The adaptive cruise control and collision warning systems are foremost comfort and safety systems, respectively. In addition, they constrain harsh acceleration and braking maneuvers, which, depending on the driver, may improve efficiency of the fuel consumption by up to 1 %. The fleetboard telematics system records the data of several vehicle parameters, which can be transmitted in real-time or at specified intervals. Typically, these include information/data on the vehicle speed, engine rpm, the rate of braking and acceleration, idling time, etc. Together with the driver ID (Identification) obtained from the digital tachograph, this information/data enable any potential correlation between the driving style and fuel consumption to be analyzed (AEA 2011).

Technical productivity

Technical productivity is expressed by the quantity of output produced by one unit of production input (in this case a single vehicle) per unit of time. For either a loaded standard or mega truck, this can be estimated as follows:

$$TP(\lambda, PL, v(d)) = \lambda * PL * v(d) \tag{3.7}$$

where

PL is the vehicle payload capacity (tons/truck);

λ is the vehicle load factor (between 0 and 1); and

$v(d)$ is the vehicle speed dependent on the freight/goods delivery distance d.

The technical productivity $TP(,PL, v(d))$ expressed in t-km/h (ton-kilometers/hour) increases in proportion with the payload, operating or commercial speed, or both simultaneously. For example, a full mega truck operating at a speed of 70 km/h has the technical productivity of 40 tons × 70 km/h = 2,800 t-km/h. A full standard truck operating at the same speed has the productivity of 26 tons × 70 km/h = 1,820 t-km/h.

3.3.2.4 Economic Performances

The main economic performances of standard and mega trucks include their operating costs. The operating costs of standard trucks include the costs of vehicle depreciation (10 %), interest (2 %), insurance (6 %), road tax (2 %), tires (1 %), fuel (30 %), wages/labor (26 %), overheads (18 %), and others (5 %) (The figures in brackets illustrate the share of particular cost component in the total operating costs for a standard truck operated by an EU road freight transport operator (AEA 2011)). The structure of operating costs of mega trucks is similar with the exception of a higher proportion of insurance, road tax, and tire costs. As expressed in averages such as €/vehicle-km and/or €/t-km, the operating costs of both vehicles become relevant for setting the prices of services. In general, these costs decrease with increasing of the delivery distance more than proportionally, thus indicating the existence of economies of distance. The typical analytical relationship obtained by using the empirical data is as follows (Janic 2007):

$$c(d) = a_0 * d^{a1} \tag{3.8a}$$

where

$c(d)$ is the average unit cost (€/vehicle-km);

a_0, a_1 are coefficients to be estimated; and

d is the distance (km).

By dividing the cost $c(d)$ in Eq. 3.8a by the vehicle payload capacity or the actual payload, the average cost per unit of payload per unit distance (€/t-km) can be estimated as follows:

$$c(d, PL, \lambda) = \left(a_0 * d^{a1}\right)/(\lambda * PL) \tag{3.8b}$$

where the particular symbols are analogous to those in the previous equations.

The payload capacity and corresponding unit costs in Eq. 3.8b can also be expressed depending on the type of goods transported per pallet and/or per unit of available space on the vehicle (m^3).

Comparison between the average unit costs of mega and standard road trucks can be carried out after introducing the following hypothesis (Fraunhofer 2009): The mega truck is more efficient than its standard counterpart mainly thanks to its greater payload capacity. Further elaboration in this section is based on the potential substitution of standard trucks with mega trucks and the possible effects. For such purpose, the generic model of the substitutive capability of mega trucks as compared to that of the standard trucks based on the equivalent quantity of freight/goods to be transported along the given distance/route is developed. Let (i) and (j) be the standard and the mega truck, respectively. The number of required vehicles of both categories, n_i and n_j, respectively, can be estimated as follows:

$$n_i(d) = Q(d)/\lambda_i * PL_i \text{ and } n_j(d) = Q(d)/\lambda_j * PL_j \tag{3.8c}$$

where

$Q(d)$ is the quantity of goods to be transported either by truck (i) or by truck (j) on the distance d (tons);

λ_i, λ_j is load factor of the truck (i) and (j), respectively;

PL_i, PL_j is the payload capacity of the truck (i) and (j), respectively (tons); and
d is the delivery distance of the freight/goods $Q(d)$ (km);

If the trucks (j) substitute the trucks (i), the factor of substitution can be estimated as follows:

$$S_{j/i}(d) = n_j(d)/n_i(d) = \lambda_j * PL_j/\lambda_i * PL_i \tag{3.8d}$$

where the symbols are as in the previous equations.

Applying Eqs. 3.8a, b, c, d to the above-mentioned data on the payload capacity of both types of trucks indicates that two full mega trucks can substitute three full standard trucks, which a priori represents savings in operators' costs. But how much do these savings amount to? The answer lies in estimating the difference between the average unit costs of transporting a given quantity of freight/goods according to the above-mentioned scenario (2 instead of 3 vehicles deployed). For such purpose, the given quantity of freight/goods is assumed to be consolidated into TEU (20-Foot Equivalent Unit) containers as the most common in Europe. Each of these containers has an average gross weight of 14.3 tons (12 tons of goods plus 2.3 tons of tare) (Janic 2007). Consequently, respecting their payload capacity, a standard truck can carry the equivalent of two TEUs and a mega truck the equivalent of 3 TEUs. Thus, for six TEUs, 3 standard or 2 mega trucks are needed. Then, applying the regression technique based on (Eqs. 3.9a, b) to the empirical data, the average costs for a standard truck loaded with 2 TEUs is estimated as follows: $c_i(d) = 5.456d^{-0.277}$ (€/vehicle-km) ($\lambda = 0.85$; $N = 26$; $R^2 = 0.781$; $25 \leq d \leq 1{,}600$ km) (Janic 2007). By taking into account the difference in the payload capacity and the number of required trucks for the above-mentioned task, the average costs for the mega truck are estimated similarly as follows: $c_j(d) = 6.913d^{-0.277}$ (€/vehicle-km). Both regressions indicate economies of distance at both standard and mega trucks. At the same time, for the given distance, the average unit cost of a mega truck is greater by about 27 %, However, by dividing these costs by the corresponding payload capacity and assumed load factor, the following average unit costs are obtained for a standard truck and for a mega truck, respectively: $c_i(d, 26, 0.85) = 0.201d^{-0.277}$ (€/t-km) and $c_j(d, 40, 0.77) = 0.167d^{-0.277}$ (€/t-km). In calculating the mega truck average unit costs, its lower load factor, higher fuel efficiency and share of fuel costs in the total operating costs are taken into account. Consequently, it seems realistic to expect that, for a given distance, the average mega truck unit costs will be by about 20 % lower than those of standard trucks under the above-mentioned substitution scenario.

3.3.2.5 Environmental Performances

The environmental and social/policy performances of mega trucks relate to their energy/fuel consumption and related emissions of GHG (Green House Gases), and land use.

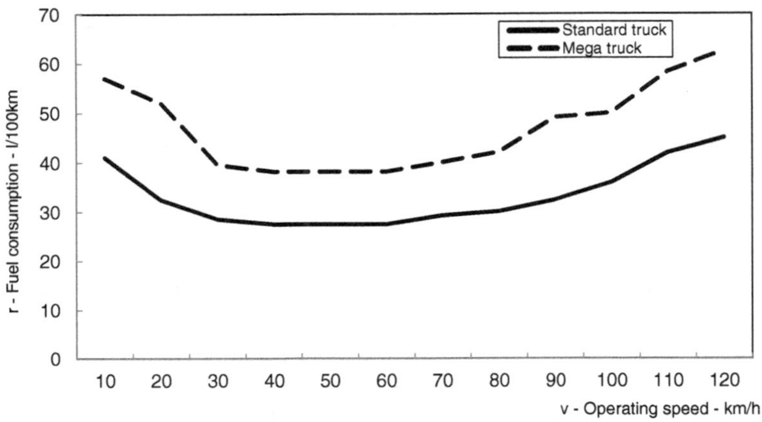

Fig. 3.6 Relationship between the operating speed and the average fuel consumption of the road standard and mega trucks (AEA 2011)

Fuel consumption

The engines of both standard and mega trucks consume diesel fuel as a derivative of crude oil. If this consumption is at the rate r (l/100 km) for each truck, the total consumption of the given convoy of n trucks operating along the distance d can be estimated as follows:

$$FC(n) = n * r * (100 * d) \tag{3.9a}$$

The main operational factors influencing the fuel consumption are the operating speed, and the aerodynamic and the rolling resistance, the latter being higher in the case of a mega truck compared to a standard truck due to the greater number of axes (7/8 vs. 5/6). With increasing speed, the total resistance linearly increases mainly thanks to an increase in the aerodynamic resistance on the account of the rolling resistance (AEA 2011). Figure 3.6 shows an example of the relationship between the average fuel consumption and the operating speed of a mega and a standard truck with 70 % of the maximum payload of 40 and 25 tons, respectively.

As can be seen, the average fuel consumption of both categories of trucks first decreases with increasing of the operating speed, then remains relatively constant for a range of speeds between 40 and 60 km/h, and then increases again. In most European countries, the maximum speed of these trucks is and will be limited to 90 km/h on motorways and 70 km/h on other roads, which implies that the corresponding fuel consumption will be 29–32 l/100 km for standard and 40–45 l/100 km for mega trucks.

The stronger engines of road mega trucks will enable performances similar to those of its standard counterpart. For example, if the engine power/gross weight ratio remains the same as that of standard trucks, mega trucks will need more powerful engines by about 50 %. Despite having stronger engines, mega trucks are expected to be more fuel efficient than standard trucks by about 8, 10/11, 10, and

3 % with respect to the gross weight and payload capacity in terms of weight, the number of pallets, and the volumes of goods, respectively. At the same time, however, mega trucks will certainly consume more fuel in absolute terms (by about 40 % per unit of distance—l/100 km; l—liter) than their standard counterparts (UIC 2008; http://lipasto.vtt.fi/yksikkopaastot/tavaraliikennee/tieliikennee/tavara_tiee.htm).

Emissions of GHG

Similarly as fuel consumption, emissions of GHG (Green House Gasses) can be considered for a single vehicle and for a convoy of vehicles. For example, the emissions of GHG in terms of CO_{2e} (Carbon-Dioxide equivalents) by a convoy of standard or mega trucks operating along the distance d can be estimated as follows:

$$EM = FC * e = n * r * (100 * d) * e \tag{3.9b}$$

where

FC is the total fuel consumption of the convoy of n trucks/vehicles (l; l-liter));
r is the rate of fuel consumption per a vehicle/truck (l/100 km); and
e is the emission rate per unit of fuel consumed ($kgCO_{2e}$/l of fuel)

The other symbols are as in the previous equations.

The rate e usually relates to on-wheel emissions including emissions from manufacturing fuel and emissions from the direct burning of fuel.

Mega and standard trucks consume diesel fuel at the rate r depending on the operating speed and other conditions as shown in Fig. 3.6. As mentioned above, mega trucks consume by about 40 % more fuel than standard trucks under the same conditions. However, if convoys are considered, then a convoy of two mega trucks consumes about 6–8 l/100 km less fuel that a convoy of three standard trucks operating at a speed of between 50 and 90 km/h. Similar applies to emissions of GHG. For example, when both mega and standard trucks operate at an average speed of $v_j = v_i = 70$ km/h, their fuel consumption rate will be: $r_j = 40$ l/100 km, and $r_i = 29$ l/100 km, respectively. The specific gravity of diesel fuel is: 0.82–0.95 kg/l, and its calorific value: 12.777 kWh/kg (EC 2005). This produces an average rate of energy consumption of about: $SEC_j = 4.59$ kWh/km for mega trucks, and: $SEC_i = 3.33$ kWh/km for standard trucks. The emission rate of GHG is: $e_j = e_i = 0.324$ $kgCO_{2e}$/kWh (CO_{2e} includes CO (Carbon Oxide), HC (Hydro Carbons), NO_x (Nitrogen Oxides), and PM (Particulate Matters)). This gives emissions of GHG by a convoy of two mega trucks of: $EM_j = 2 * 1.487 = 2.974$ $kgCO_{2e}$/km and for a convoy of three standard trucks of: $EM_i = 3 * 1.079 = 3.237$ ($kgCO_{2e}$/km). Thus, despite the emissions of GHG by a single mega truck (1.487 $kgCO_{2e}$/km) are higher than that of a single standard truck (1.079 $kgCO_{2e}$/km), which is in proportion to the differences in the rates of fuel consumption, they appear lower for a convoy of mega trucks substituting a convoy of standard trucks under the assumed 2/3 scenario. In addition, the average emission intensity is lower for the convoy of mega trucks than that of standard trucks, i.e., 53.11 gCO_{2e}/tkm compared to 61.66 gCO_{2e}/tkm, respectively (payload rate is 70 %).

Land use

As compared to the already operating standard trucks, mega trucks will not require additional roads and consequent land use. They will nevertheless need bigger parking and maneuvering spaces, which is greater for a mega truck than for a standard truck by about 35 %, i.e., in proportion to the difference in their length. However, if mega trucks substituted standard trucks at the 2/3 rate, the total additional parking space would likely increase by about 23 %.

3.3.2.6 Social/Policy Performances

The social/policy performances of standard and/or mega trucks relate to their noise, congestion, and traffic/incidents (safety).

Noise

The noise of standard and/or mega trucks generally depends on the following factors:

- Intensity of noise at source;
- Distance of the noise source from the receiver(s)/observer(s);
- Duration of the noise event(s);
- The number of successive noise events during a given period of time; and
- Noise mitigating measures.

The intensity of noise at source depends on the constructive (design) characteristics of the vehicles (trucks) and their speed while passing the given receiver(s), i.e., the potentially affected inhabitants.

The distance of noise source(s) from the receiver plays an important role since the volume of sound at the receiver's location generally decreases in line with the square of the distance from the noise source(s). This implies that highways with intensive truck traffic are preferred to be sufficiently far away from populated areas.

The duration of noise event implies the time of exposure of a receiver to noise from passing truck(s). In general, this time is proportional to the length of truck and its speed.

The number of successive noise events—passing trucks—during a given period of time reflects the persistency of noise over time.

Noise mitigating measures include diminishing causes of noise through truck design and construction, limiting their operating speed, constraining operation during particular periods of the day (night), and/or installing noise protecting barriers at particularly "noise-sensitive" locations. In order to estimate the level of noise a receiver near a given motorway/highway is exposed to, a convenient model is developed based on the generic scheme shown in Fig. 3.7.

The shaded polygon represents a convoy of mega trucks of the length S_j moving at an average speed v_j along the distance d. A receiver represented by the small

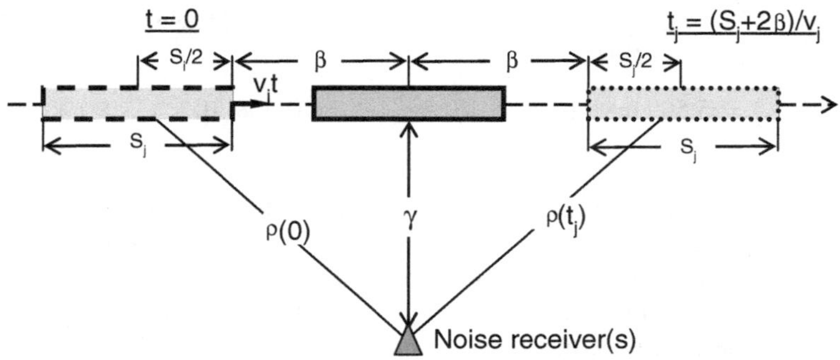

Fig. 3.7 A scheme for estimating the noise exposure of a receiver by passing road truck(s) (Janic and Vleugel 2012)

triangle starts to consider an approaching convoy of trucks when it is at distance β from the point along the line, which is at the closest right angle distance γ from him/her. The consideration stops after the convoy moves behind the above-mentioned closest point again to the distance β. Under such circumstances, the distance between a receiver and passing trucks changes over time as follows:

$$\rho^2(t) = (S_j/2 + \beta - v_jt)^2 + \gamma^2 \quad \text{for} \quad 0 < t < = (S_j + 2\beta)/v_j \tag{3.10a}$$

If each truck is of length s_j and if the distance between successive trucks in the convoy moving along the segment d is δ (v), the length of a convoy will be as follows:

$$S_j = n * s_j + (n - 1) * \delta(v_j) \tag{3.10b}$$

where the number of trucks in the convoy n_i can be estimated from Eq. 3.8c. In general, the distance between the trucks in the convoy δ (v_i) is an increasing function of the operating speed v_i.

If the level of noise received from the r-th truck passing the receiver at the speed v_i at the shortest distance γ is $L_{eq}(\gamma, v_i)$, the level of noise during time t_j can be estimated as follows:

$$L_{eq}\left[\rho(t), v_j\right] = L_{eq}\left(\gamma, v_j\right) - 8.6562 * \ln\left[\rho\left(t_j\right)/\gamma\right] \tag{3.10c}$$

The second term in Eq. 3.10c represents the noise attenuation with distance between the noise source and a receiver over an area free of barriers.

The total noise exposure of a receiver(s) from n_j successive trucks passing by during the period of time T can be estimated as follows:

$$L_{eq}\left(n_j\right) = 10 \log \sum_{r=1}^{n} 10^{\frac{L_{eq}\left[\rho(t), v_j\right]}{10}} \tag{3.10d}$$

The noise of an individual event is commonly measured as the standard at the right-angle distance of 25 m (m–meter) from the source (i.e., passing trucks) and at height of 3 m. Typical values depending of the standard truck's operating speed are as follows (EC 2001a):

$$L_{eq}(25,\ v_j) = 5.509 \ln v_j + 25.36\,\mathrm{dB(A)},\ \left(R^2 = 0.988; 10 < v_j < 90\,(\mathrm{km/h})\right) \quad (3.10\mathrm{e})$$

The noise from mega trucks is expected to be higher mainly because of their stronger engines and increased rolling resistance due to the greater number of axles. In such case the above-mentioned expression needs to be slightly modified. In addition, the exposure to noise of a single mega truck will last longer, likely in proportion to the difference in its length and the length of a standard truck (25.25/18.75, which is about 35 %), However, the exposure to noise from a convoy of three standard trucks will be longer than the exposure to that of a convoy of two mega trucks by about 11 %. Furthermore, the equivalent noise level will be higher for the former than for the later convoy of trucks, i.e., $L_{eq}(i) = 55.25$ dB(A) and $L_{eq}(j) = 53.89$ dB(A), respectively, if both move at the same speed of 90 km/h and pass the noise receiver at the same distance.

The above-mentioned estimates of noise exposure by standard and mega trucks can be further generalized by including the frequency of noise events and the number of population located close and consequently exposed to these events. In considering the actual exposure of population located close to a highway/motorway to noise by passing convoy(s) of trucks, it is necessary to take into account the influence of noise protective barriers. They are usually set up to protect particularly noise-sensitive areas by absorbing the maximum level of noise of about 20 dB (A) (single barrier) and 25 dB (A) (double barrier).

Congestion

Congestion caused by a convoy of standard and/or mega trucks is expressed by the time losses of the individual passenger cars and other smaller vehicles running behind while waiting to overtake them. Figure 3.8 shows a simplified scheme of a traffic scenario.

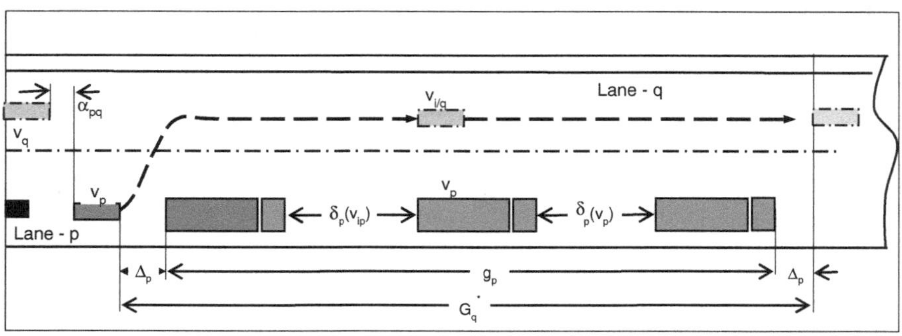

Fig. 3.8 Scenario for estimating time losses of smaller vehicles waiting to overtake a convoy of mega or standard road trucks (Janic and Vleugel 2012)

The convoy moves along the distance d along the far right line of the given motorway/highway. The potentially affected vehicles use this (right) lane just after entering and before leaving a given segment of the highway/motorway. In particular, in the former case, these vehicles follow the convoy by staying behind in the right lane until the opportunity for overtaking it arises. As this happens, they pass onto the left lane and move in parallel with the convoy until passing it and finishing overtaking, or even longer. While running behind the convoy at approximately the same speed, which is generally lower that it would be otherwise, these vehicles spend extra time between their origins and destinations, which could be considered as their time losses. In order to quantify these losses, let the right and left lane be denoted by p and q, respectively, as in Fig. 3.8. The length of a convoy of trucks can be estimated as in Eq. 3.10b. Each vehicle following the convoy in the same lane starts and finishes its overtaking at the distance Δ_p from the last and from the front truck in the convoy. In addition, at the moment of entering the lane q, the minimum distance from the vehicle already being there should be equal at least α_{pq}. Consequently, a gap between the vehicles in the traffic flow along the lane q needed for the safe overtaking of the convoy can be estimated as:

$$G_q^* = \alpha_{pq} + 2 * \Delta_p + S \tag{3.11a}$$

If the convoy of trucks moves at the average speed v_p and a vehicle overtaking it at the average speed of v_q, the average time the vehicle needs to safely overtake the convoy can be estimated as follows:

$$\bar{t}_{pq} = G_q^* / \left(v_q - v_p\right) \text{ where } v_q > v_p \tag{3.11b}$$

The vehicles behind the convoy try and overtake it after each other, i.e., the next vehicle does not start overtaking before the preceding one finishes. If the intensity of vehicles intending to overtake the convoy is Λ_{pq} (veh/h), the average waiting time of the first vehicle in the queue following the convoy along the distance x_{pq} before starting to overtake it can be estimated analogously as in the theory of steady-state queues as follows:

$$\overline{w}_{pq} = x_{pq} * \left(1/v_q - 1/v_p\right) \tag{3.11c}$$

The total time losses of all vehicles Λ_p queuing behind a convoy of trucks while waiting for the first one to overtake it can be estimated as follows:

$$W = \left(\Lambda_{pq} * w_{pq} - 1\right)\left(w_{pq} + \bar{t}_{pq}\right) \tag{3.11d}$$

where all symbols are analogous to those in the previous equations and Fig. 3.8. In addition to the above-described scenario of moving convoys of mega and standard trucks, the inputs are specified to illustrate their influence on congestion and delays imposed on other traffic. Thus, the convoy of mega trucks consists of $n_j = 2$ and the convoy of standard trucks of $n_i = 3$ vehicles, both moving at an average speed of: $v_j = v_i = 80$ km/h. The flow of vehicles with the average intensity of $\Lambda_{pq} = 1$ veh/min queues behind these convoys at an average distance of $x_{pq} = 1$ km before starting to overtake any of them. The free speed of these vehicles is assumed to be: $v_q = 120$ km/h. The length of a standard and mega truck is: $s_i = 18.75$ m, and $s_j = 25.25$ m, respectively. The distance between particular trucks in both convoys is: $\delta_j(80) = \delta_i(80) = 80$ m. In addition, the distance between the overtaking vehicles, and the last and the first truck in the convoy is assumed to be: $\Delta_p = 100$ m. The distance between the overtaking vehicle and the first incoming vehicle in the overtaking lane is: $\alpha_{pq} = 100$ m. Consequently, the waiting time or delay of a vehicle before starting to overtake either convoy is estimated to be: $w_{j/pq} = w_{i/pq} = 1.5$ min. The time for overtaking the convoy of mega trucks is estimated to be: $t_{j/pq} = 0.5$ min, and that of the convoy of standard trucks: $t_{i/pq} = 0.77$ min. The corresponding total times of passing are: 2.0 and 2.27 min, respectively. Thus, the waiting time for overtaking the convoy of mega trucks is shorter by about 13.5 % that that of the convoy of standard trucks, which seemingly indicates benefits of the 2/3 substitution scenario.

Safety

The traffic incidents/accidents of standard and/or mega trucks reflect their safety, which is usually measured by the number of actual and/or potential fatalities and injuries occurred on a given segment of the road. For example, for a convoy of either trucks, the number of incidents/accidents along the distance d can be estimated as follows:

$$A = n * \lambda * PL * d * a \tag{3.12}$$

where

a is the rate of fatalities/severe injuries per unit of output of either truck (events/t-km).

The other symbols are analogous to those in the previous equations.

The rate of fatalities and severe injuries in traffic incidents/accidents of standard trucks has been estimated by using some figures from the EU27 Member States. These show that the number of fatalities in traffic accidents in which these vehicles have been involved has generally decreased from 4,586 in 1996 to 3,350 in 2006, which is a reduction of about 27 % (ERSO 2007). Since the volume of transport output was 1,528 billion t-km in 2006, the average fatality rate was: $a = 2.191 \times 10^{-9}$ (fatalities/t-km). It is reasonable to expect that mega trucks will be at least equally if not even safer than their standard counterparts. In order to meet such safety targets, the following operational issues for mega trucks need to be considered:

- Impacts of accidents could be more serious in terms of the scale of damage and fatality rate;
- In tunnels, particularly in Alpine regions in Europe, cross-sections, parking niches/breakdown bays, and ventilation ducts will need to be reconstructed;
- The psychological impacts of mega trucks on the drivers of lighter vehicles should be taken into account;

- The existing design standards for guardrails/crash barriers would need to be strengthened in order to absorb impacts of mega trucks;
- The access space/lanes to the emergency parking in cases of breakdowns would need to be reconstructed; and
- Retrieving mega trucks in cases of breakdown would require special currently non-standard equipment and emergency procedures applicable to fire brigades or the breakdown services.

3.3.3 Evaluation

Mega trucks possess both advantages (strengths and opportunities) and disadvantages (weaknesses and threats) as compared to their standard truck counterparts, which can be considered for a single vehicle and/or for a convoy of vehicles substituting each other.

Advantages

Single vehicle

- Enabling higher utilization of the capacity of the available road infrastructure;
- Operating at the same speed as standard trucks, which in combination with the greater payload capacity provides higher technical productivity; and
- Providing lower average cost per unit of output (t-km) due to economies of scale/size given an adequate load factor.

Convoy of vehicles

- Reducing the number of vehicles required for transporting the given volumes of freight/goods;
- Preventing uncontrolled escalation of operator costs and impacts on the environment and society thanks to the lower number of vehicles engaged for transporting the same volumes of freight/goods; and
- Bringing benefits to almost all actors/stakeholders involved when substituting a convoy of standard trucks.

Disadvantages

- Extending time for logistics operations such as loading and unloading of freight/goods shipments, which increase the vehicle's turnaround time, and thus in turn: (i) requires a greater number of vehicles in the operator's fleet and (ii) increases the inventories of freight/goods and the related costs at both ends of the logistics chain(s);
- Increasing the vehicle's operating costs with likely a higher share of fuel costs in the total costs;
- Consuming more fuel per service and consequently emitting higher quantities of GHG (Green House Gases);

- Requiring adaptation/modification of the road infrastructure, which could be rather expensive (bridges, tunnels, parking lots, access lanes (paths), freight terminals, emergency routes/paths, etc.);
- Generating higher noise due to stronger engines and longer exposure times due to their greater length;
- Contributing to increasing congestion due to their greater length;
- Graver consequences of traffic incidents/accidents due to their larger size and weight; and
- Possibly affecting the market share of rail and rail-based intermodal freight transport if commercialized on a wider scale.

Finally, road mega trucks, although technically/technologically maturated, can be considered, regarding their size, related operations, and economics, as qualified components of advanced transport systems.

3.4 Long Intermodal Freight Train(s)

2003/2006	The feasibility of launching longer trains (up to 1,000 m) in the Paris-Amsterdam corridor is examined as part of the European project led by the SNCF (French National Railways) (France, The Netherlands)

3.4.1 Definition, Development, and Use

The LIFTs (Long Intermodal Freight Train(s)) have been initiated by some rail freight operators in France (SNCF FRET), Belgium (B-Cargo), Germany (DB), and the Netherlands (PRORAIL) as an innovative and presumably more efficient competitive product than the current CIFTs (Conventional Intermodal Freight Trains). These trains, longer than conventional ones, are supposed to operate as the main rail services within intermodal or multimodal freight transport networks including the above-mentioned advanced collection/distribution networks. This implies that they transport freight/goods shipment consolidated into loading units such as containers, swap-bodies, and semi-trailers. The specific objectives of launching LIFTs are to: (i) improve the internal efficiency within the rail freight sector; and (ii) increase the competitiveness of rail compared to road freight transport in the medium- to long-distance corridors between the North and South of Europe. Initial trials were carried out between the Netherlands (port of Rotterdam) and France (Paris), and further to the south toward Lyon and other Mediterranean ports (EC 2007; Janic 2008). In addition, the enlargement of the EU (European Union) from 15 to 25 Member States in 2004 and to 27 Member States in 2007 (when Bulgaria and Romania joined the EU) has seemed to open up new opportunities for such rail services in the long-distance markets (corridors) such as those between West (Germany) and East (Poland) Europe, as well as between Northwest (North of Germany, Netherlands, Denmark) and Southeast Europe (Greece, and later on Turkey).

The preconditions for the success of the concept are perceived to be, on the one hand, maintaining the compactness of LIFTs as block trains in order to control freight delivery time and costs at a reasonable (competitive) level, and on the other, the sufficient and time regular volumes of freight demand in both directions along the corridors.

3.4.2 Analyzing Performances

3.4.2.1 Background

LIFTs (Long Intermodal freight Train(s)) are characterized by their infrastructural, technical/technological, operational, economic, environmental, social, and policy performances. These performances are elaborated in the scope of the rail/road intermodal freight transport network in which they operate as the main mode instead of CIFTs (Conventional Intermodal Freight Train(s)), and the road freight transport network used for comparative purposes.

3.4.2.2 Infrastructural, Technical/Technological and Operational Performances

Rail/road intermodal freight transport network

Components and operations

Rail/road intermodal freight transport network delivers loading units from shippers to receivers in five steps: (i) collection in the origin "zone" and transportation to the "origin" rail-road intermodal terminal located in the "shipper" area by road trucks; (ii) transshipment at the "origin" rail-road intermodal terminal from road trucks to CIFTs or LIFTs; (iii) rail haulage between the "origin" and "destination" intermodal terminal; the CIFTs or LIFTs running between two terminals are trains with a fixed composition,[5] which use the same technology (infrastructure, traffic control and signaling system, and rolling stock) but have distinct technological/operational characteristics, such as braking systems and operating speed; the former trains have been operating for a long time while the latter are as a concept still under preparation (EC 2006); (iv) transshipment at the "destination"

[5] This excludes the shunting and marshaling of trains while operating between their origin and destination intermodal terminal at both ends of a given corridor (market) (EC 2006).

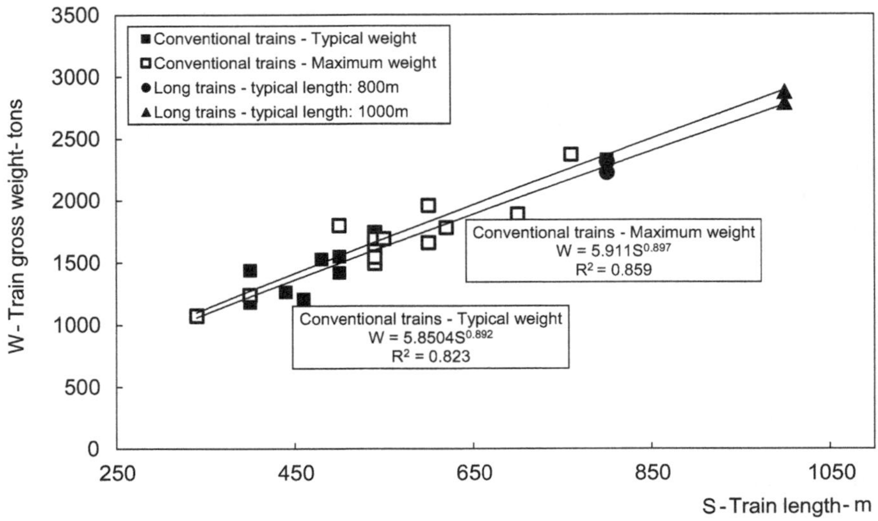

Fig. 3.9 Relationships between the gross weight and the length of CIFTs and LIFTs in Europe (EC 2002, 2006)

intermodal terminal in the "receiver" area from the rail to the trucks; and (v) distribution from the "destination" terminal to the destination "zone" by trucks.

The main mode: CIFTs (Conventional Intermodal Freight Train(s))

CIFTs (Conventional Intermodal Freight Train(s)) already operate in many national/country and Trans-European corridors (EC 2001a, b, 2007). Typically, each train consists of 25–30 four-axle rail flat wagons of an approximate length of 20 m, which gives a typical length of a CIFT of about 500–600 m. The empty weight of a flat wagon depends on its type and varies from about 24 tons (for carrying containers and swap bodies) to 32 tons (for carrying semi-trailers). The carrying capacity of each flat wagon is up to 50 tons. This wagon can carry an equivalent of three TEU (20-Foot Equivalent Unit), each of a length of 20 ft or about 6 m (the most common in Europe). An empty TEU weighs 2.3 tons (EC 2002). Goods in the TEUs and their tare represent the CIFTs' payload. The length of CIFTs depends on the number of wagons. Its weight depends on the weight of empty flat cars and the locomotive, as well as the payload, i.e., the load factor of each wagon and/or of the entire train. Figure 3.9 shows the relationship between the length and typical and maximum gross weight of the CIFTs and LIFTs in particular European corridors.

The prospective length and weight of LIFTs are shown for comparative purposes. The length of CIFTs, typically varying between 450 and 650 m, is within the current UIC (International Union of Railways) regulation on the pneumatic braking distance of 750 m. The constraints of the current braking system[6] as well as commercial reasons, i.e., finding sufficient and regular (stable) quantities of goods flows, are the main factors influencing the above-mentioned length of CIFTs. In addition, their weight increases at a decreasing rate with increasing of the length in both typical (load factor 75 %) and maximal configuration (load factor 100 %) (EC 2001a, 2002, 2006). CIFTs have a fixed composition, which a priori excludes additional shunting and/or marshaling along the route(s). Due to difference in the electric power supply in particular European countries (1.5 kV DC, 3 kV DC, 25 kV 50 Hz AC), multisystem 6 MW ALSTOM PRIMA 6000B or SIEMENS CLASS 189 locomotives weighing 86–89 tons are usually used to power the CIFTs. Thus, the necessary interoperability is provided by eliminating cross-border delays due to locomotive changing. CIFTs can run at the maximum operating speed of about 120 km/h, but the average commercial speed tends to decrease with distance resulting in an average of about 40–50 km/h between the short- and medium distance terminals (300–600 km). This happens mainly due to the relatively frequent speed changes and intermediate stops. CIFTs are usually dispatched once to three times per week during the weekdays, depending on the volume and time pattern of demand. Such frequency and speed make these trains relatively competitive alternative to road haulage with respect to the door-to-door delivery time over longer distances (Janic 2007).

The main mode: LIFTs (Long Intermodal Freight Train(s))

LIFTs (Long Intermodal Freight Train(s)) actually represent extended CIFTs composed of 38 or 48 flat wagons giving a train length of 800 m and 1,000 m, respectively. LIFTs are also supposed to have a compact-fixed composition implying no-intermediate marshaling and/or shunting along the route(s). At an empty weight and load factor of 75 %, the gross weight of a LIFT can vary from about 2,400 to 2,800 tons (see Fig. 3.9). The length and weight of LIFTs require a breaking system to be designed, which is, in addition to be the major technical/ technological advancement, also the most significant barrier to their immediate implementation[7] (EC 2006). Two instead of one multisystem 6 MW electric

---

[6] The conventional braking system operates as follows: when the braking command is initiated from the locomotive (front side of the train), there is a propagation delay of about a half a minute per rail wagon before the breaking becomes effective after the air pressure stabilizes within the system. This delay increases the train's braking distance, which is, for example, standardized to about 950 m for freight trains running at a maximum speed of about 100 km/h and to those trains of 690 m running at a maximum speed of 120 km/h. New rail wagons for both categories of trains have a braking distance of about 890 m (Railtrack 1998).

[7] Freight trains longer than 750 m do not comply with the current UIC braking regulation. However, installing a remote controlled ABD (Additional Braking Device) at the rear end of the train in order to distribute pressure more symmetrically towards the middle of the train can shorten the pressure stabilization delay and thus enable shortening of the braking distance(s) according to the prescribed standards, and consequently enable trains of up to 1,000 m in length to be operated safely. This additional braking device will be controlled by means of a radio-based communications system and related network, whose main features are autonomy without interference with those of other trains and interoperability across different European countries (EC 2006).

locomotive are supposed to pull LIFTs of both lengths (weights) in order to achieve maximum operating speeds of about 90 km/h and commercial speeds of about 45 km/h (EC 2006). The length and speed of LIFTs can interfere with the shorter but faster CIFTs and other passenger trains on the same line(s), thus causing delays of the latters. These delays can be further extended upstream of LIFTs due to their intermediate stops on short sidings, passing loops, marshaling, and shunting yards, currently with the maximum length of 775 m, which is shorter than the LIFTs' length of 800 or 1,000 m. Nevertheless, despite the above-mentioned shortcomings, in order to be sufficiently competitive, LIFTs should be dispatched similarly as CIFTs—at least three times per weekday, which implies that sufficient volumes of freight/goods need to be provided.

Road freight transport network

The road freight transport network enables delivering loading units between particular shippers and receivers in three steps: (i) collection in the origin "zone" by truck; (ii) road haulage from the shipper area to the receiver area by the same vehicle without intermediate storage and associated unloading/loading, which would also include exchanging vehicles; and (iii) distribution in the destination zone by the same vehicle (EC 2000a, 2006; Janic 2007, 2008).

3.4.2.3 Economic Performances

The economic performances of intermodal freight transport networks relate to the operational (internal) costs and revenues of the particular actors involved. Specifically, the operational costs of intermodal transport and terminal operators generally depend on the service frequencies, delivery distances, the volumes of transshipment activities in the intermodal terminals, and the prices of basic inputs such as labor (staff), material (means), and energy (fuel). The costs of loading units mainly relate to the costs of their owners–freight/goods shippers and receivers and/or intermodal system operators (EC 2001a, b; Levison et al. 1996). The operational (internal) costs are constant in the short-term for a given volume and intensity of system activities (Janic 2007).

Revenues are obtained by charging users/shippers of loading units for door-to-door services. These are then distributed to the particular actors in proportion to their operating costs for providing the given services.

3.4.2.4 Environmental Performances

The environmental performances of intermodal freight transport networks relate to their energy consumption and related emissions of GHG (Green House Gases) and land use/take (EC 2001a; Levison et al. 1996).

Energy consumption and emissions of GHG

The trucks carrying out the collection and distribution of loading units usually burn diesel fuel, thus causing emissions of GHG (Green House Gases) with local and global impacts, the latter if transported and deposited into the atmosphere (Janic 2007; Levison et al. 1996).

Emissions of GHG from CIFTs and/or LIFTs are indirect, depending on the primary sources for producing the electric energy for their (electric) locomotives. In any case, the energy consumption and related emissions of GHG associated with LIFTs and their external costs are higher than those of CIFTs (The potential use of diesel locomotives and their associated impacts is not considered). Large contemporary intermodal terminals use electric cranes for transshipment of loading units, thus generating indirect emissions of GHG (The use of diesel-powered reachtackers is not considered). The costs of GHG emissions usually reflect the marginal expense of maintaining and/or recovering people's health, as well as the flora and fauna in the affected areas (EC 2001a, 2006).

Land use

Rail/road intermodal freight transport networks operated by CIFTs or LIFTs do not require taking additional land use for modification of rail infrastructure. Some seemingly very limited exceptions can be modifications for extensions of tracks for handling LIFTs. However, some additional land could be taken by intermodal terminals for expanding storage space for the increasing number of loading units to be transported by LIFTs.

3.4.2.5 Social/Policy Performances

The social/policy performances of intermodal freight transport networks operated by LIFTs or CIFTs relate to their noise, congestion, and traffic incidents/accidents (safety) (EC 2001a; Levison et al. 1996).

Noise

Trucks collecting and distributing loading units in the shipper and receiver area, respectively, and the CIFTs and LIFTs operating between them generate noise, which can affect nearby populated areas. Truck noise depends on its weight and operating speed, while train noise depends on its length, weight, and speed, which implies that the bigger but rather slower LIFTs are supposed to generate more persistent but not necessarily louder noise than CIFTs. In most cases, direct noise from the transshipment of loading units in the intermodal terminals is not significant as it remains inside the terminal.

The costs of noise generated by trucks can include the costs of additional noise insulation of the affected houses and the costs of constructing protecting barriers along the road lanes. The costs of noise from trains can be considered as the marginal cost of setting up noise protection barriers along the rail lines (EC 2006; Janic 2007).

Congestion

Trucks performing collection and distribution of loading units usually operate in densely urbanized and/or industrialized "zones," which frequently experience costly congestion and delays. They also impose delays on other vehicles whose consequent costs of time losses need to be taken into account as an externality.

CIFTs and LIFTs are assumed to be free of congestion under regular (non-disturbing) conditions, which is provided through their scheduling. This a priori prevents the mutual interference with other trains, both freight and passenger, operating on the same lines. Thus, the costs of delays that such trains may impose on each other are negligible. Nevertheless, the slower LIFTs may impose delays on those CIFTs and other passenger trains behind them even under regular operating conditions. The costs of these delays as externalities depend on the number of affected CIFTs, the value of their time, and the length of imposed delays (EC 2006). The loading of units usually bundled in the storage areas of intermodal terminals or onto the trains do not interfere and thus do not impose delays and related costs on each other (EC 2006; Janic 2007).

Safety

Traffic incidents/accidents cause injuries and loss of human life, as well as damage and loss of goods and property affecting intermodal transport system operators, third parties, and neighboring people. In particular, trucks performing collection and distribution of loading units have an accident rate comparable to that of the overall road freight transport of the same category. LIFTs are expected to have the very low, if any, incidents/accidents similarly to CIFTs. Accidents at intermodal terminals are shown to be very rare events. The total costs are expressed as the direct and indirect costs of impacts (EC 2002, 2006; Levison et al. 1996).

3.4.3 Modeling Performances

Modeling performances of the intermodal freight transport networks operated by CIFTs or LIFTs implies developing a methodology consisting of models for estimating the full costs of the particular steps—collection of loading units from their shippers by road haulage, rail-line hauling, and distribution of loading units to their receivers, again by road haulage. The full costs consist of the operating (internal) and external costs, the latter reflecting the cost of impacts of the system on the environment and society (EC 2001a, b; Levison et al. 1996; Janic 2007, 2008). This actually means modeling the economic and environmental performances of such networks respecting their close relationship with the technical/technological and operational performances. Modeling includes assumptions and the basic structure of the methodology for estimating the full costs.

3.4.3.1 Assumptions

The methodology for estimating the full costs of an intermodal freight transport network operated by either CIFTs or LIFTs is based on assumptions related to the general conditions, collection and distribution of loading units by trucks, and the rail haulage between intermodal terminals.

General conditions

- The demand is always sufficient for dispatching a train of a given category, either a CIFT or LIFT, with a reasonable frequency and utilization (load factor).

Collection and distribution of loading units

- Trucks of the same capacity and load factor collect and/or distribute loading units from/to shippers/receivers in a given origin/destination zone/region around the corresponding intermodal terminal(s) by making tours of the approximately same length at a constant average speed;
- The collection step starts at the vehicle's initial position, which can be anywhere within the "shipper's" area and ends at the origin's intermodal terminal. The distribution step starts at the destination intermodal terminal where the vehicles may be parked and ends at the "receiver's" area at the doors of the last receiver (Morlok et al. 1995; Janic 2007, 2008).

Rail haulage between intermodal terminals

- The service frequency of either CIFTs or LIFTs between a given pair of intermodal terminals follows the practice of many rail operators in Europe, which dispatch few regular weekly services (EC 2001a, 2006);
- CIFTs and/or LIFTs are block trains performing "shuttle" or "direct" train services. They consist of a fixed number of rail flat wagons of the same capacity, which, on the one hand, implies their constant capacity, and on the other, the absence of any additional shunting and/or marshaling along the line (Janic 2007, 2008);
- The average speed and anticipated arrival/departure delays of particular train services provided either by CIFTs or LIFTs are approximately constant and equal.

3.4.3.2 Structure of the Methodology

The full costs of the intermodal transport system

The full costs of the given intermodal freight transport network operated by either CIFTs or LIFTs can be estimated in generic terms as follows (Janic 2008):

- **Operational (internal) cost**

– Transport cost

$$\begin{aligned}\text{Transport Cost} &= (\text{Frequency}) \times (\text{Cost per Frequency}) \\ &= [(\text{Demand})/(\text{Load Factor} \times \text{Vehicle Capacity})] \\ &\quad \times (\text{Cost per Frequency})\end{aligned} \tag{3.13a}$$

– Handling cost

$$\text{Handling Cost} = (\text{Demand}) \times (\text{Cost per Unit of Demand}) \tag{3.13b}$$

- **External cost**

$$\text{External cost} = (\text{Frequency}) \times (\text{External cost per Frequency}) \tag{3.13c}$$

The variables in Eq. 3.13a are specific to particular steps of the intermodal freight transport network. The "Frequency" variable in the collection and distribution step relates to the number of truck runs in collecting and/or distributing a given volume of load units. In zone (k), "Frequency" f_k is directly proportional to the volume of loading units Q_k, and inversely proportional to the product of the truck capacity M_k and load factor λ_k. The "Cost/Frequency" variable relates to the operating cost of a truck and is usually expressed in relation to distance (i.e., length of the tour) as $c_{ok}(d_k)$. The distance d_k includes the segments between the truck's initial position and the first stop x_k, the average distances between successive stops δ_k, and the distance between the last stop and given intermodal terminal r_k. Analogous reasoning for the trip frequencies and distances is used for the distribution step. In the rail line-hauling step, the "Frequency" variable f is directly proportional to the total volume of loading units Q and inversely proportional to the product of the train carrying capacity Q_t and the load factor λ. The variable "Cost per Frequency" implies the internal (operating) costs per train. It is modeled as dependent on the train's gross weight W, payload q, and distance d, i.e., as $c_o(W, q, d)$.

The variables in Eq. 3.13b have the following meaning: the handling costs in the collection step in zone/region (k) are proportional to the quantity of loading units q_k, the unit handling time and costs, t_{hk} and c_{hk}, respectively. The calculation of these costs is the same for the distribution step in zone/region (l). In the rail line-hauling step, the handling cost is proportional to the total quantity of loading units q and the unit handling cost at both intermodal terminals, c_{h1} and c_{h2}, respectively. In many cases, these costs can be considered as the costs of loading/unloading a train.

The variables in Eq. 3.13c are as follows: the external cost in the collection step in zone/region (k) is proportional to the frequency of truck trips f_k, which depends of the quantity of loading units q_k, the truck's capacity and load factor, i.e., m_k and λ_k, respectively. The external cost per frequency is given at the aggregate level and for a given truck type depends mainly on the operating distance (i.e., route length) d_k, i.e., as $c_{ek}(d_k)$. Calculation of the external cost (Eq. 3.13c) is the same for the distribution step in "zone" (l). In the rail line-hauling step, the external costs are proportional to the total quantity of loading units q, the unit aggregate external cost of each intermodal terminal c_{e1} and c_{e2}, and the unit aggregate external cost per train service, i.e., $c_e(w, q, d)$. The detailed analytical expressions for particular cost components are given as follows: (Janic 2007).

1. **Transport operating (internal) cost**

(a) Collection/distribution step:

$$C_{1/k} = (Q_k/\lambda_k * M_k) * c_{ok}(d_k) \tag{3.14a}$$

(b) Line-haul step:

$$C_{1/lh} = f * c_o(W, q, d) = (Q/q) * c_o(W, q, d) \tag{3.14b}$$

2. **Handling cost**

(a) Collection/distribution step:

$$C_{2/k} = Q_k * t_{hk} * c_{hk} \tag{3.15a}$$

(b) Line-haul step:

$$C_{2/t} = Q * (c_{h1} + c_{h2}) \tag{3.15b}$$

3. **External cost**

(a) Collection/distribution step:

$$C_{4/k} = (Q_k/\lambda_k * M_k) * c_{ek}(d_k) \tag{3.16a}$$

(b) Line-haul step:

$$\begin{aligned} C_{3/lh} + C_{4/lh} &= Q * (c_{e1} + c_{e2}) + f * c_e(W, q, d) \\ &= Q * (c_{e1} + c_{e2}) + (Q/q) \cdot c_o(W, q, d) \end{aligned} \tag{3.16b}$$

4. **Subtotal**

(a) Collection/distribution step:

$$\sum_{i=1}^{3} \sum_{k=1}^{K} C_{c/i/k} \tag{3.17a}$$

(b) Line-haul step:

$$C_{lh} = \sum_{i=1}^{4} C_{i/lh} \tag{3.17b}$$

5. **The full (total) cost**

$$C_F = C_c + C_{lh} + C_d$$
$$= \sum_{i=1}^{3}\sum_{k=1}^{k} C_{c/i/k} + f[c_o(W, q, d) + c_e(W, q, d)] + \sum_{i=1}^{3}\sum_{k=1}^{k} C_{c/i/k} \quad (3.18)$$

where

Q	is the quantity of goods to be transported between given origin and destination intermodal terminals (tons);
n_l	is the number of locomotives per train;
w_l	is the locomotive weight (tons);
n_w	is the number of flat-wagons;
w_w	is the weight of an empty wagon (tons);
$n_{c/w}$	is the carrying capacity of a flat-wagon (loading units/wagon);
q_c	is the average weight (tare + goods) of a loading unit (tons/unit);
λ	is the train load factor;
T	is the period of time in which the transport of goods between two terminals is considered (h);
d	is the distance of the given rail line connecting two intermodal terminals (km);
v	is the commercial speed of a train along a given line (km/h);
D	is the anticipated delay of a train running between two intermodal terminals (hours);
α	is the value of time of goods while waiting for and during transportation, respectively (€/h-ton);
f	is the train frequency along a given line (departures/T);
W	is the gross weight of a train including rolling stock and payload (tons);
w	is the weight of the train's rolling stocks (wagons and locomotive(s)) (tons);
q	is the payload on the train (tons);
$c_o(W, q, d)$	is the operational cost of a train of the gross weight W over the distance d (€);
n_d	is the number of drivers;
t_{dp}	is the driver's preparation and finishing time before and after the trip (h);
$c_e(W, q, d)$	is the external cost of a train of the gross weight W along the distance d (€);
N	is the number of different staff categories serving and operating a given train;
$n_{s/i}$	is the number of staff of category (i) serving and operating the train;
$c_{s/i}$	is the cost of labor of the staff of category (i) serving and operating the train (€/h-staff);

$t_{s/i}$ is the time of engagement of staff of category (i) needs to serve and operate the train (h);
P_j is the price of rolling stock of type (j) ($j = 1$ for locomotives; $j = 2$ for wagons) (€);
r_j is the interest rate on loans for acquiring rolling stock of type (j)(%);
n_j is the life cycle period of rolling stock of type (j) (years);
$n_{r/j}$ is the number of rolling stock of type (j);
$m_{r/j}$ is the utilization of rolling stock of type (j) during the life cycle (km);
c_{mji} is the unit maintenance cost of rolling stock of type (j) (€/km);
e is the unit energy consumption of a train (kWh/ton-km);
c_e is the cost per unit of energy consumed (€/kWh);
c_a is the unit charge of using the railway infrastructure (€/ton-km);
C_i is the cost of train insurance (€);
a is the quantity of emissions of GHG per unit of energy consumed by a train ($kgCO_{2e}$/kWh);
c_{ap} is the unit cost of damage by emissions of GHG (€/kg);
C_n is the cost of additional barriers along a given rail line to protect people from the noise of a given train (€);
U is the utilization of a given train along the line during its life cycle (km);
a_r is the train's accident rate (events/ton-km);
c_{ac} is the cost per train accident (€/event);
D_e is the delay the a given train imposes on other trains (h);
c_{ed} is the average cost per unit of delay of other trains (€/h);
$d_{k/l}$ is the tour of a road vehicle (truck) collecting/distributing loading units in the zones (k) and (l), respectively;
$M_{k/l}$ is the capacity of road vehicle collecting/distributing loading units in the zone (k) and (l), respectively;
$\lambda_{k/l}$ is the capacity of road vehicle carrying out collection/distribution of loading units in zone (k) and (l), respectively.

The average operational (internal), external, and full cost per unit of system output (t-km) can be estimated by dividing the full (total) cost (Eq. 3.18) by the volume of demand and door-to-door delivery distance. This can be used for comparisons either within the intermodal freight transport system where exclusively CIFTs or LIFTs operate or between these and road transport systems operating either standard or mega trucks. In the case of road transport systems, Eqs. (3.17a, b) and (3.18) can be modified using the door-to-door distance between the zones/regions (k) and (l).

The full cost of rail haulage

The full costs of a given rail haulage, i.e., train service, consist of the operational (internal) and external costs, $c_o(W, q, d)$ and $c_e(W, q, d)$ and given in Table 3.4 and Table 3.5, respectively.

Table 3.4 Structure of the train's operational (internal) cost c_o (W, q, d) (Janic 2008)

Cost component	Equation	
(i) Investments in rolling stock (rail flat wagons + locomotive(s))	$C_{01} = \sum_{j=1}^{2} n_{r/j} * P_j * \left(\frac{r_j(1+r_j)^{nj}}{(1+r_j)^{nj}-1}\right)$	(3.19a)
(ii) Maintenance of rolling stock (rail flat wagons + locomotive(s))	$C_{02} = \sum_{j=1}^{2} n_{m/j} * c_{m/j} * d$	(3.19b)
(iii) Accessing and using infrastructure	$C_{03} = c_a * W * d$	(3.19c)
(iv) Energy consumption	$C_{04} = c_e * e * W * d$	(3.19d)
(v) Labor (staff)	$C_{05} = \sum_{j=1}^{N} n_{s/i} * c_{s/i} * t_{ai}$	(3.19e)
(vi) Loading/unloading	$C_{06} = (c_{h1} + c_{h2}) * q$	(3.19f)

Table 3.5 Structure of the train's external costs c_e (W, q, d) (Janic 2008)

Cost component	Equation	
(i) Emissions of GHG	$C_{e1} = (c_{e1} + c_{e2}) + q * c_{ap} * e * a * W * d$	(3.20a)
(ii) Noise	$C_{e2} = (C_n/U) * d$	(3.20b)
(iii) Congestion	$C_{e4} = c_{ed} * D_e$	(3.20c)
(iv) Traffic incidents/accidents	$C_{e3} = c_{ar} * a_r * W * d$	(3.20d)

Operational (internal) cost

The operational (internal) cost $c_o(W, q, d)$ in Eq. 3.19 consists of six components: (i) investments in rolling stock—wagons and locomotive(s); (ii) the costs of maintaining rolling stock–wagons and locomotives; (iii) the costs of using the railway infrastructure, i.e., the infrastructure charge; (iv) the costs of energy consumption; (v) labor costs for assembling/decoupling and driving the trains; and (vi) the costs of loading/unloading the trains at two intermodal terminals. The corresponding equations are given in Table 3.4.

Consequently, the costs $c_o(W, q, d)$ are equal to the sum of the particular cost components (Eq. 3.19a–f) in Table 3.4.

External cost

The external cost $c_e(W, q, d)$ (Eq. 3.16) consists of four cost components: (i) emissions of GHG from assembling/decoupling, running, and loading/unloading the trains; (ii) noise; (iii) congestion; and (iv) traffic incidents/accidents. The corresponding equations are given in Table 3.5.

Consequently, the external cost $c_e(W, q, d)$ is equal to the sum of the particular external cost components (Eq. 3.20a–d) in Table 3.5). The train's gross weight W consisting of the weight of the empty train (flat-wagons + locomotive(s)) and payload (loading units) in (Eq. 3.19c, d) and Table 3.4 and 3.5 can be calculated as follows:

$$W = w + q = n_l * w_l + n_w * w_w + \lambda * Q_t) = n_l * w_l + n_w * \left(w_w + \lambda * n_{c/w} * q_c\right) \tag{3.21}$$

where all symbols as in previous equations.

3.4.3.3 Application of the Methodology

The proposed methodology for estimating the full costs is applied to the simplified European intermodal system operated by LIFTs and CIFTs, and its road truck counterpart for comparative purposes (EC 2000a; 2001a, b, 2006; Janic 2007, 2008).

Input

Loading units

Both systems deliver loading units of 20 ft or about 6 m (TEU-Twenty Foot Equivalent Unit), the most common in Europe. Each load unit has an average gross weight of 14.3 metric tons (12 tons of goods plus 2.3 tons of tare) (EC 2001a; Janic 2007).

Collection/distribution by road trucks

In each zone of the intermodal transport system, the average length of tour and the speed of each vehicle, which is assumed to make only one stop during the collection and distribution step, are taken to be as in Table 3.6.

The average collection/distribution distance is assumed to be longer for LIFTs than for CIFTs because the LIFTs' loading units are collected and distributed over a wider area, implying the constant spatial concentration of the shippers and receivers and their generating and attracting potential. The load factor of 0.60 reflects the possibility that the train will be partly or completely empty during returning trips (EC 2001a, b).

The truck's operational cost during the collection/distribution step, based on the full load equivalent of two 20 foot loading units, is determined as the regression equation, which uses the empirical data: $c_0(d) = 5.456d^{-0.277}$ €/vehicle-km ($N = 26$; $R^2 = 0.78$; $25 \le d \le 1{,}600$ km). The average load factor is: $\lambda = 0.60$,

Table 3.6 Characteristics of the collection/distribution step in the given example (Janic 2007; 2008)

Parameter	Train category		
	CIFT	LIFT—800 m	LIFT—1,000 m
• Collection/distribution distance by road $d_{k/l}$ (km)	50	75	75
• Truck's carrying capacity $M_{k/l}$ (tons)	2 * 14.3	2 * 14.3	2 * 14.3
• Load factor $\lambda_{k/l}$	0.60	0.60	0.60
• Average speed during a tour $u_{k/l}$ (km/h)	35	35	35

Table 3.7 Characteristics of CIFTs and LIFTs in the given example (EC 2006; Janic 2008)

Parameter	Train category		
	CIFT	LIFT—800 m	LIFT—1,000 m
Train load factor λ	0.75	0.75	0.75
Payload q (tons)	14.3 * 3 * 26 * 0.75 = 837	14.3 * 3 * 38 * 0.75 = 1223	14.3 * 3 * 48 * 0.75 = 1544
Empty weight w (tons)			
1 locomotive	24 * 26 * 89 = 713	24 * 38 + 89 = 1001	24 * 48 + 89 = 1241
2 locomotives		24 * 38 + 2 * 89 = 1090	24 * 48 + 2 * 89 = 1330
Total weight W (tons)			
1 locomotive	1550	2224	2785
2 locomotives (tons)	–	2313	2874
Operating speed v (km/h)			
1 locomotive	110	70	65
2 locomotives	–	95	90
Commercial speed v (km/h)			
1 locomotive	60	50	45
2 locomotives	–	50	45
Anticipated delay D (h)	1	1	1

implying the possibility that the truck will be partly or even completely empty during the returning trip(s). The costs $c_0(d)$ already include the handling costs of loading units (EC 2001a, b; Janic 2007). The same equation is used for determining the operational costs of haulage by standard trucks between particular shippers and receivers with respect to an average load factor of: $\lambda = 0.85$. This load factor has been observed for the most long-distance road operators in Europe reflecting, in many cases, full or semi-full returning trips (EC 2001a, b).

From the same sources of data, the externalities comprising local and global air pollution, congestion, noise, and traffic accidents are determined in the following aggregate regression form: $c_e(d) = 9.884d^{-0.6235}$ €/vehicle-km (N = 36; $R^2 = 0.70$; $25 \leq d \leq 1{,}600$ km) (EC 2001a, b; Janic 2007, 2008).

Rail hauling

- The train composition, weight, speed, and anticipated delay

The characteristics of CIFTs and LIFTs running between two intermodal terminals are given in Table 3.7 (EC 2006; Janic 2007, 2008):

The load factor of a train of 0.75 reflects operating not always completely full train(s) in both directions along a given route (market-corridor) (EC 2006; Janic 2007).

- The train's operational (internal) cost

The train's operational (internal) cost $c_o(W, q, d)$ expressed in €/train is estimated as follows (AEAT 2005; Baungartberm 2001; EC 1996a, 1997a, 2000a, 2001a, 2006; INFRAS 2000; Janic 2008):

$$c_o(W, d, q) = (4.60 * n_l + 0.144 * n_w + 0.3) * d + 12.98(n_l + n_w) + 5.6q + 0.0019 * W * d \\ + \sum_{l=1}^{L} \left[0.227 * 10^{-6} v_l^2 / \ln d_l + 0.000774\right] * W * d + 33 n_d * \left(t_{dp} + d/v + D\right)$$

The coefficients of particular variables have the following meaning: the first represents the unit cost of depreciation and maintenance of the rolling stock (flat wagons and locomotives) and the monitoring cost of a train while running along the line (€/km); the second represents the unit cost of assembling/decomposing the train at both ends of the corridor (€/train unit); the third expresses the unit cost of loading and unloading the train at both ends of the rail-line, i.e., the transshipment cost of loading units at two intermodal terminals (€/ton); the fourth represents the unit cost of using the rail infrastructure (i.e., the infrastructure charge) (€/t-km); the fifth represents the unit cost of the energy consumption along the line with L segments, i.e., intermediate stops (€/t-km); and the last sixth rate represents the unit cost of the train's driver(s) (€/h).

- The train's external cost

The train's external cost $c_e(W, q, d)$ expressed in €/train is estimated as follows (EC 1996a, 1997a, 2000a, 2001a, 2006; INFRAS 2000; Janic 2008):

$$c_o(W, d, q) = 0.000128 * W * d + 0.0549q + \sum_{l=1}^{L} \left[1.889 * 10^{-7} v_l^2 / \ln d_l + 0.00064\right] * W * d \\ + 0 W * d + 5.6 \overline{d * D_m}$$

The coefficients of particular variables have the following meaning: the first represents the unit cost of noise (€/t-km); the second expresses the unit external cost of the train loading/unloading, i.e., transshipment of loading units at two intermodal terminals (€/ton); the third represents the cost of emissions of GHG due to energy consumption (€/t-km); the fourth expresses the cost of traffic incidents/accidents (€/t-km); and the fifth rate represents the unit external cost of congestion (€/h-km).

Handling cost

The handling cost of a given loading unit at each intermodal terminal is already included in the cost of loading/unloading the train (the third term in the train's internal cost function), i.e., at both terminals: $c_{h1} = c_{h2} = 2.8$ €/ton. This is taken as the average value for a given level of utilization of intermodal terminals (EC 2001b, c; Janic 2007).

The external costs of intermodal terminals include only the costs of local and global air pollution imposed by production of electricity for moving the cranes used for the transshipment of loading units, as follows: $c_{e1} = c_{e2} = 0.0549$ €/ton (EC 2001a).

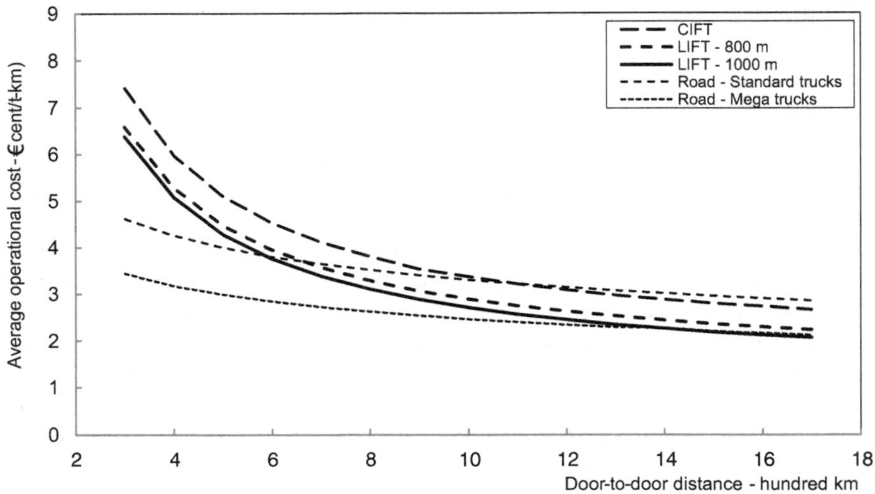

Fig. 3.10 Relationship between the operational (internal) cost and door-to-door distance of intermodal CIFTs, LIFTs, and road trucks (Janic 2008)

Results

The results from applying the methodology based on the above-mentioned inputs are shown in Figs. 3.10 and 3.11. For the purpose of the sensitivity analysis, the length of rail-line hauling distance (i.e., the door-to-door distance) is varied as a parameter for both CIFTs and two categories of LIFTs. The volume of demand is considered to be always sufficient for dispatching one train of either category per day, which is, and will be, the preferred departure frequency for many existing and prospective users- customers (and consequently train operators) in Europe (EC 2001a, b, 2006; Janic 2007).

Figure 3.10 shows the relationship between the average operational (internal) cost and the door-to-door distance of an intermodal transport system operating a CIFT and different categories of LIFTs. The corresponding cost figures for the road transport system operating standard and mega trucks are also provided.

The average operational (internal) cost decreases more than proportionally with increasing the door-to-door distance for both intermodal rail/road and road transport, thus indicating the existence of economies of distance. In general, the operational cost of intermodal transport decreases at a higher rate than that of the road transport. If CIFTs are used, the costs of intermodal transport equalizes and becomes increasingly lower than the cost of road transport beyond distances of about 1,100 km. If LIFTs of either length are used, this cost-breakeven distance shortens to 600–650 km. This occurs because the average internal cost of LIFTs of 800 m and 1,000 m are lower by about 12–18 % and 18–27 %, respectively, than that of CIFTs for the range of door-to-door distances between 300 and 1,300 km. As an illustration, if road mega trucks are used instead of standard trucks, this cost-

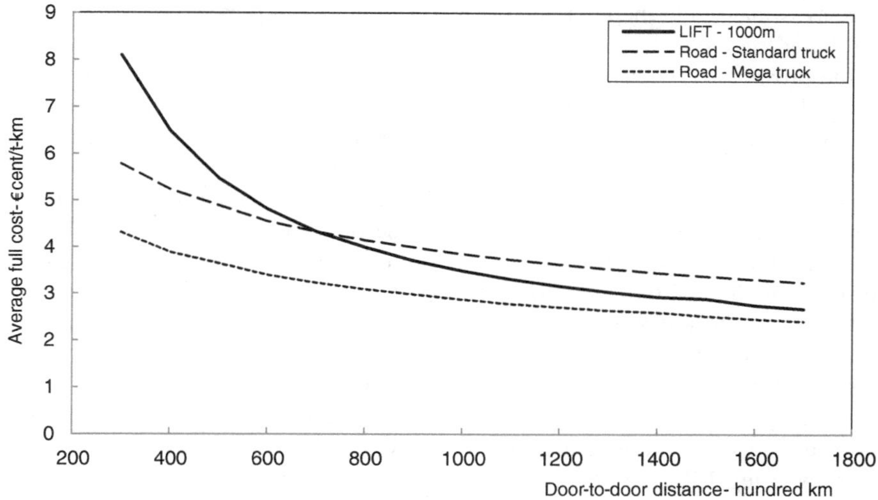

Fig. 3.11 Relationship between the full (operational and external) cost and door-to-door distance of intermodal LIFTs and road trucks (Janic 2008)

breakeven distance for both categories of LIFTs extends to about 1,300–1,400 km. The above-mentioned differences between the operational costs indicate the existence of the economies of scale for LIFTs as compared to CIFTs.

The relationships between the average operational (internal) costs of both systems indicate that intermodal transport networks operating CIFTs or LIFTs are and can be competitive alternatives to long-haul road transport operating standard trucks beyond the above-mentioned "break-even" distances. Therefore, introducing LIFTs under the above-specified conditions and circumstances could seemingly improve the competitiveness of the intermodal transport by decreasing the cost-based prices over a wider range of even shorter distance(s) where more voluminous demand actually exists. Thus, LIFTs could generally improve the efficiency of intermodal (rail) transport operators and eventually contribute to changing the current modal split, both of which are amongst the important objectives of the concept/product. A real threat to such developments could come in the form of more intense introduction of road mega trucks.

Figure 3.11 shows the relationships between the average full cost, as the sum of operational (internal) and external cost, and the door-to-door distance for the intermodal transport system operating LIFTs (category of 1,000 m) and road transport operating standard and mega trucks.

In the above example, the share of external cost in the full cost of the intermodal transport system increases by about 20–23 % as the door-to-door distance extends from about 300 to 1,300 km. The share of the external cost in the full cost of road haulage decreases from about 20–13 % for the same range of distances independently of the truck type. The external cost of both modes appears to be relatively low as compared to the internal (operating) cost. One of the main reasons is that while the internal cost is based on real figures, the external cost is still based on the above-mentioned caveats including the slight (under) estimates of the prospective impacts, rather than on their real (market-recognized) values. The full cost of both modes decreases more than proportionally with increasing of the door-to-door distance(s). The rate of decrease is again higher for the intermodal transport system, thus equalizing its full cost with that of its road counterpart operating standard trucks at the "break-even" distance of about 700 km. This is longer than in the case where only the operational (internal) cost is considered (about 600 km). In addition, the full cost of road mega truck transport is lower than that of the intermodal transport operating LIFTs of the length of 1,000 m for the entire range of the door-to-door delivery distances. Since the volume of demand generally decreases with increasing of the door-to-door delivery distance, basing the prices on the (higher) full costs may generally affect the already low (though still present) price-sensitive demand, and thus make it more difficult for intermodal transport to eventually gain higher market shares even by using LIFTs. Introducing road mega trucks under the given conditions would additionally worsen the market position of intermodal transport operating LIFTs. This again raises the question of the efficiency of EU policies, which expect that internalizing transport externalities could strengthen the market position of the entire, but particularly rail/road intermodal transport systems in Europe, also despite more widespread use of road mega trucks (EC 2001a; Janic 2008).

3.4.4 Evaluation

LIFTs (Long Intermodal Freight Train(s)) possess both advantages (strengths and opportunities) and disadvantages (weaknesses and threats) as compared to CIFTs (Conventional Intermodal Freight Train(s)) and their road counterparts as follows:

Advantages

- The internal and full costs of both intermodal and road transport systems decrease more than proportionally with increasing of the door-to-door distance, thus indicating economies of distance; and
- The average operational (internal) and full cost of intermodal transport operated by LIFTs are lower than those operated by CIFTs, thus indicating economies of scale;
- The full costs decrease faster with increasing of the door-to-door distance at both intermodal than at the road transport system; the costs of both types of intermodal trains equalize at the "break-even" distance(s), thus indicating conditions for competition by distance.

Disadvantages

- Contributing to improving the internal efficiency and intra- and inter- competitiveness of the rail freight transport sector only if there is sufficient demand around the "break-even" distances guaranteeing the operation of frequent/competitive services; and
- Increasing the "break-even" distances after internalizing externalities, which will generally increase and thus push LIFTs to compete with their road counterparts in longer distance markets with increasingly diminishing volumes of demand; these will not be able to justify the customer-driven services required for successful competition with the road counterpart.

Finally, it can be said that LIFTs possess sufficient advancement in both operational and economic performances enabling them to be qualified as the components of advanced transport systems.

3.5 Large Commercial Freight Aircraft

1973	FedEx (Federal Express) begins exclusive freight delivery services (U.S.). as a pure air cargo company
1988	UPS (United Parcel Service) obtains permission from the US FAA (Federal Aviation Administration) to begin operating its own services (U.S.)
2006/2007	Twenty-seven orders for the A380F-800 aircraft are cancelled or converted (France, Germany)
2011	The first B747-8F enters service with the Cargolux airline (Luxembourg)

3.5.1 Definition, Development, and Use

The large commercial freight aircraft is defined as aircraft with the greatest structural payload capacity among existing commercial freight aircraft. Except the single largest Antonov An-225,[8] two aircraft are currently the largest: the Boeing B747-8F and the forthcoming Airbus A380-800F. Their size compared to the size of other commercial freight aircraft qualifies them as advanced in the given context. The B747-8F aircraft entered service in 2011, while the entry of the latter A380-800F

[8] The USSR/Ukrainian 6-turbofan (229 kN thrust each) Antonov An-225 (Mriya—"Dream") is currently the largest commercial freight aircraft in the world with a length of 84.0 m and a wingspan of 88.4 m, a MTOW (Maximum Take-Off Weight) of 640 tons, and a payload capacity of 250 tons (http://en.wikipedia.org/wiki/Antonov_An-225_Mriya).

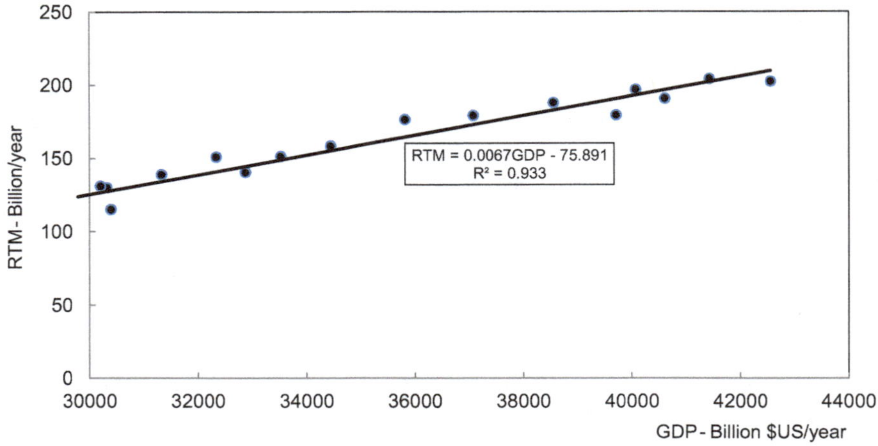

Fig. 3.12 Relationship between global volumes of air freight in RTK (Revenue Ton-Kilometer) and GDP (Gross Domestic Product) (period 1995–2011) (Boeing 2012a)

aircraft has been postponed indefinitely. Both aircraft, together with their smaller counterparts, are supposed to provide sufficient capacity for supporting the current and future expected growth of air freight transport demand, which is usually expressed in RTKs (Revenue Ton Kilometer(s)). The average annual growth rates of the world's air freight RTKs over the past three decades have been about 6.9 % (1981–1990), 6.1 % (1991–2000), and 3.7 % (2001–2011). According to some forecasts, the volume of 194.5 billion RTK in 2011 will increase to about 550–560 billion by 2031. GDP (Gross Domestic Product) will be the main driving force behind such growth (Boeing 2012). Figure 3.12 shows that the volumes of air freight and GDP have been and are very likely continue to be in strong correlation. Industrial production is an essential component of GDP, and has also shown to correlate very well with growth of the volumes of air freight demand. In addition, about 80 % of the prospective air freight demand is expected to be long-haul intercontinental demand. Furthermore, about 60 % of the total air freight demand is likely to take place on the US-Europe-Asia Pacific routes (Boeing 2012).

Air freight airlines serve the above-mentioned demand by providing scheduled and charter flights/services. The former account 90–93 % and the latter about 7 % of the total global air freight volumes. These airlines operated 1738 freight aircraft in 2011, of which 31 % were the large aircraft with a payload capacity greater than 80 tons. The proportion of these aircraft in the total fleet was 22 % in 2001. The forecasts suggest that 3198 freight aircraft of which 31 % will be large ones will be needed by 2031. Such developments will also be influenced by the decreasing role of passenger aircraft in serving air freight demand (Boeing 2012).

3.5.2 Analyzing and Modeling Performances

3.5.2.1 Background

The large commercial freight aircraft are characterized by their infrastructural, technical/technological, operational, economic, environmental, social, and policy performances, which all influence each other. The infrastructural performances mainly relate to suitability for airport operations. The technical/technological performances relate to the aerodynamic design and materials used, engines, and aircraft systems. The operational performances relate the apron-gate turnaround time, payload range characteristics, technical productivity, the service network, and fleet size. The economic performances mainly include operating costs and revenues. The environmental performances consider fuel consumption and related emissions of GHG (Green House gases), and land use. The social policy performances include noise, congestion, and safety (traffic incidents/accidents).

3.5.2.2 Infrastructural Performances

The infrastructural performances of large commercial freight aircraft such as the B747-8F and A380-800F mainly refer to airport operations. Some of these performances are given in Table 3.8.

The 747-8 and A380-800F aircraft are classified into FAA (Federal Aviation Administration) Airplane Design Group VI (Wing Span and Tail Height) and ICAO (International Civil Aviation Organization) Aerodrome Reference Code 4F (Wing Span and Outer Main Landing Gear Width) category aircraft. According to the FAA, both aircraft belong to the speed category D: 141–165 kts. However, in terms of ATC wake vortex separation rules during the final approach and landing, the B747-8F is considered a Heavy and the A380-800F a Super Heavy aircraft.

Both B747-8F and A380-800F aircraft can maneuver safely, efficiently, and effectively on existing airport infrastructure-runways, taxiways, and apron-gate complexes. Specifically, the B747-8F aircraft can use runways and taxiways 45 and 23 m wide, respectively; the external and internal radius of curvature for the turning angles of 90/135° is 52 and 26 m, respectively. The A380-800F aircraft can operate on same the runways and taxiways as B747-8F aircraft; the external and internal radius of curvature for the turning angles of 90/135° is 45.7 and 25.9 m, respectively. On runways and taxiways 60 and 30 m wide, respectively, the external and internal radius of curvature for the turning angle of 135° is 51.0 and 25.5 m, respectively. In addition, as shown in Fig. 3.13. The B747-8F and A380-800F aircraft occupy the apron-gate parking space of 6,350 and 5,840 m^2, respectively, which is larger by about 9.3 and 1.0 %, respectively, then the space occupied by their previously largest counterpart B747-400F (this occupies the space of 5,810 m^2). Furthermore, the B747-8F occupies an about 9 % larger space than its A380-800F counterpart (Airbus 2012; Boeing 2012).

Table 3.8 Infrastructural performances of the B747-8F and A380-800F aircraft—airport operations (Airbus 2012; Boeing 2012a)

Aircraft	B747-8F	A380-800F
Parameter	Value	
Length (m)	76.25	72.73
Wingspan (m)	68.40	79.75
Wing area (m^2)	554	845
Wing sweepback (°)	37.5	33.5
Height (m)	19.35	24.45
Fuselage constant diameter (m)	6.5	7.14/8.41
Maximum take-off weight (MTOW) (t)	448	590
Maximum landing weight (t)	346	427
Maximum payload (t)	133	151
Volume of cargo compartments (m^3)	874	938
Designed freight density (kg/m^3)	152	161
Take-off field length (m)[a]	3150	2900
Landing field length (m)[b]	2250–2350	2150
Maximum pavement width (m)	45	45
Effective steering angle (°)	70	70
Aircraft Classification Number (ACN)[c]	64–11 (flexible pavement) 66–102 (rigid pavement)	66–116 (flexible pavement) 60–120 (rigid pavement)

[a] *MTOW* (Maximum Take-off Weight), Sea level pressure altitude, standard day + 15 °C temperature, dry runway
[b] *MLW* (Maximum Landing Weight), sea level pressure altitude, dry runway; m meter, t ton
[c] *ACN* is the ratio between the pavement thickness required for a given aircraft and that required for the standard aircraft single wheel load
(The take-off and landing field length of the An-225 aircraft are about 3,200 m)

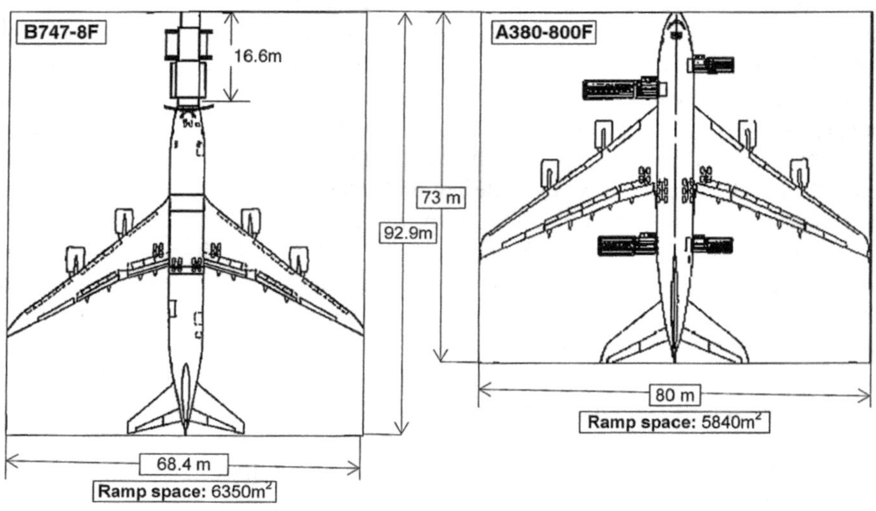

Fig. 3.13 Scheme of the required apron-gate parking area for B747-8F and A380-800F aircraft (Airbus 2012; Boeing 2012a)

3.5.2.3 Technical/Technological and Operational Performances

The technical/technological performances of the B747-8F and A380-800F large commercial freight aircraft include their aerodynamic design and materials used, their engines, and aircraft systems.

Aerodynamic design and materials

B747-8F

The aerodynamic design of the B747-8F aircraft has been improved as compared to that of its predecessor B747-400F. In particular, this refers to the new wing design including new airfoils and raked wingtips that replace the winglets. The wings are thicker and deeper. The new wingtip structures reduce the wingtip vortices at the lateral edges of the wings, thus decreasing wake turbulence and aerodynamic drag, and consequently improving fuel efficiency. Some composites such as carbon fiber-reinforced plastic in addition to next generation alloys are used for parts of the airframe primarily in order to reduce the weight. In addition, using fly-by-wire technology for the majority of the lateral controls has additionally contributed to reducing the weight. Nevertheless, the above-mentioned design and structural modifications can be considered mainly as evolutionary, rather than revolutionary, as compared to the B747-400F aircraft (Boeing 2012a).

A380-800F

The A380-800F is expected to have stronger wings (and structure) than its current passenger version A380. This implies that the most of its fuselage will be made of the regular aluminum (7010) rather than of composites, which comprise about 20 % of the airframe of the A380 passenger version. In this latter case, composites such as carbon-, glass, and quartz-fiber reinforced plastic are used extensively for the wings, the sections of fuselage such as the undercarriage and the rear end, the tail surfaces, and the doors. However, it remains to be seen what materials will be used for the final version of the A380-800F. What remains pretty certain is that the smoothly contoured wing cross-section and the central wing box made of carbon-fiber reinforced plastic will stay in place in order to improve aerodynamic efficiency, and consequently reduce fuel consumption. Nevertheless, very limited use of composites can be expected in order to preserve the required strength of the structure, thus making the aircraft design and structure again evolutionary rather than revolutionary, this time, as compared to the B-747-8F (Airbus 2012).

The designed freight density of both aircraft is closely adjusted to the medium- to long-term relatively stable average density of the world's air freight shipments of about 155 kg/m^3 (Table 3.8). (For comparison, the designed freight density of the largest An-225 aircraft is 250 tons/1,300 m^3 = 192 kg/m^3, which is about 24 % greater than the world's average of 155 kg/m^3 (http://www.antonov.com/aircraft/transport-aircraft/an-225-mriya)).

Engines

The B747-8F and A380-800F large commercial freight aircraft can use two types of engines—Engine Alliance GP7200 and RR (Rolls-Royce) Trent 900—whose main characteristics are given in Table 3.9.

It is evident that although both engines have very similar characteristics, this is no guarantee of similar efficiency and effectiveness of the powered aircraft.

Table 3.9 Characteristics of GP7200 and RR Trent 900 aircraft engines (ICAO 2012; Ryck 2008)

Characteristic	Engine type	
	GP7200	RR Trent 900
Type	Dual rotor, axial airflow, high bypass ratio turbofan engine with a single stage fan with large chord hollow blades, 5-stage LPC (Low Pressure Compressor), 9-stage HPC (High Pressure Compressor), annular combustion chamber, 2-stage HPT (High Pressure Turbine), 6-stage LPT (Low Pressure Turbine), dual channel FADEC (Full Authority Digital Engine C), and EOS (Electronic Over-speed Protection)	Three shaft high bypass ratio, axial flow, turbofan engine with LP (Low Pressure), IP (Intermediate Pressure and HP (High Pressure) Compressors driven by separate turbines through coaxial shafts
Dimension/ weight		
Length (m)	4.92	5.48
Fan diameter (m)	2.96	2.95
Maximum diameter (m)	3.14	3.94
Dry weight (tons)	6.718	6.246
Performances		
Maximum thrust (kN)	330	310–360
Bypass ratio	8.7	8.02–8.15
Overall pressure ratio	36.92	37–39
SFC (Specific fuel consumption) (kg-fuel/ kg-thrust/h)[a]	0.518	0.518
Trust-to-weight ratio (kN/ ton)	49.12	49.43–57.40

[a] *Cruise; m* meter, *kN* kilo-Newton, *kg* kilogram

GP7200

The GP7000 engine is developed by Engine Alliance comprised of two aircraft engine manufacturers—Pratt and Whitney and General Electric. This new engine is intended to be used for both the passenger and freighter version of B747-8 and A380 aircraft. The engine is derived from the GE90 and PW4000 engine families. As such, it is expected to fulfill stricter standards in terms of in-service reliability and performance, simplicity of maintenance, and environmental and social impacts such as fuel consumptions and related emissions of GHG (Green House Gases) and noise, respectively.

RR (Rolls-Royce) Trent 900

The RR (Rolls-Royce) Trent 900 engine is the largest and one of the most powerful engines developed by the manufacturer. It is derived from the RR Trent 1,000 engine specifically developed for powering B787-8 aircraft (see Chap. 2). The Trent 900 engine is a three shaft high bypass ratio, axial flow, turbofan with LP (Low Pressure Compressor), IPC (Intermediate Pressure Compressor), and HPC (High pressure Compressor) driven by separate turbines through coaxial shafts. The combustion system consists of a single annular combustor. The LPC's and IPC's assemblies rotate independently in an anti-clockwise direction, and the HPC's assembly rotates clockwise, when viewed from the rear of the engine. The engine control system utilizes an EEC (Electronic Engine Controller) which has an airframe interface for digital bus communications. The engine is considered to have the lowest noise and emissions of GHG per unit of thrust as compared to other large turbofan engines powering other commercial passenger and freight aircraft.

Aircraft systems

The aircraft systems mainly include the flight deck and the systems for actuating the aircraft control surfaces such as the flaps, slats, allerons, spoilers, elevators, rudder, and stabilizers.

B747-8F

The flight deck of the B747-8F aircraft is improved as compared to that of the B747-400F aircraft, but at the same time preserving their operational similarity, which contributes to reducing the costs of training crew and the number of required crews. The flight deck architecture is similar to that of the B777 aircraft including the new FMS (Flight Management System) with larger memory and increased functionality, the new VSD (Vertical Situation Display), the integrated moving map display, and built-in EFB (Electronic Flight Bag).

System improvements include a new RAT (Ram Air Turbine) for additional hydraulic/electric power, improved interior and cargo handling equipment, and an improved fire suppression system (SSG 2012).

A380-800F

The A380-800F aircraft is expected to have the same IMA (Integrated Modular Avionics) architecture as its passenger version. This technological innovation is based on networked computing modules, which support different applications. Data communication networks based on fast-Ethernet reduce the required amount of wiring and minimize latency. The cockpit layout is similar to that of other Airbus aircraft, which certainly contributes to reducing the cost of training crews and the overall number of required crews as well. Eight displays including two MFDs (Multi-Functional Display(s)) provide an easy-to-use interface to the flight management system. The NSS (Network Systems Server) as a critical component of the paperless cockpit eliminates need for hard-copy manuals and navigational charts and backup paper documents. The MFDs enable keyboard access to the NSS with the stored data and electronic documentation, required equipment list, navigation charts, and performance calculations.

In addition, the primary hydraulic actuators are backed by power-by-wire flight control actuators with self-contained hydraulic and electrical power supplies. EHA (Electro-Hydrostatic Actuators) are used in the aileron and elevator, electric, and hydraulic motors to drive the slats. EBHA (Electrical Backup Hydrostatic Actuators) are used to drive the rudder and some spoilers. Reductions in the weight and size of pipelines (made of titanium), actuators, and related components is achieved through innovative high-pressure hydraulics. Variable frequency electrical generators and aluminum power cables are used as the main components of the completely computerized electric power system (Airbus 2012).

Operational performances

The operational performances of the B747-8F and A380-800F aircraft include the apron-gate turnaround time, payload range, technical productivity, the service network, and fleet size.

Apron-gate turnaround time

The apron-gate turnaround time depends on the quantity of freight to be unloaded/loaded, and the aircraft's suitability for such operations, usually expressed by the position, number and size of doors, and the loading/unloading facilities and equipment. For example, for the B747-8F aircraft, this maximum time is 91 min if the nose cargo door is used, 98 min if the side doors are used, and 54 min if both the nose and side doors are used (Boeing 2012a). The turnaround time of the A380-800F is projected to be about 120 $\pm$ 20 min depending on the layout of the ULD (Unit Load Device) (ULDs appear in two forms: pallets and containers of different sizes) (Airbus 2012; Boeing 2012a).

Payload range

As also applies to other passenger and freight aircraft, one of the most important operational performances of the B747-8F and A380-800F aircraft are the payload range characteristics. This is influenced by the other aircraft operational performance characteristics given in Table 3.10.

Table 3.10 Operational performances of the B747-8F and A380-800F aircraft (Airbus 2012; Boeing 2012a)

Indicator/measure	Aircraft type	
	B747-8F	A380-800F
Cockpit crew	2	2
MTOW (tons)	448	590
MLW (tons)	346	427
MZFW (tons)	330	402
OEW (tons)	197	251
Payload (tons)	133	151
Capacity		
Containers[a]	2LD1 + 12 LD6 s	59–71 LD3 s
Pallets[b]	46	66
Cruising speed at 35,000 ft/11,000 m (Mach/kts)	0.845/490	0.890/510
Maximum speed at 35,000 ft/11,000 m (Mach/kts)	0.92/533	0.96/551
Range, fully loaded (nm/km)	4150/7685	5643/10450
Maximum fuel capacity (000 *l*)	226	310
Engines (×4)		
Type	GEnx2B67	GP7177/Trent 977B
Thrust (kN)	4 × 296	4 × 340

[a] *Contoured full-width container: 1 LD6 is equivalent to 2 LD3* (1 *LD3—volume* 4.9 m^3 , *tare weight* 123 kg, *maximum gross weight* 1588 kg); *Contoured half-width containe—LD1*
[b] *Contoured pallet—width/length* 2.44 × 3.18 m; *volume* 10.8–11.52 m^3 ; *MTOW* Maximum Take of Weight, *MLW* Maximum Landing Weight, *MZFW* Maximum Zero Fuel Weight, *OEW* Operating Empty Weight, *l* liter, *ft* foot, *kt* knot, *kN* Kilo Newton, *GE* General electric, *GP* Engine Alliance of General Electric and Pratt and Whitney Aircraft Engine Manufacturers

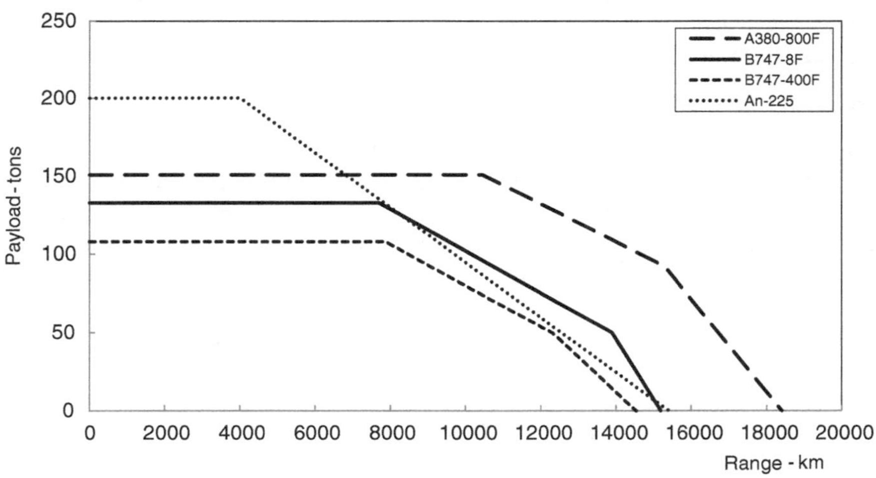

Fig. 3.14 Payload-range characteristics of the selected commercial freight aircraft (Airbus 2012; Boeing 2012a; http://www.antonov.com/aircraft/transport-aircraft/an-225-mriya)

In particular, the freight density expressed as the ratio of freight weight carried and the volume of aircraft indoor space occupied by such weight can be relevant for setting up the payload range relationship. In such context, the payload can be expressed in terms of weight, volume, or freight density. Figure 3.14 shows the payload range diagrams for the B747-400F, B747-8F and A380-800F aircraft (Diagrams for the An-225 and B747-400F aircraft are shown for comparative purposes).

As can be seen, the A380-800F aircraft appears superior as compared to the other two counterparts from Boeing thanks to it being able to carry the greatest payload over the longest range. The B747-800F aircraft is superior in a similar way to the B747-400F aircraft. In addition to the cancellation of orders for A380-800F aircraft, this superiority can also explain the dynamism of orders for the B747-7F aircraft: 27 by 2005, 53 by 2006, 66 by 2007, 68 by 2011, and 70 by 2012 (http://active.boeing.com/commercial/orders).

Technical productivity

Technical productivity is defined as the product of the aircraft payload and cruising speed (see Chap. 2). For the B747-8F and A380-800F aircraft, the payload versus range is shown in Fig. 3.14, and the cruising speed is given in Table 3.10. The calculated relationship between the technical productivity and the range affecting the maximum payload for both aircraft is shown in Fig. 3.15.

As can be seen and as intuitively expected, the technical productivity of the A380-800F aircraft exceeds that of the B747-8F aircraft since both influencing factors—payload and cruising speed—are greater. This difference decreases from about 42 % for the corresponding range with the maximum payload to about 20 % for the corresponding range with 60 % of the maximum payload. At the maximum range, which can be achieved without any payload, the technical productivity becomes equal to zero by definition at both aircraft.

Service network

The B747-8F and A380-800F large commercial freight aircraft are expected to operate in the global air freight service network, which will presumably be of the multi hub-and–spoke configuration. The primary network nodes will be large hubs located on different continents (one or two per continent). Each hub will have spoke airports as the secondary network nodes for feeding and distributing traffic located in the regions of the same and/or different continents. Air freight services scheduled between particular airports-nodes will act as the links. Figure 3.16 shows the simplified scheme of such a network. Long-haul intercontinental services between hub airports will be carried out by large commercial aircraft including the B747-8F and A380-800F. The short- to medium-haul connecting usually daily services will be scheduled between particular hubs and the associated spokes and carried out by medium size freight aircraft. This ensures the efficient interlining of services on a global scale. In addition, both the hub and spoke airports of the global network will have to have efficient and effective inland road (and in some instances rail) networks for the collection and distribution of freight

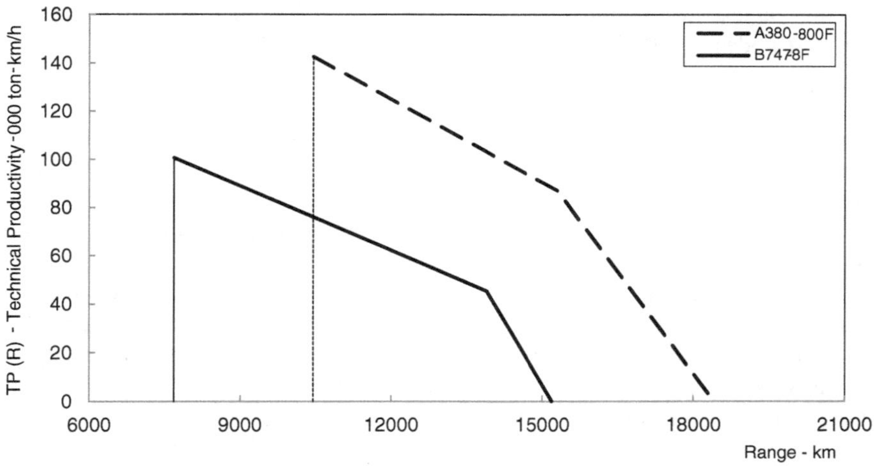

Fig. 3.15 Relationship between the technical productivity and range for the B747-8F and A380-800F aircraft (Airbus 2012; Boeing 2012a)

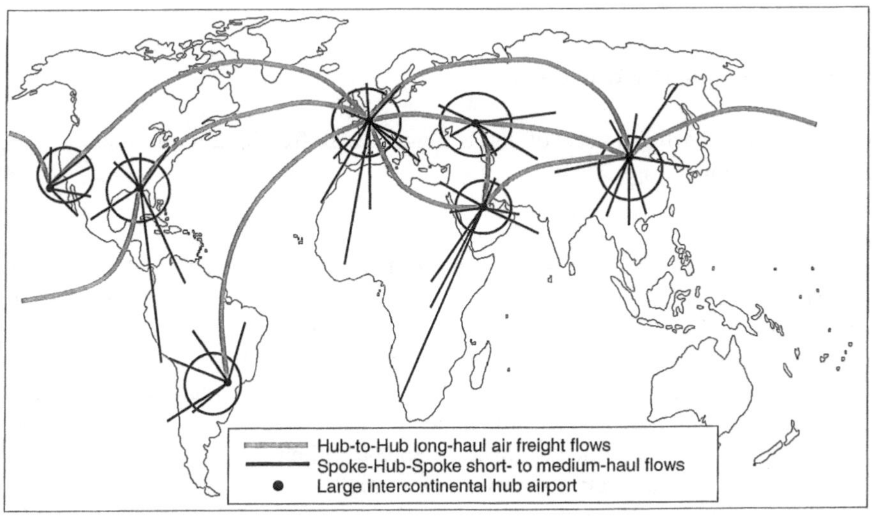

Fig. 3.16 The potential future global air freight transport network (Cargolux 2011)

shipments at/to the doors of shippers and receivers, respectively. For example, the European feeder road network of the Cargoulx airline from and/to its hub—Luxembourg airport—spreads over 17 countries with 63 origin/destination cities. The airline also has six road feeder networks from other European airports with total of 50 associated origin/destination cities (Cargolux 2011).

This makes the involved airports air/road or air/rail/road freight intermodal transport nodes.

Fleet size

The fleet size of air freight airlines is generally proportional to the product of the service frequency during any given period of time and the aircraft turnaround time on the network. For example, LHC (Lufthansa Cargo AG) operates a fleet of 18 MD-11F aircraft and has 5 new B777F aircraft on order. The airline network with its main hub at Frankfurt-Main airport includes 211 destinations/cities in 94 countries on all continents except Australia served per week (about two destinations/cities/country on average). The average service frequency is more than 20 flights/day/route. The Cargolux airline with its hub at Luxembourg airport currently operates a fleet of 11 B747-400F and B747- 400BCF, and 2 B747-8F aircraft, and has 11 new B747-8F aircraft on order. Its network includes 62 destinations/cities on all continents except Australia. It is evident that the fleet of both airlines is much smaller and more unified in terms of the number and type/size of aircraft, than that of, for example, some passenger airlines. This is mainly influenced by the characteristics of their rather global intercontinental networks and the type (scheduled, charter) and frequency of the offered services (Cargolux 2011; LHC 2011).

3.5.2.4 Economic Performances

The economic performances of large commercial freight aircraft include their costs and revenues.

Costs

As with other freight and passenger aircraft, the operating costs of large commercial freight aircraft such as the B747-8F and A380-800F aircraft consist of the costs of aircraft depreciation, flight crew, fuel, maintenance, and other flight costs including navigational and airport charges (Morrell 2011). In general, the evidence to date does not indicate the existence of economies of scale regarding the aircraft size, but in general, economies of distance do exist as shown in Fig. 3.17.

In addition, the economies of distance appear to be relatively weak in the case of long-haul freight aircraft as shown in Fig. 3.18.

It should be mentioned that US air freight carriers operate the B747-400F, B747- 400BCF, B747F-100/200F, B767-200F, and MD-11F aircraft. As mentioned above, Cargolux operates 11 B747-400F and B747- 400BCF, and 2 B747-8F aircraft, while LHC (Lufthansa Cargo AG) operates a fleet of 18 MD-11F aircraft.

The share of fuel costs in the total operating costs of long-haul and particularly the B747-400F aircraft operated by most freight airlines amounts to about 50–70 %. Therefore, if the B747-8F aircraft is expected to have lower fuel consumption by about 7–15 % (Table 3.11) than the B474-400F aircraft, its operating

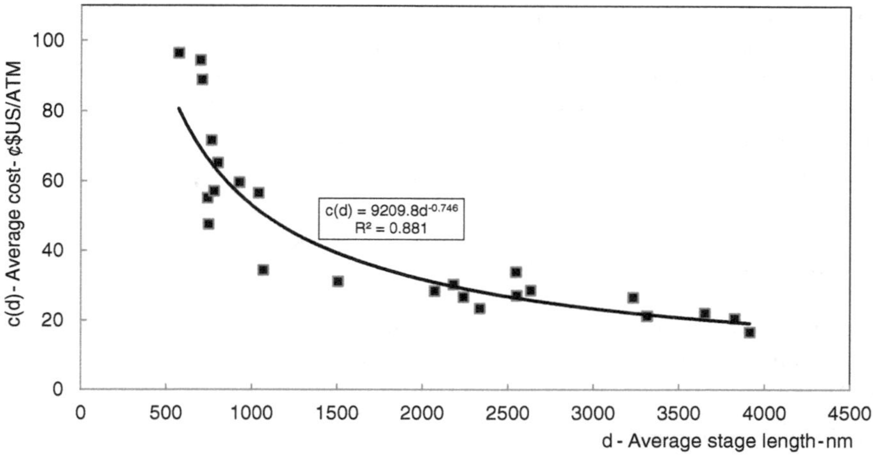

Fig. 3.17 Relationship between the average operating cost and the average stage length of U.S. commercial freight aircraft (AM 2011)

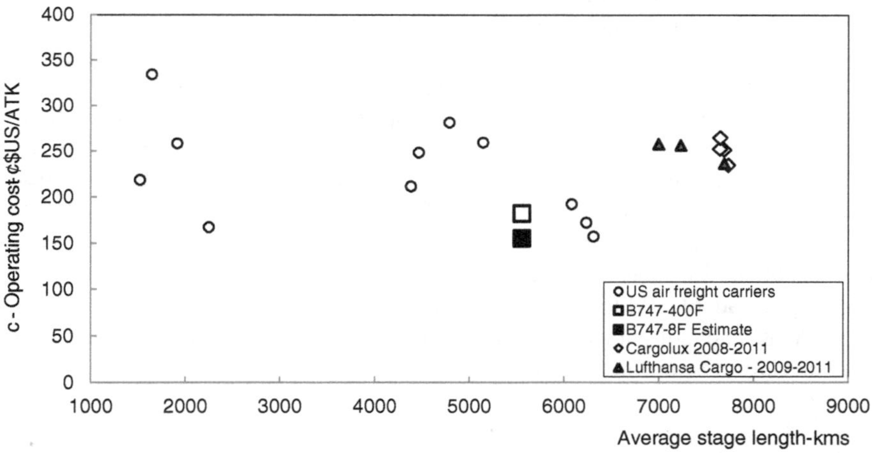

Fig. 3.18 Relationship between the average operating costs and the average stage length of commercial long-haul freight aircraft (Cargolux 2011; LHC 2011; Morrell 2011)

costs will be lower by about 7–15 %. As fuel prices increase, so will the cost savings. The information about the cost of the A380-800F aircraft are not sufficiently reliable in order to make more detailed analysis of their potential effects.

Revenues

Revenues are obtained by charging for air freight services. For example, during the 2008–2011 period, the above-mentioned Cargolux freight airline recorded average revenues (i.e., yield) of about 339 €$US/RTK and the Lufthansa Cargo AG freight airline of about 319 €$US/RTK (Cargolux 2011; LHC 2011). Despite there not being any strong evidence on the economies of scale relating to the aircraft size, it is reasonable to expect that the fuller and more fuel efficient B747-8F and A380-800F aircraft will enable freight airlines to offer lower rates while still providing sufficient revenues for covering their operational and other costs.

3.5.2.5 Environmental Performances

The environmental performances of large commercial freight aircraft include fuel consumption and related emissions of GHG (Green House Gases) and land use/take.

Fuel consumption and emissions of GHG

The main characteristics of the fuel consumption and related emissions of GHG (Green House Gases) of the B747-8F and A380-800F aircraft powered by different engines are given in Table 3.11.

Specifically, the fuel consumption and related emissions of GHG of the B747-8F aircraft are lower than those of the B747-400F aircraft by about 15–22 % (0.117 kg/ATK). The fuel consumption of the A380-800F aircraft derived from its passenger version of 0.096 kg/ATK appears to be by about 5–6 % higher than the lowest of B747-8F aircraft. Nevertheless, despite such differences and the inherent uncertainty of the figures for the A380-800F aircraft, the average fuel consumption and related emissions of GHG (Green House Gases) in terms of CO_{2e} (Carbon Dioxide Equivalents) of both large aircraft appear to be quite comparable (ICAO 2009). For comparison, the fuel consumption of the largest An-225 aircraft amounts about 23.5 tons/h while flying at the average cruising speed of 800 kmh with a payload of 200 tons. These give an average fuel consumption of about 147 g/ATK and emissions of GHG of about 639 g/ATK (http://www.antonov.com/aircraft/transport-aircraft/an-225-mriya)

Table 3.11 Environmental performances of B747-400F, B747-8F and A380-800F aircraft—fuel consumption and emissions of GHG (Green House Gases) (Airbus 2012; Boeing 2012a; various websites)

Indicator/measure	Aircraft/Engines		
	B747-400F/PW, GE, RR	B747-8F/GP7200	A380-800/RR Trent 900
Fuel consumption (g/ATK)	117	91–101	88–103/113
Emissions of GHG (gCO_{2e}^{a}/ATK)	510	399–438	383–449/492

[a] CO_{2e} Carbon-Dioxide equivalents, *PW* Pratt and Whitney, *GE* General Electric, *GP* Engine Alliance, *RR* Rolls Royce, *ATK* Available Ton Kilometer

Land use

The B747-8F and A380-800F large commercial freight aircraft are designed to operate on runways and taxiways of ICAO reference code F airports safely, efficiently, and effectively. At the apron-gate gate complex in front of the airport cargo terminal(s), parking stands need to be appropriately designed within the existing airport area (Fig. 3.13). This implies that these aircraft are "neutral" regarding additional land use.

3.5.2.6 Social/Policy Performances

The social/policy performances of large commercial freight aircraft include noise, congestion, and traffic incidents/accidents (safety).

Noise

Aircraft noise is discussed in Chap. 2. Table 3.12 gives the noise level at noise-certified locations for the selected freight aircraft.

It is evident that both the B747-8F and A380-800F aircraft are much quieter than the current long-haul freight aircraft including their closest counterpart—the B747-400F aircraft, namely by about 2.9–4.6 dB, depending on the noise certification location. In addition, Fig. 3.19 shows that both large aircraft are also superior regarding their SN (Specific Noise) during arrivals and departures.

Consequently, replacing current aircraft with either the B747-8F and/or A380-800F aircraft will significantly contribute to mitigating the aircraft noise around airports.

Congestion

The B747-8F and A380-800F large commercial freight aircraft are categorized as Heavy aircraft since their take-off weight is greater than 136 tons (ICAO) and 162 tons (CAA UK). Consequently, ATC (Air Traffic Control) separation rules are set up for the arriving and departing sequences in which these aircraft are the leading ones in order to avoid hazards arising from their wake vortices. In such context, the A380-800F aircraft is differentiated from other Heavy aircraft as a Supper Heavy aircraft. As the leading aircraft in the arriving sequence, it is separated from other Supper Heavy, Heavy, Upper and Medium Heavy, Small and Light following aircraft at 4, 6, 7, 7, and 8 nm, respectively. At the same time, the B747-8F aircraft as the leading aircraft in the arrival sequence is separated from other Supper Heavy, Heavy, Upper and Medium Heavy, Small and Light following aircraft at 4, 4, 5, 6, and 7 nm, respectively. This indicates that the presence of an A380-800F in the arriving stream increases the average ATC separation rules by about 20 % as compared to the B747-8F aircraft, and thus almost proportionally reduces the airport runway landing capacity under given conditions. Similarly, the presence of an A380-800F in the departure stream increases ATC separation rules by about 20 % as compared to those applicable to B747-8F aircraft, and thus reduces the corresponding runway capacity (Janic 2012).

Table 3.12 Social performances of the selected commercial freight aircraft—noise (EASA 2011)

Aircraft type	Aircraft weight (tons)	Noise level (EPNLdB)		
	MTOW/MLW	Lateral	Flyover	Approach
B787-4F	448/346	94.0	94.0	100.9
A380-800F	590/427	94.2	95.6	98.0
B747-400F	386/296	98.3	98.6	103.8
MD-11F	286/223	96.1	95.8	104.4
A330-200F	233/187	97.4	90.7	97.1
MD10-30F	263/198	97.9	97.4	106.3
B777F	287/221	98.7	87.0	99.7

MTOW (Maximum Take-Off-Weight)*; MLW* Maximum Landing Weight
EPNLdB Effective Perceived Noise Level in decibels (typical engines)

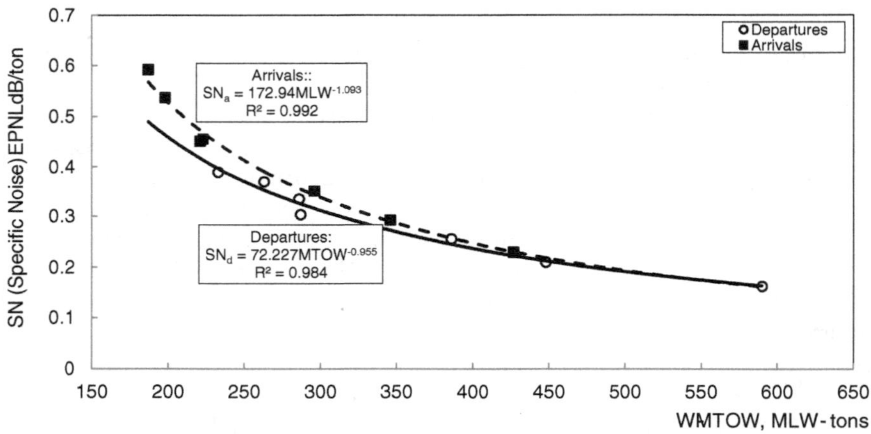

Fig 3.19 Relationship between the SN (Specific Noise) and weight of the commercial freight aircraft (MTOW—Maximum Take-Off Weight; MLW—Maximum Landing Weight) (EASA 2011; FAA 1997)

However, operating large commercial freight aircraft rarely has such an impact on the runway capacity since freight flights are usually scheduled during off-peak periods and/or on the separate runways.

Safety

The traffic incidents/accidents (safety) of large commercial freight aircraft can be considered as those of other commercial freight and passenger aircraft (Chap. 2). An additional aspect can be the safety of staff involved in ground operations, particularly in loading and unloading the aircraft. This can be measured similarly as the safety of aircraft operations in terms of the number of deaths and injuries per quantity of freight handled or per an aircraft kilometer carried out, both for the specified period of time, which is usually 1 year. Similarly, the safety of freight onboard can be measured by the quantity/value lost per aircraft kilometer. Regarding the B747-8F and A3380-800F aircraft, no judgment on their safety can be made at this moment since the former has just started and the latter is expected to begin operations sometime in the future.

3.5.3 Evaluation

The large commercial freight aircraft possess both advantages (strengths and opportunities) and disadvantages (weaknesses and threats) as compared to their closest smaller counterparts such as the B747-400F aircraft.

Advantages

The main advantages are as follows:

- The flight deck design enables flight operational commonality implying cross-crew qualification (i.e., qualification for different aircraft types), and consequently reducing the crew training time, the number of required crews, and providing for simpler and cheaper maintenance;
- The designed freight density fits very well with the current and prospective density of air freight shipments around the world;
- The lower fuel consumption and related emissions of GHG (Green House Gases) in relative and noise in absolute terms (per ATK (Available Ton-Kilometer)) are lower as compared to those of their closest counterpart(s);
- The designed range enables non-stop flights on long-haul intercontinental routes connecting large hub airports on different continents, thus consolidating the global air freight multi hub-and-spoke network(s) on the one side and fulfilling requirements of global companies in terms of just-in-time delivery of shipments and made-to-order call frequencies on the other; and
- The required fleet and its number of flights for transporting a given volume of air freight are smaller, contributing to reducing the related total fuel consumption and emissions of GHG, and congestion and delays at airport(s) as compared to those of their closest smaller counterparts.

Disadvantages

The man disadvantages are as follows:

- The actual behavior of composites used for making the structure—airframe and engines—is uncertain during the stressful life-cyclecycle, which consequently may increase maintenance cost;
- The time for collecting, loading, unloading, and distributing the larger volumes of freight is longer, which inherently increases the level of inventories and related costs at both ends of the given supply chain(s);
- The fuel consumption and related emissions of GHG per individual flight are grater that those of their closest counterparts;

- Airport infrastructure, particularly at the apron-gate complex where parking stands and related maneuvering space of the required size need to be provided, need to be modified;
- Adapting to strong and fast fluctuations in the volumes of air freight transport is inherently weak, i.e., the inflexibility in such adaptation is very high; and
- Congestion and delays can increase due to longer ATC separation rules applied to these and the other aircraft in the arrival and departure streams at a given airport (that said, the flights of freight aircraft are usually scheduled outside of the main peaks and/or on separate runways).

Finally, large commercial freight aircraft can be considered as advanced mainly because of their size, operational, and economic performances, which all enable forming and consolidating the global air freight transport network supporting the further internationalization and globalization of the world's economy.

References

AEA. (2011). *Reduction and testing of greenhouse gas (GHG) emissions from heavy duty vehicles—Lot 1: Strategy,* Didcot, Oxfordshire, UK: Final Report to the European Commission, DG Climate Action Ref: DG ENV. 070307/2009/548572/SER/C.

AEAT. (2005). *Structure of the costs and charges review—Environmental costs of rail transport,* Final Report to the Office of Rail Regulation, London, UK: AEA Technology Rail.

Airbus. (2012). *A380 airplane characteristics for airport planning AC*, Cedex, France: Airbus Blagnac.

AM. (2011). Block hour operating costs by airplane type for the year 2010: Executive summary. *The Airline Monitor*, *24*(3), 1–49.

ASECAP. (2010). *Introduction of longer and heavier vehicles: Impacts on road infrastructure*, Brussels, Belgium: Association des Europeennes des Cpncessionaires d'Autoroutes et d'Ouvrages a Peage.

Baungartberm, J. P. (2001). *Prices and costs in the railway sector.* Lausanne, Switzerland: EPEL—Ecole Politechnique Federale de Lausanne.

Boeing. (2012). *World air cargo forecast 2012–2013.* Seattle, WA, USA: Boeing Commercial Airplanes.

Boeing. (2012a). *747-8 airplane characteristics for airport planning*. Seattle, WA, USA: Boeing Commercial Airplanes.

Cargolux. (2011). *Annual Report 2011*, Luxembourg, Luxembourg: Cargolux Airlines International S.A., Luxembourg Airport.

CNT. (2006). *Transport/Europe, bulletin of the observatory on transport policies and strategies in Europe*, Paris, France: Michel Savy, December.

EASA. (2011). *Type-certificate data sheet for noise A380, TCDSN EASA, Issue 6*. Koln, Germany: European Aviation Safety Agency.

EC. (1996). *Council Directive 96/53/EC*, Brussels, Belgium: Official Journal of European Communities, No. L235/59.

EC. (1996a). *Towards fair and efficient pricing in transport: Policy options for internalizing the external costs of transport in European Union*, Supplement 2/96, Brussels, Belgium: European Commission.

EC. (1997a). *Innovative bundling networks in europe*, TERMINET, Deliverable D1, WP1, The Transport RTD Program of the 4th Framework Program, European Commission, (TERMINET Consortium).

EC. (1997b). *New Generation Terminals and Terminal-Node Concepts in Europe*, TERMINET, Deliverable D2, WP2, The Transport RTD Programme of the 4th Framework Programme, European Commission, (TERMINET Consortium).

EC. (1998a). *Building the blocks for preferable layout of innovative networks*, Deliverable D4, WP 3, Final Report, The Transport RTD Programme of the 4th Framework Programme, European Commission, (TERMINET Consortium).

EC. (1999). *The common transport policy—Sustainable mobility: Perspectives for the future*, Brussels, Belgium: European Commission, Economic and Social Committee and Committee of the Regions, Directorate General DG VII.

EC. (2000a). *Intermodal quality-IQ project*, Brussels, Belgium: European Commissions, Directorate General DG VII, RTD 4th Framework Program.

EC. (2000b). *The way to sustainable mobility: Cutting the external cost of transport*, Brussels, Belgium: Brochure of the European Commission.

EC. (2001a). *Real cost reduction of door-to-door intermodal transport—RECORDIT*, Brussels, Belgium: European Commission, Directorate General DG VII, RTD 5th Framework Programme.

EC. (2001b). *Improvement of pre- and end- haulage—IMPREND*, Brussels, Belgium: European Commissions, Directorate General DG VII, RTD 4th Framework Programme.

EC. (2001c). *Towards a new generation of networks and terminals for multimodal freight transport—TERMINET*, Brussels, Belgium: European Commission, Directorate General DG VII, RTD 4th Framework Programme.

EC. (2002). *EU intermodal transport: Key Statistical data 1992–1999*, Luxembourg: European Commissions, Office for Official Publications of European Communities.

EC. (2005). *Energy and fuel consumption from heavy duty vehicles*, COST 346, Final Report on the Action, Brussels, Belgium: European Cooperation in the Field of Scientific and Technical Research.

EC. (2006). *Long innovative intermodal interoperable freight trains*, Brussels, Belgium: European Commission, INTERREG IIIB NWE, C041.

EC. (2007). Customer-*driven rail-freight services on a european mega corridor based on advanced business and operating models (CREAM)*, Integrated Project, Brussels, Belgium: European Commission, Directorate General DG VII, 6th EU Framework Programme.

EIRAC. (2007). *Strategic intermodal research agenda 2020*. Brussels, Belgium: European Intermodal Research Advisory Council.

ERSO. (2007). *Traffic safety: Basic figures 2007—Heavy goods vehicles and buses*, European Road Safety Observatory, Safety Net, Transport, http://ec.europa.eu/transport/wcm/road_safety/erso/index.html

FAA. (1997). *Noise level for U.S. certified and foreign aircraft*, Washington, USA.: AEE-110Federal Aviation Administration, U.S. Department of Transportation.

Fraunhofer. (2009). *Long-term climate impacts of the introduction of mega-trucks: Study for the Community of European Railway and Infrastructure Companies (CER)*. Karlsruhe, Germany: The Fraunhofer-Institute for Systems and Innovation Research.

Hay, W. W. (1977). *An introduction to transportation engineering*. New York, USA: Wiley.

Hall, R. W. (1987). Direct versus terminal freight routing on a network with concave costs. *Transportation Research B, 21*(4), 287–298.

Hall, R. W. (1993). Design for local area freight networks. *Transportation Research Part B, 27B*(2), 70–95.

Hwang, L. C., & Yoon, K. (1981). *Multi attribute decision-making: Methods and applications, lecture series in economics and mathematical systems*. Berlin, Germany: Springer.

ICAO. (2009). *ICAO carbon emissions calculator version 2*. Montreal, Canada: International Civil Aviation Organization.

ICAO. (2012). *ICAO engine exhaust emissions data bank: Subsonic engines*. Montreal, Canada: International Civil Aviation Organization.

INFRAS. (2000). *External cost of transport: Accident, Environmental and Congestion Costs in Western Europe*. Zurich, Switzerland: INFRAS Consulting Group for Policy Analysis and Implementation.

Janic, M., Reggiani, A., & Nijkamp, P. (1999). Sustainability of the European freight transport system: Evaluation of the innovative bundling networks. *Transportation Planning and Technology, 23*(2), 129–156.

Janic, M. (2005). Modelling performances of intermodal freight transport networks. *Logistics and Sustainable Transport, 1*(1), 19–26.

Janic, M. (2007). Modelling the full costs of an intermodal and road freight transport network. *Transportation Research D, 12*(1), 33–44.

Janic, M. (2008). An assessment of the performance of the European long intermodal freight trains (LIFTs). *Transportation Research—A, 42*(10), 1326–1339.

Janic, M. (2012). Modeling effects of different air traffic control operational procedures, separation rules, and service disciplines on runway landing capacity, *Journal of Advanced Transportation*, August 24, 2012. doi:10.1002/atr.1208

Janic, M., Vleugel, J. (2012). Estimating Potential Reductions in Externalities from Rail-Road Substitution in Trans-European Transport Corridors. *Transportation Research D, 17*(2), 154–160

Larsson, S. (2009). *Weight and dimensions of heavy commercial vehicles as established by directive 96/53/EC and the European modular system (EMS)*. Brussels, Belgium: Workshop on LHVs.

Levison, D., Gillen, D., Kanafani, A., & Mathieu, J. M. (1996). *The full cost of intercity transportation—A comparison of high-speed rail, air and highway transportation in california*, USA: Institute of Transportation, University of California, Berkeley, Research Report, UCB-ITS-RR-96-3.

LHC. (2011). *Thinking ahead: 2011 Annual report*, Frankfurt, Germany: Lufthansa Cargo AG, Frankfurt Airport.

Manheim, M. L. (1979). *Fundamentals of transportation system analysi:Basic Concepts* (Vol. 1). Cambridge, Massachusetts, USA: The MIT Press.

Morlok, E. K., Sammon, J. P., Spasovic, L. N., & Nozick, L. K. (1995). Improving productivity in intermodal rail-truck transportation. In P. Harker (Ed.), *The service productivity and quality challenge* (pp. 407–434). The Netherlands: Kluwer Academic Publishers.

Morrell, S. P. (2011). *Moving boxes by air: Economics of air cargo*. Farnham, UK: Ashgate Publishing Company.

Railtrack, (1998). *Braking system and performance for freight trains*. London, UK: Railtrack PLC.

Ryck De, H. (2008). *Turbofan design for the commercial aircraft*. Warsaw, Poland: University of Technology, Faculty of Power and Aeronautical Engineering.

SSG. (2012). *747-8 Quick reference pilot guide*, Supercritical Simulations Group. http://www.supercritical-simulations.com

Tarski, I. (1987). *The time factor in transportation processes, developments in civil engineering 15*. Amsterdam, The Netherlands: Elsevier.

UIC. (2008). *Mega trucks versus rail freight*. Paris, France: International Union of Railways.

Chapter 4
Advanced Transport Systems: Technologies and Environment

4.1 Introduction

This chapter deals with the performances of advanced passenger cars, large advanced container ships, and LH_2 (Liquid Hydrogen)-fuelled commercial air transportation. The prime objective is to show the potential effects of such advanced technologies on the environment in terms of energy/fuel consumption and related emissions of GHG (Green House Gases).

Man-made GHG emissions, particularly those from using nonrenewable energy sources, have become an increasing burden on the industry, society, and politics all around the world. This is because these emissions and particularly their CO_2 component (Carbon Dioxide) are perceived to remain in the atmosphere for prolonged periods of time (presumably hundreds of years) and are proven to contribute to global warming and consequent climate change (Archer 2008). In order to mitigate or even diminish these impacts, both national and international policy makers, industrial organizations, and associations have undertaken a range of different measures. For example, in Europe, the EU (European Union) 27 Member States have fully institutionalized the problem by introducing national and international legislations and conventions, in addition to setting up specific targets for the absolute and relative reduction in emissions of particular GHG. These targets are expected to be achieved by a range of advanced technical/technological and operational improvements and by monitoring and reporting developments throughout particular air polluting sectors of the economy and society (EEA 2010). The most recent evidence indicates that some results have already been achieved: the total emissions of GHG have decreased by about 20 % over the 1990–2009 period, from 5,589 in 1990 to 4,674.5 million tons of CO_{2e} (Carbon Dioxide equivalents) in 2009 (CO_{2e} include CO (Carbon Oxide), CO_2 (Carbon Dioxide), SO_4 (Sulfur Oxides), NO_x (Nitrogen Oxides), H_2O (water vapor), and particles). However, at the same time, the share of transport sector in the total emissions of CO_{2e} has increased from about 17 % in 1990 to about 26 % in 2009, which is an equivalent of about 951 and 1,225 million tons of CO_{2e}, respectively (EC 2010a, b).

M. Janić, *Advanced Transport Systems*,

DOI: 10.1007/978-1-4471-6287-2_4,

In particular, road transport and specifically passenger cars have substantially contributed to the above-mentioned increase in the total emissions of GHG, as their share in the total volumes of vehicle-kilometers by road amounted to about 73 % during the 2005–2009 period. In absolute terms, these total volumes have increased from about 2,433 million in 1995 to about 3,061 million vehicle-kilometers in 2009 (EC 2010a, b; EEA 2010).

The world's economic development and international trade have been strongly supported by maritime transport whose freight ship fleet has been permanently growing aiming to satisfy growing demand. The fleet consists of different types of ships such as bulk, container, general cargo ships, oil tankers, and other ships. A substantial fast-growing part of this fleet consists of large advanced container ships. These and other freight ships use diesel fuel, which, in combination with higher demand, has resulted in an increase of GHG emissions. Consequently, maritime transport, which is currently not part of the Kyoto Protocol, accounts for about 3.3 % of the global man-made GHG emissions with the share of its international part in this total of about 2.2 %. Unless global policies aimed at controlling these emissions are put into place, they will likely increase by about 200–300 % by 2050 as compared to the figures in 2009, mainly driven by the expected continued growth in international trade. For example, in the EU-27 Member States, international maritime transport is the second largest source of GHG emissions by the transport sector. In 2007, international maritime transport emitted about 176 million tons of CO_{2e} mainly on account of transportation of 3,934 million tons of freight/goods. As compared to the emitted volumes in 2002, this represents an overall increase of about 18 % (EEA 2012; UNCDAT 2012).

In order to prevent the above-mentioned negative developments in the EU27 and the rest of the world, the shipping industry has undertaken a variety of technical/technological and operational measures to improve the sector's efficiency. The aim is two-fold: to reduce operational costs on the one hand and improve energy efficiency by reducing fuel consumption and related emissions of GHG on the other. Particularly relevant for large advanced container ships, the former measures include building fuel-efficient and environmentally friendly ships and promoting a switch to alternative/cleaner fuels. The latter measures imply adopting slow steaming. In addition, the international community, including the IMO (International Maritime Organization), has undertaken some measures to influence the energy efficiency of all, including large advanced container ships, aiming at reducing the rates of emissions of GHG (CO_2/ton-mile) below the current level. By combining the technical/technological and operational improvements, it seems possible to reduce these emissions by about 15–20 % by 2020 and by about 30 % by 2025 and beyond (IMO 2011; MEPC 2012).

The commercial APT (Air Passenger Transport) system mainly driven by economic/GDP (Gross Domestic Product) growth has grown over the past decades contributing to both globalization of the world's and national economies and overall social welfare on the one hand, and increasing energy consumption of nonrenewable sources (crude oil), related emissions of GHG and local noise, on the other. For example, the number of RPK (Revenue Passenger Kilometers) has increased from

0.5 trillion in 1971 to about 4.25 trillion in 2006. Some long-term forecasts by international air transport organizations (IATA, ICAO, ACI), and in particular by the two main manufacturers of commercial aircraft Boeing and Airbus, predict the rather stable long-term growth of RPKs at an average annual rate of 4.6–5 % over the next 20 years, mainly on account of average annual GDP growth of about 3.5 %. This will increase the total volumes of the world's traffic to about 10.545 trillion RPKs (Airbus 2006) and 11.4 trillion RPKs (Boeing 2007) by 2025/26. At the same time, the number of passengers is predicted to rise at an annual rate of 4.5 %, which will result in a total of about 6.8 billion in 2025/26 (Boeing 2007). The above-mentioned growth of air traffic will require an increasing number of aircraft, from the current 18,230 (of which 16,250 are passenger aircraft) in 2006 to about 36,420 (of which 32,440 will be passenger aircraft) in 2025/26 (Boeing 2007). Since all these aircraft are assumed to use conventional jet fuel as a derivative of crude oil, the total fuel consumption and related emissions of GHG will continue to increase, contributing to global warming and climate change (Airbus 2006; Boeing 2007; IPCC 1999). Some estimates indicate that the air transport sector emitted about 513 MtCO_2 in 1992. This is expected to increase to about 1,468 MtCO_2 in 2050. The latter quantity will likely continue to account for between 3–5.5 % of the total man-made emissions of CO_2 (ICAO 2008; IPCC 1999, 2001).

4.2 Advanced Passenger Cars

1901	Electric cars-taxi cabs appear in New York (U.S.)
1911	The first gasoline–electric hybrid car is released by the Woods Motor Vehicle Company of Chicago (U.S.)
1997	The world's first mass-produced Toyota Prius electric–gasoline hybrid car is released (Japan)
1999	The first Honda Insight hybrid car since the little-known Woods hybrid of 1917 is sold in North America; sales of the hybrid Toyota Prius substantially increase; many car makers release hybrid models and several began to produce new electric car prototypes (U.S.)
2008	The first Tesla Roadster all-electric car developed by Tesla Motors in serial production is sold to customers (U.S.)
2009/2010	The Mitsubishi i-MiEV electric car is launched (Japan)

4.2.1 Definition, Development, and Use

At present, the majority of passenger cars use petrol and diesel fuel as a derivative of crude oil and natural gas, the burning of which contributes to the above-mentioned emissions of GHG. Consequently, under the assumption that volumes of passenger car use will continue to grow, that the reserves of crude oil and natural gas will become depleted and eventually vanish, and that the emissions of GHG will remain in the atmosphere for prolonged periods, thus indicating their continuous increase in cumulative amounts, various improvements of existing and developing advanced passenger car technologies have been undertaken. Among other, they aim, together with other operational, social, and policy measures, to mitigate the above-mentioned emissions of GHG over the medium- to long-term future (IPTS 2008).

4.2.2 Analysis and Modeling Performances

4.2.2.1 Background

Both existing conventional and advanced forthcoming passenger car technologies are characterized by their technical/technological, operational, economic, environmental, and social/policy performances. In the given context, the technical/technological performance mainly relates to the vehicle size (the number of seats, weight), and the engine type characterized by its volume/power and energy/fuel efficiency. The operational performances include the maximum and the most fuel/energy efficient speed. The economic performances include the purchase price and operational costs (fuel, maintenance). The environmental performances mainly include the emissions of GHG, which depend on the energy/fuel used. These could also include land use for operating—maneuvering and parking. The social/policy performances relate to noise, congestion, and general public acceptance.

4.2.2.2 Analyzing Performances

At present, the following passenger car technologies based on the type of energy/fuel use can be distinguished: conventional petrol/diesel/gas ICEVs (Internal Combustion Engine Vehicles), HYVs (Hybrid Vehicles), BEVs (Battery Electric Vehicles), HVs (Hydrogen Vehicles), and HFCVs (Hydrogen Fuel Cell Vehicles). The last three categories of cars, and particularly the latter two, are expected to more intensively penetrate the EU27 and other world markets over the forthcoming decades. However, this can only be expected if they are able to provide an equivalent overall convenience to their users—at least at the level provided by today's conventional ICEVs and/or if they become exclusive alternatives due to the depletion of reserves of crude oil, making more convenient ICE cars practically unusable.

ICEVs (Internal Combustion Engine Vehicles)

Conventional ICEVs (Internal Combustion Engine Vehicle(s)) are considered as relatively low energy/fuel efficient due to the fact that as a result of converting fuel into propulsion, most of the energy is emitted as heat. Typical petrol ICEVs engines effectively use only 21 % of the fuel energy content to move the vehicle and their diesel ICE counterparts are efficient up to 25 %. This WTW (Well-to-Wheel) efficiency includes the energy consumed to produce and deliver fuel to the station (WTT (Well-to-Tank)) and the energy used to fill and consume it in the car (TTW (Tank-to-Wheel)) (Bodek and Heywood 2008).

Currently, in the EU (European Union)-27 Member States, conventional ICEVs are categorized into three categories depending on the engine volume: Small <1.4 l, Medium >1.4 and ≤2.0 l, and Big >2.0 l (l—liter). Regardless of the fuel used, Small cars are most numerous and their Big counterparts the least. The typical engine power of these cars is about 60–80 kW. The engine volume is correlated to the car weight, which is related to the fuel efficiency as follows: $FE = 0.004 + 5.249\ W$ ($R^2 = 0.839$) (*FE* is fuel efficiency, i.e., the average fuel consumption (l/100 km)); *W* is the car weight (kg)). In addition, the fuel consumption of an average car using petrol, diesel, and/or gas amounts to 6.7 l/100 km (0.683 kW-h/km). Specifically, the average fuel consumption of an average petrol car is 7.3 l/100 km (0.706 kW-h/km) (this is expected to decrease to 5.8 l/100 km (0.561 kW-h/km) by 2020), and that of an average diesel/gas car 5.8 l/100 km (0.594 kW-h/km), which is expected to decrease to 4.6 l/100 km (0.493 kW-h/km) by 2020. The average age of a passenger car in the EU-27 is 7.5 years (this is expected to increase to about 11–13 years by 2020) (IPTS 2003; ICG 2010).

Emissions of GHG by conventional ICEVs are usually considered as closely related to their WTW energy/fuel efficiency. In many cases, both can be standardized and as such become country or region specific. For example, the standards set up for the EU-27 Member States in 2007–2008 were 6–8 l/100 km (0.612–0.760 kW-h/km) of energy/fuel consumption and 165–200 gCO_2/km. The newly proposed standards are around 6.2 l/100 km (0.632 kW-h/km) and 140 g CO_2/km. The targets to be achieved by 2030 are energy/fuel consumption of about 3.5 l/100 km (0.357 kW-h/km) and emissions of about 82–84 gCO_2/km (CO_2—Carbon Dioxide) (IPTS 2008).

HYVs (Hybrid Vehicle(s))

HYVs (Hybrid Vehicle(s)) can be considered an advanced passenger car technology. They are powered by conventional petrol or diesel ICEs and an electromotor. While the former uses petrol or diesel fuel, the latter uses electric energy stored in on-board batteries, which are charged by the energy from the ICE engine. This means that recharging batteries by plugging in at street stations and/or at home is not possible. In general, the electromotor is used for driving at low speeds predominantly in urban areas, while the power switches to ICE when driving at higher speeds requiring greater engine power. The WTW energy/fuel efficiency of these cars is about 40 % (Toyota Prius) and is expected to improve to about 55 % in the mid-term future. For example, the most efficient hybrid car in 2005 was the Honda Insight whose WTW energy/fuel efficiency was 0.64 km/MJ (0.391 kW-h/km) followed by the Toyota Prius with 0.56 km/MJ (0.491 kW-h/km), and the petrol

ICE Honda Civic VX with 0.52 km/MJ (0.534 kW-h/km) (MJ—Mega Joule; kW-h—kilowatt hour).

In general, in 2010, the fuel consumption of an average hybrid electric-petrol car amounted to about 5.4 l/100 km (0.799 kW-h/km) and that of an average hybrid electric–diesel car to about 4.51 l/100 km (0.483 kW-h/km). The corresponding emissions of GHG were 125 and 90 gCO_2/km, respectively. Some improvements particularly to the fuel supply systems in these cars lead to expectations that their consumption will decrease to about 3.4 l/100 km (0.329 kW-h/km) in the former and to about 2.45 l/100 km (0.251 kW-h/km) in the latter by 2035. The corresponding emissions of GHG will be 52 and 47 gCO_2/km, respectively. This implies that in terms of energy/fuel efficiency and related emissions of GHG, electric/petrol and electric/diesel HYVs are more efficient than their conventional ICE counterparts by about 25 and 30 %, respectively (Bodek and Heywood 2008).

BEVs (Battery Electric Vehicles)

BEVs (Battery Electric Vehicle(s)) can be considered as an advanced passenger car technology. They are propelled by electromotors using the electric energy stored in batteries on-board the vehicle. The batteries are recharged from the power grid (at home or at street/shop charging stations). The WTW energy efficiency of electric cars is expected to reach up to about 80 %. This can be achieved, among other factors, also thanks to converting the stored energy into propelling the car, not consuming energy while stopping, and regenerating some (about 20 %) through regenerative braking. For example, the Tesla Roadster BEV has a WTW energy efficiency of about 1.14 km/MJ (0.235 kW-h/km). Other typical electric cars are expected to have a WTW energy efficiency of about 1.125 km/MJ (0.247 kW-h/km) (Hamilton 1980) and 1.583 km/MJ (0.175 kW-h/km) (Toyota Rav4EV) (ICG 2010). It should be mentioned that about 20 % of this energy consumption is due to inefficiencies in recharging the on-board batteries. These are the most sensitive parts of electric cars in terms of their specific energy capacity versus the weight, replacement, durability, and the short and full charging time. With a single charge, they need to provide sufficient energy for the car to cover a reasonable distance at a reasonable speed as compared to conventional ICE petrol/diesel cars. Contemporary lithium batteries usually have a specific energy capacity of about 130 W-h/kg, which is one of the reasons for their frequent use despite their rather limited lifespan. Modified lithium iron phosphate and lithium–titan batteries have an extended life span of up to several thousand cycles and are relatively easily replaced. Their recharging time also needs to be reasonable. This is not particularly important if recharging takes place at home during off-peak hours (Koyanagi and Uriu 1997); however, it becomes very important if recharging takes place at street stations. Depending on the car's charger and battery technology, the recharging time can be 10–30 min to fill the batteries to about 70 % of their capacity. For example, the forthcoming models in the EU-27 market in 2011 such as Nissan Leaf, Renault Fluence Z. E. and Hyundai Blue have ranges between 140 and 170 km, top speeds between 130 and 145 km/h, full charging times of 6–8 h, and rapid charging times (up to 80 %) of about 25–30 min. The above-mentioned characteristics make these cars particularly convenient for use in urban and suburban areas with rather short daily driving distances (ICG 2010).

Electricity for BEVs can be obtained from different primary nonrenewable and renewable primary sources (EEA 2008; OI 2011). The former include coal, crude oil, natural gas, biomass, and nuclear energy, and the latter solar, wind, and hydro energy. The shares of the above-mentioned primary sources (usually country or region specific) make GHG emissions by BEVs exclusively dependent on their WTT (Well-To-Tank) energy/fuel efficiency.

HVs (Hydrogen Vehicles) and HFCVs (Hydrogen Fuel Cell Vehicle(s))

Hydrogen passenger vehicles (cars) are powered by hydrogen fuel. Two categories of these vehicles can be distinguished. The first are slightly modified conventional ICEs that use hydrogen instead of petrol/diesel/gas as fuel—HVs (Hydrogen Vehicle(s)). In order to cover a reasonable distance, hydrogen is highly compressed in the fuel storage tanks of these vehicles, mainly thanks to its low density. HFCVs (Hydrogen Fuel Cell Vehicle(s)) represent an advanced technology in passenger cars. They consist of five components which distinguish them from their HV counterparts: their fuel cell stack, electric motor, power control unit, hydrogen storage tank, and high-output batteries. Specifically, the fuel cell stack consists of individual fuel cells whose number depends on their size and the required electric energy. Each fuel cell uses either pure hydrogen from hydrogen-rich sources, or oxygen to generate electric energy. Fuel is used to feed the electric motor that actually propels the car. The intensity of electric energy delivered from the fuel cells to the electric motor is regulated by the power control unit. Hydrogen as the source of electricity is stored in the hydrogen storage tank either as a liquid or as a highly compressed gas. In addition, high-output batteries are installed to accumulate the electric energy from the regenerative braking, thus providing additional power to the electric motor.

Hydrogen as a prospective fuel exists in nature as a component of natural gas (CH_4) and water (H_2O). This means that in order to provide hydrogen as fuel for hydrogen fuel cell cars, it needs to be extracted from the above-mentioned sources. This can be carried out by reforming natural gas or through the water electrolysis either at large plants or at local fuel supply stations. In the former case, distribution from the producing plants to local supply stations needs to be provided either by truck or an underground pipeline network. Hydrogen has more energy per unit of mass than other crude oil-based fuels including natural gas. On the other hand, it is much less dense (Janic 2010). The design of the fuel tanks of HFCVs will have to take the above-mentioned facts into consideration. Nevertheless, the volume of these tanks should not be much greater than that of conventional ICEVs as more energy per unit of mass of hydrogen is expected to compensate its lower density to a large extent. In addition, this will enable a similar pattern of utilization of HFCVs compared to their modern conventional ICEV counterparts.

The primary sources for obtaining hydrogen heavily influence the energy/fuel efficiency of HFCVs. At present, in practice, the WTW energy efficiency of HFCVs reaches about 50–60 % (i.e., 0.85 km/MJ or 0.327 kW-h/km) if hydrogen is obtained from reforming natural gas, and to only about 22 % (i.e., 0.30 km/MJ or 0.926 kW-h/km) if it is obtained through water electrolysis. However, the theoretical overall efficiency of HFCVs can be nearly 100 % (i.e., 1.39 km/MJ or 0.198 kW-h/km and 2.78 km/MJ or 0.102 kW-h/km, respectively) (http://www.fueleconomy.gov/FEG/fuelcell.shtml).

If hydrogen is derived from water electrolysis, the emissions of GHG by HFCVs will mainly depend on the primary sources of the electric energy used for this electrolysis. This can be from both nonrenewable and renewable sources, which influences the total WTW emissions of GHG. In the WTT segment, these will be zero if electricity is obtained exclusively from renewable sources and much higher otherwise. In the TTW segment, the emissions will be zero except for those of water vapor (H_2O), which will increase by about three times as compared to those from conventional ICEV and HYV fuels (Janic 2010).

4.2.2.3 Modeling Performances

Modeling the performances of different passenger car technologies is focused on their environmental performances in terms of the energy/fuel consumption and related emissions of GHG. When and where necessary, other infrastructural, technical/technological, operational, economic, and social policy performances are also considered. Modeling includes an analysis of the previous efforts, objectives, and assumptions, as well as the basic structure of the methodology for assessing the above-mentioned impacts/effects on the environment.

Previous efforts

Alterative passenger car technologies and particular dimensions of their performances (technical/technological, operational, economic, environmental, social, and policy) have been intensely investigated over the past decade. In general, efforts can be classified into the following five segments:

- *Market demand*: This aspect has focused on investigating the demand for alternative energy/fuel passenger cars and the consequent prospective market structure, while respecting the various operational, economic, environmental, and institutional (policy) conditions/constraints. These particularly relate to demand for HYVs and BEVs in specified regions. In such context, different supporting tools mainly based on the logit modeling approach have been developed for estimating the relative market share and the dynamics of introducing particular passenger car technologies, as well as their absolute demand

and supply with both and particularly the latter respecting the long-term plans of major car manufacturers (Ewing et al. 1998; Hörmandinger and Lucas 1996; Heffner et al. 2007; Higgins et al. 2007; ICG 2010; IPTS 2003; Kurani et al. 1996; Mabit and Fosgerau 2011).

- *Impacts and effects of energy consumption and emissions of GHG in urban areas*: This aspect has focused on investigating the characteristics of energy/fuel consumption by conventional ICEVs, HYVs, and new BEVs and HFCVs, the overall logistics for the energy/fuel supply, estimating the demand for the particular energy/fuel type, and the impacts of this demand on the eventual depletion of these energy/fuel primary sources. In the above-mentioned cases, the related emissions of GHG affecting the environment at the specified urban, suburban, and wider regional scale have also been estimated (Chi and Stone 2005; Coelho and Luzia 2010; DeLuchi 1989; Georgakellos 2008; Hamilton 1980; IPTS 2008; Johansson and Åhman 2002; Kang and Recker 2009; Kempton and Letendre 1997; Koyanagi and Uriu 1997; Lave and MacLean 2002; Nakata 2000; Rienstra and Nijkamp 1998; Schock et al. 1995; Wang and DeLucchi 1991; Wang et al. 2008).
- *Design and performance of new passenger car technologies including infrastructure for energy/fuel supply*: This aspect has dealt with the technical/technological solutions (material, design, safety requirements) influencing the operational, economic, and safety performances of innovative (HYV) and new (BEV and HFCV) technologies. In addition, the characteristics and needs for energy/fuel supply infrastructure for new passenger car technologies have been investigated (Chen and Ren 2010; Eberhard and Tarpening 2006; Ogden 1997; Schwoon 2007; Spiegel 2004).
- *Social costs and benefits*: This has focused on assessing the overall social and environmental costs and benefits from using innovative HYVs and new BEVs and HFCVs including the economy of providing and using energy/fuel by these cars in specified regions (Funk and Rabl 1999; Haller et al. 2008; Johansson 1999); and
- *Policy implications due to introducing alternative passenger car technologies*: This research has analyzed energy/fuel economy in various regions of the world and related emissions of green house gases, and compared them with the standards set up for passenger cars and other transport vehicles (An and Sauer 2004).

Assumptions

The methodology for assessing the prospective medium- to long-term effects of the above-mentioned advanced passenger car technologies on energy consumption and related emissions of GHG (CO_{2e}—Carbon Dioxide equivalents) has been developed respecting the following facts:

- The fuel/energy consumption and related emissions of GHG depend on the volumes of passenger car use, which is mainly driven by the overall socio-economic development;

- The energy/fuel used by currently predominant conventional ICEVs is obtained from exhaustive resources (crude oil), the burning of which generates emissions of GHG, which after being deposited tend to remain in the atmosphere for a long time;
- The structure of the passenger car market is expected to gradually change due to the more intensive use of advanced HYVs, BEVs, and HFCVs, in addition to permanent improvements of conventional ICEVs; and
- Production of electric energy used by BEVs and indirectly by HFCVs is expected to generally shift toward more intensive use of renewable sources.

In addition, the methodology is based on the following assumptions:

- The particular passenger car technologies remain and/or penetrate the market of a given region according to the specified "what-if" scenarios, which remain stable (constant) over a given period of time; this approach has been adopted since it is at present practically impossible to precisely predict the structure of this market based on the users' acceptance of BEVs and HFCVs even in the short- and especially in the medium- and long-term future;
- The average (typical) values of the WTW energy/fuel efficiency for particular passenger car technologies and related emissions of GHG and their gradual improvements are implicitly taken into account; energy consumption is estimated only in relative terms (per unit of car output—veh-km) and used as an input for calculating the GHG emissions;
- The emissions of GHG by passenger car use over the specified future period of time are exclusively considered; this implies that the lifetime and related rates of dissipation of GHG have not been taken into account mainly due to the very high level of uncertainty and diverging expert opinions on these issues; and
- Advanced passenger car technologies are assumed to use energy/fuel mainly obtained from GHG-neutral primary sources whose shares are also specified according to the "what-if" scenarios remaining stable (constant) over the given period.

Structure of the methodology

The methodology consists of three components/models: (i) the model for estimating the volumes of passenger car use; (ii) the model for estimating the fuel/energy consumption by the above-mentioned passenger car use; and (iii) the model for estimating emissions of GHG from the above-mentioned fuels/energy consumption.

The model for estimating the passenger car use

The existing and future passenger car use in terms of the vehicle kilometers carried out during a given period of time (year) in a given region can be estimated by two types of submodels using empirical data: (i) the time series model; and (ii) the causal model based on multiple regression analysis.

- **The time series submodel**

The time series model uses the empirical data on the passenger car use for the period and establishes their relationship in dependence of time as follows:

$$v_k = f(t^n) \tag{4.1a}$$

where

v_k is the volume of passenger car use in the k-th year of the observed period ($k = 1,2,\ldots, N$);

t^n is the variable representing time (years of the observed period); and

n is the coefficient estimated by establishing the given relationship.

The submodel (Eq. 4.1a) can be estimated using past data, thus enabling extrapolation of the passenger car use as the dependent variable into future period of time (the independent variable). In this case, it is implicitly assumed that the main forces influencing the dependent variable in the past period will continue to similarly drive future development.

- **Causal submodel**

This submodel implies that the passenger car use in a given region over a given period (the dependent variable) depend on a set of influencing factors considered as the independent (explanatory) variables. One of the possible generic forms in the given context is as follows:

$$v_k = f(t^n) \tag{4.1b}$$

where

GDP_k is the Gross Domestic Product of a given region in the k-th year of the observed period (billion currency units);

MR_k is the motorization rate of a given region in the k-th year of the observed period (cars/thousand inhabitants); and

P_k is the population of a given region in the k-th year of the observed period (million).

The other symbols are analogous to those in Eq. 4.1a.

In Eq. 4.1b, the past data of one or more independent variables can be simultaneously taken into account to estimate the dependent variable. By specifying the future values of particular independent variables, the future volumes of passenger car use in the given region can be estimated (as the dependent variable). Similarly as in Eq. 4.1a, the driving forces from the past are expected to act in a very similar way in the future.

The model for estimating energy consumption by passenger car use

Let the period of time considered be N years long. During this period, M different passenger car technologies are expected to be used. The average fuel/energy consumption of the passenger car technology (i) in the k-th year of the observed period can be estimated as follows:

$$E_{ki} = v_k * p_{ki} * (1 - r_{ki}) * e_{ki} \tag{4.2a}$$

where

p_{ki} is the proportion (i.e., market share) of passenger car technology (i) in the k-th year of the observed period;

r_{ki} is the average rate of improvement of the fuel/energy efficiency of passenger car technology (i) in the k-th year of the observed period; and

e_{ki} is the average energy/fuel consumption of passenger car technology (i) in the k-th year of the observed period (units of energy/fuel/car-km).

From Eq. 4.2a, the total cumulative energy/fuel consumption from the beginning to the year (n) of the observed period can be estimated as follows:

$$E(n) = \sum_{k=1}^{n} \sum_{i}^{M} E_{ki} = \sum_{k=1}^{n} \sum_{i}^{M} v_k * p_{ki} * (1 - r_{ki}) * e_{ki} \tag{4.2b}$$

where all symbols are analogous to those in the previous equations.

The model for estimating emissions of GHG by passenger car use

Based on Eqs. 4.2a and 4.2b, the emissions of GHG by the passenger car technology (i) in the k-th year of the observed period can be calculated as follows:

$$GH_{ki} = E_{ki} * g_{ki} \tag{4.3a}$$

where

g_{ki} is the average emission rate of GHG of the passenger car technology (i) in the k-th year of the observed period (emitted quantity/unit of energy/fuel consumed)

In Eq. 4.3a, the variable g_{ki} takes into account both direct and indirect (WTW) emissions of GHG emitted by the given passenger car technology in the given year. In addition, it depends on the structure of primary sources for obtaining energy/fuel for a given car technology in the given year as follows:

$$g_{ki} = \sum_{j=1}^{L} q_{kij} * \gamma_{kij} \tag{4.3b}$$

where

q_{kij} is the share of the primary source of type (j) for producing fuel/energy for the car technology (i) in the k-th year of the observed period;

γ_{kij} is the average rate of GHG emissions from producing the energy/fuel from the primary source (j) for the car technology (i) in the k-th year of the observed period (emitted quantity/unit of energy/fuel consumed); and

L is the number of different primary sources for producing the energy/fuel for passenger car technologies.

The cumulative GHG emissions from energy/fuel consumption due to the passenger car use from the beginning until the year (n) of the observed period are estimated as follows:

$$G(n) = \sum_{k=1}^{n} \sum_{i}^{M} G_{ki} = \sum_{k=1}^{n} \sum_{i}^{M} E_{ki} * g_{ki} \tag{4.3c}$$

where all symbols are as in the previous equations.

Application of the methodology

Input

- **"What-if" scenario approach**

The "what-if" scenario approach is characterized by specifying the particular variables of the proposed above-mentioned methodology for the specified medium- to long-term period, in this case from 2010/15 to 2065 (Rienstra and Nijkamp 1998). This includes GDP growth as the main driving force of the volumes of passenger car use, the composition of the passenger car fleet in terms of the share of particular car technologies, the characteristics of particular car technologies in terms of the energy/fuel consumption and related emissions of GHG respecting gradual improvements over time, and the composition of primary sources for obtaining the energy/fuel to be used by the particular car technologies. Three scenarios are considered:

- *Scenario 0* is a rather hypothetical one specified mainly for comparative purposes. It is characterized by the continuing exclusive use of conventional petrol/diesel/gas ICEVs whose WTW energy efficiency and related rates of emissions of GHG continuously improve by 2030/2035. Their hybrid counterparts are not expected to penetrate the market substantially. In addition, depletion and vanishing crude oil reserves and increased related emissions of GHG do not particularly affect the expected use of these permanently technical/technologically improved conventional passenger cars.
- *Scenario 1* is a semi-hypothetical scenario containing some remaining uncertainty regarding the extent to which advanced, conventional petrol/diesel/gas ICEVs, and their HYV versions with constantly improving WTW energy/fuel efficiency and related emissions of GHG are assumed to be exclusively used. This again implies that further depletion of crude oil reserves will not significantly affect the use of these passenger car technologies. In addition, the emissions of GHG will be controlled by the technical/technological improvements of advanced innovative HYVs and existing ICEVs. Also, advanced BEVs and HFCVs are not assumed to substantially penetrate the market.
- *Scenario 2* can be considered realistic as it implies depletion of crude oil reserves and the subsequent strong market penetration of advanced passenger

car technologies (BEVs and HFCVs). In addition, there will be a constant increase in the use of renewable primary sources to produce the energy/fuel for these cars. According to some forecasts, the known reserves of crude oil will be mostly exhausted by the end of the specified period (around 2065) (Cambell 2002; Greene and Hobson 2003; Pfeiffer 2004; IEA 2009). The petrol and diesel fuels obtained from biomass and synthetic fuels will not be sufficient and will consequently be too expensive and thus practically unavailable for mass passenger car use. This implies that conventional petrol/diesel/gas ICEVs and their hybrids (HYVs) will gradually but certainly disappear from the market, which will be exclusively shared between BEVs and HFCVs.

During the observed period, which is divided into three subperiods, a set of factors are assumed to individually or in combination influence the rate of market penetration of each advanced passenger car technology as follows:

- The technical/technological maturity of the advanced car technology and availability of the energy supply infrastructure in the given region (EU27);
- The familiarity of prospective users with advanced car technologies as compared to existing conventional technologies;
- Comparable/competitive purchasing prices and operating costs of advanced car technologies as compared to existing conventional technologies;
- Despite improvements in the TTW (Tank-To-Wheel) efficiency, continuously raising operating costs of conventional petrol/diesel/gas ICEVs and their hybrids mainly due to raising prices of the increasingly depleting/vanishing crude oil reserves;
- Gradual saturation of the market with advanced car technologies implying the market presence of competitive advanced technologies; and
- The overall increasing public awareness of the depletion of crude oil reserves as the fuel source for conventional ICEVs and their hybrids on the one hand, and of BEVs and HFCVs as the only available alternatives on the other.

In the first subperiod, the annual market penetration rate of new passenger cars is assumed to be rather modest due to the predominant influence of factors 1 and 2. In the second period, this rate is assumed to be reasonably high, stable, and influenced mainly by the above-mentioned factors 3 and 4. In the final period, this rate is assumed to again decrease mainly due to the influence of factors 5 and 6.

During the same period, energy for the transport sector and passenger cars is assumed to be increasingly produced from renewable primary sources on account of coal, crude oil, and natural gas exhaustion (OI, 2011).

- **The volumes of passenger car use**

The volumes of passenger car use in the given region are estimated by using the past GDP data as the independent variable in the expression (4.1b) as shown in Fig. 4.1. Implicitly, the average car occupancy rate is adopted to be 1.6 persons/vehicle.

The growth of volumes of passenger car use in the EU27 between 1995 and 2009 closely followed GDP growth, albeit at a decreasing rate. Consequently, the

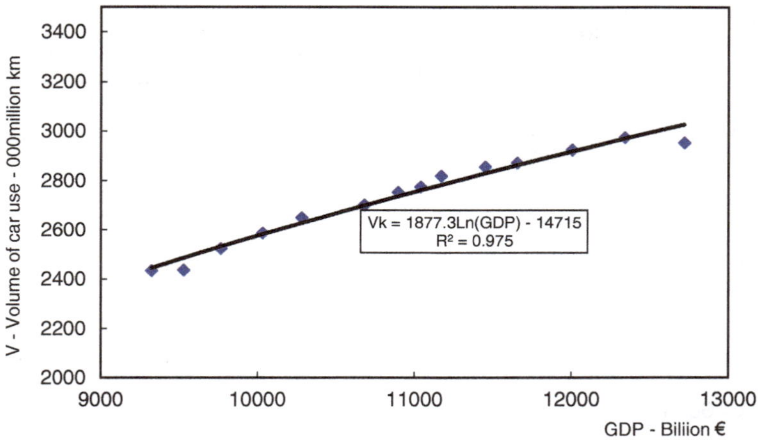

Fig. 4.1 Relationship between the annual volumes of passenger car use and GDP in the EU27 (1995–2009) (EC 2010)

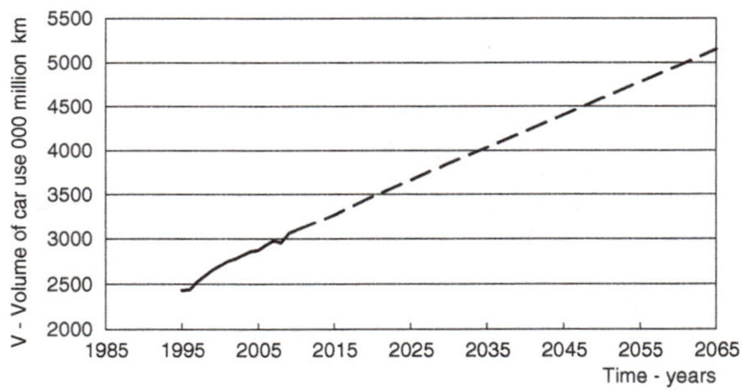

Fig. 4.2 Development of the volumes of passenger car use over time in the EU27 (2010–2065)

volumes of passenger car use over the forthcoming 2010/15–2065 period are estimated using the constant average annual growth rate of GDP of 2 % (as in the past period) as shown in Fig. 4.2.

The volumes of passenger car use in the EU27 are expected to continue to increase during the observed future period at a slightly decreasing rate mainly driven by the rather constant and stable GDP growth.

- **Market share of different passenger car technologies**

Considering the above-mentioned factors, the annual rates of market penetration by the particular advanced passenger car technologies in each of the three sub-periods of the observed period are determined for Scenario 0, 1, and 2, as shown in Fig. 4.3a, b, c (ICG 2010).

In all the above-mentioned scenarios, the proportion of conventional petrol ICEVs is expected to decrease to a modest 10 % in Scenario 0 and 0 % in Scenarios 1 and 2. The proportion of conventional diesel ICVs is expected to increase

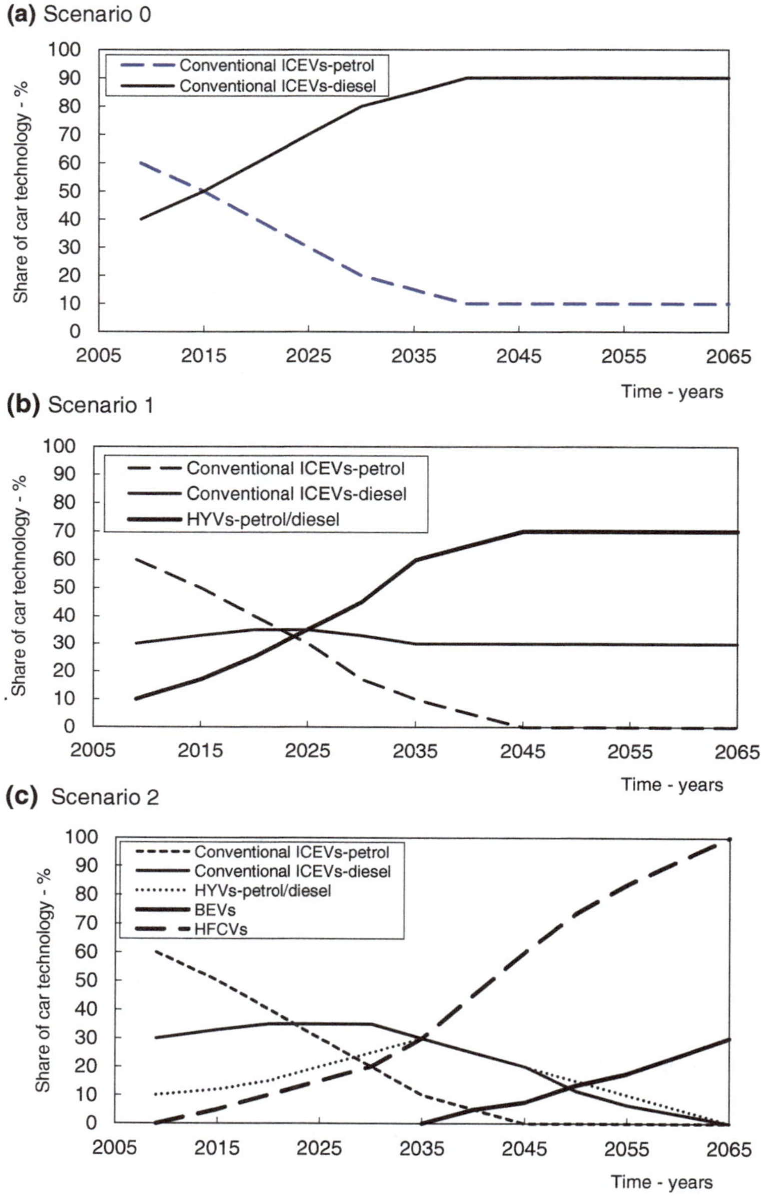

Fig. 4.3 Scenarios of the market penetration by different passenger car technologies in the given example (EU27–2010/15-2065)

up to 90 % in Scenario 0, increase and then decrease and stabilize at about 30 % in Scenario 1, and increase and then continuously decrease to 0 % until the end of the observed period in Scenario 2. The proportion of HYVs is expected to increase in

Scenario 1 during the first two-thirds of the observed period to 70 %, and to 35 % during the first half of the observed period in Scenario 2. This proportion remains at the achieved level in Scenario 1 and decreases to 0 % in Scenario 2 during the remaining observed period. From the time of entering the market, the proportion of BEVs and HFCVs is expected to increase in Scenario 2 and reach about 70 and 30 %, respectively, by the end of the observed period.

- **Energy/fuel consumption and emissions of GHG**

The average rates of energy/fuel consumption and emissions of GHG of different passenger car technologies expected to be used in the EU27 during the observed period are given in Table 4.1.

In order for conventional ICEVs and their hybrid versions to fulfill the above-mentioned targets, improvements in both WTT (Well-To-Tank) and particularly TTW (Tank-To-Wheel) efficiency will be needed. In the former case, such improvements will be rather difficult to achieve, while in the latter, improvements will mainly stem from advanced car design including increased use of generally lighter (composite) materials. In any case, the average annual rate of improvements of the WTW efficiency and related emissions of GHG by 2020/2035 will need to be about 4–5 % for conventional petrol ICEVs, 4.5–5.5 % for conventional diesel ICEVs and 4.7–5.7 % for HYVs. The WTW (Well-To-Wheel) energy efficiency of BEVs and HFCVs will also improve while their emissions of GHG will strongly depend on the primary sources for the production of electric energy (EU-27- 2010/15-2065).

Table 4.1 Environmental performances of the different passenger car technologies—energy/fuel consumption and emissions of GHG (Bodek and Heywood 2008; IPTS 2003; ICG 2010; IEA 2009; http://www.fueleconomy.gov/FEG/fuelcell.shtml)

Technology	Basic energy/fuel	Efficiency 0	WTW energy efficiency (kWh/km)	WTW emissions of GHG (gCO_{2e}/km)
Conventional ICEVs	Petrol/diesel/	20–21	0.612–0760	165–200
	Auto gas (e)	25	0.955	
	Petrol/diesel/	20–21	0.632	140
	Auto gas (n)	–	–	–
	Petrol/diesel/	20–25	0.357	82–84
	Auto gas (f)	–	–	
HYVs	Petrol (e)	40	0.799	125
	Diesel (e)	55	0.483	90
	Petrol (e)	50	0.329	52
	Diesel (e)	65	0.251	47
BEVs	Electricity	80	0.175–0.235	(A)
HVs	Hydrogen	50	0.926	(A)
HFCVs	Hydrogen	95–100	0.010–0.327	(A)

e existing standards; *n* new proposed standards to be in place by the year 2020–2035; *(A)* depending on the primary sources for producing electricity and/or hydrogen

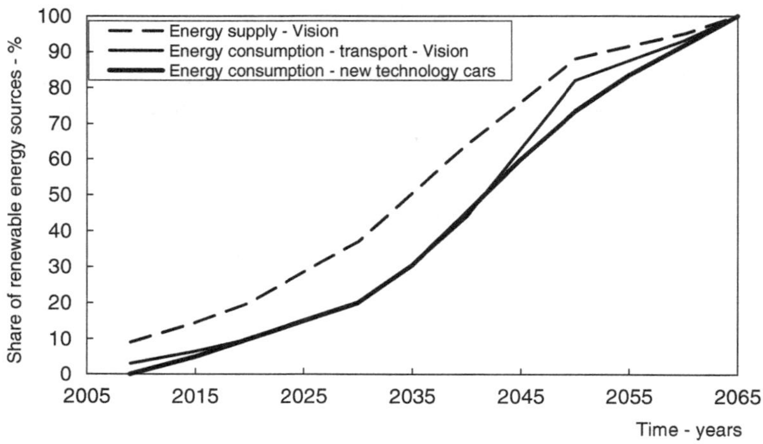

Fig. 4.4 Development of the share of renewable primary sources in the energy supply and consumption by the transport sector and advanced passenger car technologies (EU27–2010/15-2065) (EEA 2008; OI 2011)

- **Primary sources for energy/fuel supply and consumption**

A scenario of developing the share of renewable primary sources in the energy/fuel supply and the energy consumption by the transport sector and particularly by advanced passenger car technologies is shown in Fig. 4.4.

The share of renewable primary sources in the energy supply is expected to continuously grow during the given period according to the "S-curve power law." This implies relatively modest growth rates at the beginning, higher in the middle, and again lower at the end of the observed period. Such dynamism is reasonable if the EU-27 Member States intend to mitigate their currently increasing dependency on imported and increasingly expensive depleting crude oil sources. Since these reserves are expected to be exhausted by the end of the observed period, renewable sources will remain the exclusive primary energy supply sources in the region. A substantial proportion of such energy will be consumed by the transport sector—in proportion to that of the overall supply. This electricity will be mainly consumed by the advanced passenger car technologies dominating the transport sector during the observed period according to Scenario 2.

Results

The results in terms of the annual energy consumption and related emissions of GHG in particular Scenarios of passenger car use in the EU27 states during the observed period are shown in Figs. 4.5, 4.6, and 4.7.

- **Energy consumption**

The annual energy consumption by passenger car use in the given example is shown in Fig. 4.5. The energy consumed is expressed in terms of crude oil equivalents for comparative purposes.

As intuitively expected, the annual energy consumption differs in different scenarios. In addition, it changes over the observed period driven mainly by the volumes of passenger car use on the one hand, and particular passenger car technologies in combination with improvements of their WTW energy efficiency on the other. In particular, in Scenario 0, the annual energy consumption decreases at the beginning of the observed period despite growing volumes of passenger car use mainly thanks to improvements in the energy efficiency of conventional (petrol/diesel/auto gas) ICEV passenger cars (Table 4.1). When these improvements are exhausted, the annual energy consumption begins and continues to increase until the end of the observed period mainly driven by growing volumes of passenger car use. In Scenario 1, the trend of changing the annual energy consumption during the observed period is generally similar to that in Scenario 0. The differences are as follows: the annual energy consumption is always lower than that in Scenario 0 mainly thanks to more intensive use of energy efficient HYV cars; the period in which the energy consumption decreases despite growth in the volumes of passenger car use is longer due to the longer period of exhaustion of improvements in the WTW energy efficiency of all three car technologies (Table 4.1). In Scenario 2, the annual energy consumption is the lowest as compared to that in the other two scenarios. It continuously decreases despite the growing volumes of passenger car use during the observed period (2010/15--2065). This is achieved by the more intensive and continuous introduction of EVs on the one hand, and much more energy efficient BEVs, HVs, and HFCVs on the other (Fig. 4.3 and Table 4.1).

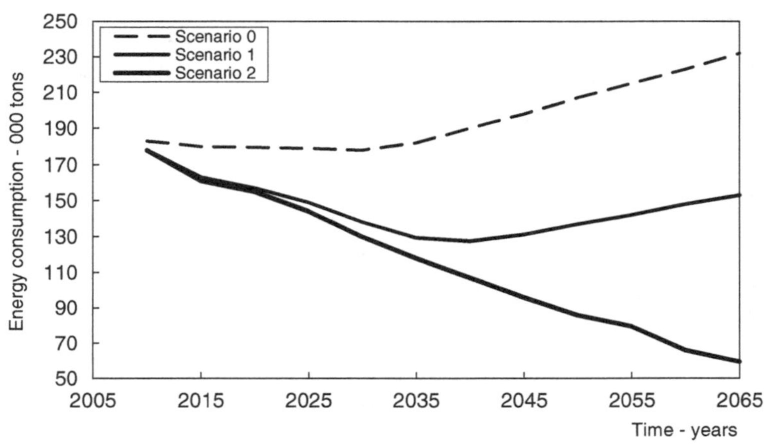

Fig. 4.5 Energy consumption over time in particular scenarios (EU27—period 2010/15–2065)

- **Emissions of GHG**

Emissions of GHG in terms of CO_{2e} are estimated using the above-mentioned inputs. The results are shown in Figs. 4.6 and 4.7.

In particular, Fig. 4.6 shows the annual emissions of CO_{2e} by passenger car use in the given example (EU27—period 2010/15–2065).

The general trends in particular Scenarios are similar to those of energy consumption. In particular, in Scenarios 0 and 1, the annual emissions of GHG (CO_{2e}) decrease during the first part and then increase until the end of the observed period. The main cause of this is the fact that during the first part of the period, the annual rate of technical/technological improvements to passenger cars (only ICEVs in Scenario 0 and both ICEVs and HYVs in Scenario 1) is higher than the annual rate of increasing volumes of their use. After the above-mentioned improvements are exhausted, these emissions again start to increase by the end of the observed period mainly driven by increasing annual volumes of passenger car use. During the entire observed period, the annual emissions of GHG (CO_{2e}) are greater in Scenario 0 than in Scenario 1, thus indicating the contribution of the HYVs to their mitigation (by about 10–15 %). In Scenario 2, the gradually increased use of advanced BEVs and HFCVs contributes to decreasing annual emissions of GHG (CO_{2e}) over the entire observed period. In addition, in each individual year of this period, these emissions are lower than those in Scenarios 0 and 1. At the same time, at the end of the observed period (in 2065), the annual emissions may be close to zero due to providing the energy/electricity for both advanced passenger car technologies prevailing in the market exclusively from renewable primary sources.

The cumulative emissions of GHG (CO_{2e}) by the given year of the observed period (EU27–2010/15–2065) are shown in Fig. 4.7.

In Scenarios 0 and 1, the cumulative emissions of GHG (CO_{2e}) increase continuously during the observed period. Specifically, they increase during the first

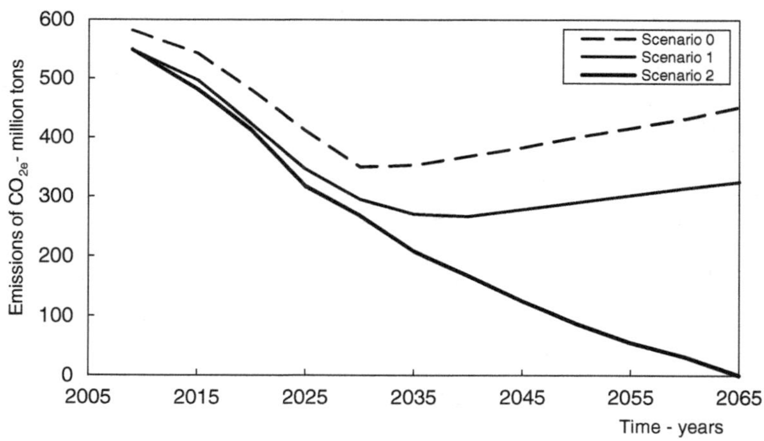

Fig. 4.6 Emissions of GHG (CO_{2e}) over time in particular scenarios (EU27–2010/15–2065)

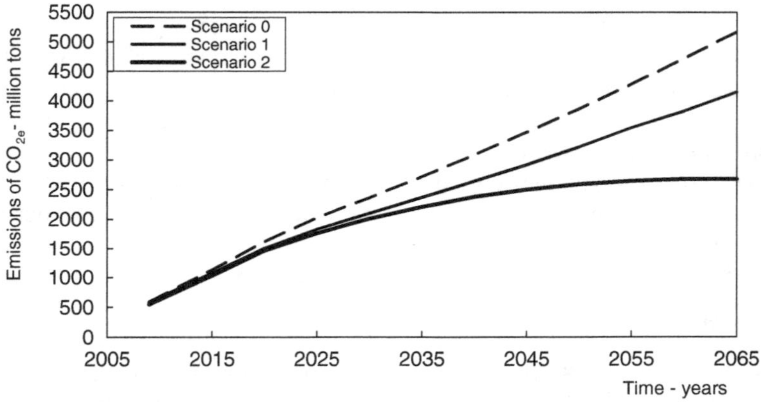

Fig. 4.7 Cumulative emissions of GHG (CO_{2e}) by given year of the observed period in particular scenarios (EU27–2010/15–2065)

part of the period at a decreasing rate mainly thanks to improvements of conventional ICEVs and both ICEVs and HYVs, respectively, despite continuous growth in passenger car use volumes. Over the remaining part of the period, when the improvements are exhausted, GHG emissions continue to grow mainly driven by and in proportion to the growing volumes of passenger car use. Again, the cumulative emissions of GHG (CO_{2e}) in Scenario 0 will always be greater than those in Scenario 1, with increasing differences over the observed period. This illustrates the increasing positive contribution of the more intensive use of HYVs over time. In Scenario 2, BEVs and HFCVs will contribute to increasing the cumulative emissions of GHG (CO_{2e}) at a decreasing rate during the entire observed period. They are mainly driven by the growth of volumes of passenger car use. Near the end of the observed period, when both advanced passenger car technologies prevail in the market, the cumulative emissions stagnate. In other words, in light of the long standing time of CO_2 in the atmosphere, these are actually emissions from ICEVs and HYVs before their replacement by their BEV and HFCV counterparts. Last but not least, the cumulative emissions of GHG (CO_{2e}) in Scenario 2 will always be lower than those in Scenarios 0 and 1. The positive differences continuously increase and reach the maximum at the end of the observed period.

4.2.3 Evaluation

Advanced passenger car technologies and their variations including electric petrol/diesel HYVs (Hybrid Vehicles), BEVs (Battery Electric Vehicles), and HFCVs (Hydrogen Fuel Cell Vehicles) possess both advantages (Strengths and Opportunities) and disadvantages (Weaknesses and Threats).

Advantages

- Decreasing total energy consumption in terms of crude oil equivalents and related emissions of GHG in terms of CO_{2e} (Carbon Dioxide equivalents) despite increasing volumes of passenger car use thanks to the technical/technological improvements of conventional ICEVs and HYVs at higher rates of growth of passenger car use volumes during the first part of the observed period; and
- Decreasing total energy consumption and related emissions of GHG during the entire observed period despite growing volumes of passenger car use after BEVs and HFCVs penetrate the market more substantially.

Disadvantages

- Contributing to increasing total energy consumption and related emissions of GHG until the end of the observed period mainly driven by continuously growing volumes of passenger car use after the potential for further improvements of ICEVs and HYVs is exhausted; the energy consumption and related emissions of GHG will be lower insofar as HYVs penetrate the market at a higher rate;
- Considering the lifespan of man (car)-made emissions of GHG in the atmosphere of several hundred years, only complete replacement of conventional ICEVs and HYVs with their BEV and HFCV counterparts under the given circumstances can actually stop their further cumulative increase in the given case;
- Achieving an energy density of batteries close to that of gasoline and diesel fuel as derivatives of crude oil, enabling the equivalent driving performances (acceleration, operating speed, and range with a single battery charge) as those of ICEVs and HYVs is going to be complex;
- Achieving a selling price comparable to that of both BEVs and HFCVs is uncertain; and
- Penetrating the market more intensively will not contribute to reducing congestion in urban and suburban areas.

Finally, advanced passenger cars can be considered a subsystem of the advanced transport system mainly regarding the techniques/technologies of the power system (engine) and the energy/fuel supply system, both of which are adapted to type of energy/fuel used.

4.3 Large Advanced Container Ships

1957	The first Sea–Land Gateway City container ship, a modified tanker loaded with 56 containers, makes its inaugural voyage between the ports of Newark Miami, Houston, and Tampa (U.S.)
1960	The first Grace Line Santa Eliana fully containerized ship begins international container shipping to Venezuela (U.S.)
2006	The then largest Emma Maersk container ship begins commercial operations (Denmark)

4.3.1 Definition, Development, and Use

Advanced freight ships are mainly characterized by improved operational, economic, and environmental performances, the latter in terms of energy consumption and related emissions of GHG (Green House Gases), as compared to those of their conventional counterparts. The improvements in performances of these ships can generally be achieved by improving existing and/or deploying completely new (advanced) technologies and operations. In particular, due to the continuously increasing use of containers for transporting freight/goods in international trade, improvements of performances will be particularly relevant for the relatively fast and continuously growing fleet of container ships.

Maritime shipping has gained a central role in global trade due to the internalization and globalization of the world's economies. During the past three decades, international seaborne trade has continuously grown, and increasing volumes of freight/goods have been transported in containers as shown in Fig. 4.8.

Despite being relatively modest, the volumes of containerized freight/goods have increased faster than the total volumes of freight/goods by their share in the totals of about 2.75 % in 1980, 10.16 % in 2000, and 15.83 % in 2011.

The above-mentioned total freight/goods volumes have been transported by a fleet composed of five types of ships respecting the categories of freight/goods such as oil tankers, dry bulk, general cargo, container, and other ships. Figure 4.9 shows development of the total world's and particularly the container ship fleet.

As can be seen, both have increased almost exponentially over the past three decades. In particular, the share of container ship capacity in the capacity of total fleet has constantly increased from about 1.61 % in 1980 and 8.02 % in 2000 to about 11.5 % in 2011.

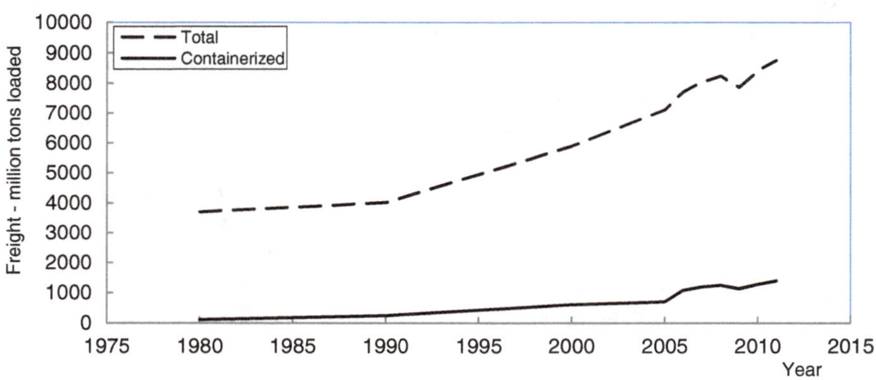

Fig. 4.8 Development of the global international seaborne trade (million tons loaded) (UNCDAT 2012)

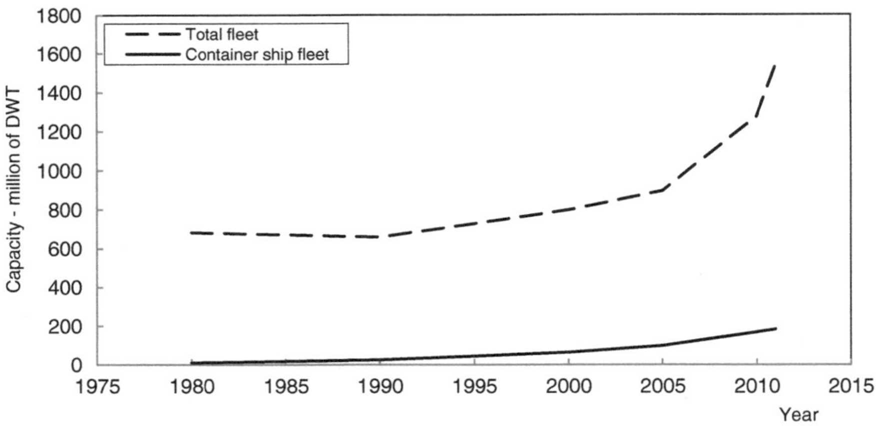

Fig. 4.9 Development of the global freight ship fleet (UNCDAT 2012)

4.3.2 Analyzing and Modeling Performances

4.3.2.1 Background

Container ships have been designed to exclusively carry containers in their holds and on the deck. The fleet of these ships is usually represented by the annual number of ships in operation, their total capacity, the average ship size expressed in TEUs (Twenty foot Equivalent Unit(s)), and DWT (Deadweight Tonnage)[1] as given in Table 4.2.

Evidently, over the past 25 years, the number of container ships has increased fivefold, their total capacity about 15-fold, and the average ship size about threefold. Thus, ships with a capacity exceeding 3,000–4,000 TEU can be considered large container ships, implying that the average ship in 2012 can be considered as a large container ship. It can also be considered advanced if it is more operationally, economically, and environmentally efficient and effective than its conventional predecessors.

Large advanced container ships possess infrastructural, technical/technological, operational, economic, environmental, and social/policy performances. Nevertheless, the aim of dealing with particular performances is primarily to emphasize the contribution of these ships to mitigating impacts on the environment in terms of the energy/fuel consumption and related emissions of GHG (Green House Gases).

[1] This is the total weight (tons) that a given ship can safely carry. It includes payload (cargo), fuel, water, supplies, crew, etc.

Table 4.2 Development of the world's container ship fleet over time (UNCDAT 2012)

Year	Number of ships	Fleet capacity (TEU)	Average capacity (TEU/ship)	Average carrying capacity[a] (DWT/ship)
1987	1,052	12,122,15	1,155	16,170
1997	1,954	3,089,682	1,581	22,134
2007	3,904	9,436,377	2,417	33,838
2008	4,276	10,760,173	2,516	35,224
2009	4,638	12,142,444	2,618	36,652
2010	4,677	12,824,648	2,742	38,388
2011	4,868	14,081,957	2,893	40,502
2012	5,012	15,406,610	3,074	43,036

[a] Based on the standard assumption: 1 TEU = 14 DWT (1 TEU = 2.3 tons of tare weight + 10 tons of average payload; the rest is allocated to the ship's fuel, fresh water, spares and other supplies)

Table 4.3 Development of the container ships over time—milestones in size (MAN Diesel 2011; Rudolf III 2007; www.worldslargestship.com)

Year/generation	Capacity (TEU)	Length (m)	Beam (m)	Draught (m)	Number of engines/power (MW)
1968	750	180	25	9.0	1/6.7
1972	1,500	225	29	11.5	1/14
1980	3,000	275	32	12.5	1/25
1987	4,500	275	39	11.0	1/40.1
1998	7,900	347	43	14.5	1/60
2006/Emma Maersk	15,000	397	56	15.5–16.0	1/80.1
2012/CMA/CGM Marco Polo	16,020	396	54	16.0	2/80
2015/Triple E Maersk	18,000	400	59	16.0	2/64

4.3.2.2 Infrastructural Performances

The main infrastructural performances of large advanced container ships are the required space and other conditions at the port container terminals used to handle them. The required space refers to the number and length of berths, which can be constructed along linear or sheltered coastline. The length of a berth is directly related to the length of a container ship implying that it can range up to 400 m. The length of a quay is then influenced by the number of berths needed to simultaneously handle container ships. The width of a seaside could be up to 60 m. These would both enable the handling of the largest forthcoming container ships such as Triple E Maersk (see Table 4.3). In addition, the sea water in the terminal accessing channels and near berths enabling access and docking needs to be sufficiently deep (in the above-mentioned example at least 17 m). Experience so far indicates that the water depth in the main ports has been continuously improved in line with deploying large advanced container ships. If the depth is adopted to be 15 m as the required minimum, the number of appropriate ports has increased from 17 in 2000 to 25 in 2003,

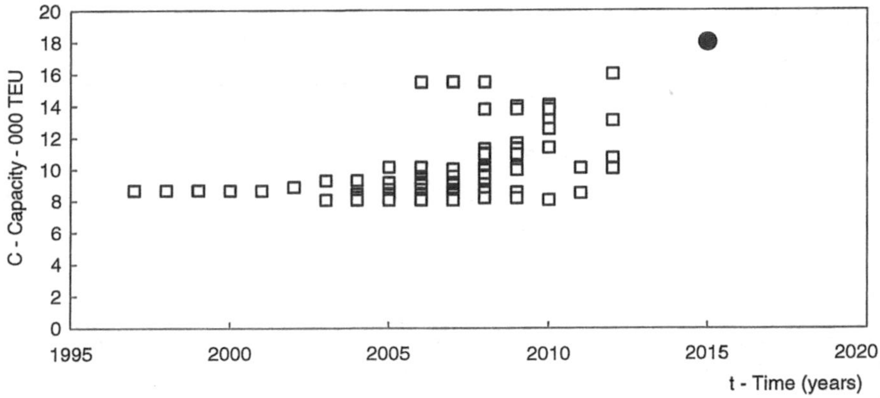

Fig. 4.10 Development of the capacity of large container ships (*period* 1997–2012) (http://en.wikipedia.org/wiki/List_of_largest_container_ships)

and to 28 in 2008. Consequently, the infrastructure of the main ports in the major trading regions in terms of the required water depth and length of berths is adequately provided for current and forthcoming large advanced container ships (Tozer 2001).

4.3.2.3 Technical/Technological Performances

Design

Large advanced container ships have been characterized by a persistent increase in size, capacity, and related engine power over time. Table 4.3 shows the milestones of such developments over the past 40 years.

In addition, Fig. 4.10 shows the general trend of increasing capacity of the 267 largest container ships over the past 15 years.

Three periods can be distinguished: the first (1997/2003) when only ships of a capacity of about 8,000 TEU were built, the second (2003/2005) when ships with a capacity of 8,000–10,000 TEU were built, and the last (2006–2012) when ships of a capacity between 8,000 and 15–16,000 TEU were built. The latest are usually called ULCS (Ultra Large Container Ships). One of them, the Triple E Maersk to be launched in 2015, will represent an advanced step in increasing the size to about 18,000 TEU/ship. However, the capacity of these ships in terms of TEU and DWT does not reflect the standard rule of 14 DWT/TEU as mentioned in Table 4.2, but rather less: from 9.9 DWT/TEU for the largest (Triple E Maersk) to 11.0 DWT/TEU to the smallest (2,800 TEU) container ship. This indicates that they are actually designed assuming that all TEUs on-board will never be completely full.

Advanced container ships have been designed for relatively constant "design" conditions. Such conditions have mainly influenced the hull form, rudder and propeller design, size, and power of the main engine, and the capacity and layout of auxiliary systems. Consequently, designing of future large advanced container ships will have to be flexible in order to be adaptable to both "design" and "off-design" conditions. The former conditions are characterized by increased resistance of the hull due to operating at higher "design" speed(s), which compromises their overall energy, economic, and environmental efficiency. The latter conditions imply operating at lower than design speeds, i.e., slow steaming, which reduces fuel consumption and improves economic and environmental efficiency. In order to avoid the negative effects of changing conditions, future large advanced container ships will have to be designed (particularly hull and propulsion system-engines) for a range of the most likely operating speeds and draughts, thus balancing between the two compromising effects: one for reducing the operating speed and the other for increasing the capacity.

For example, the forthcoming Triple E Maersk ship is designed with a wider hull in order to accommodate the specified 18,000 TEUs. Such a wider U-shaped hull creates higher propulsion resistance than the narrower V-shape hull of its closest counterpart—Emma's Maersk. However, the Triple E's operating speed is limited to 23 kts (two engines generate the required power of 65–70 MW while running at 80 rpm (revolutions per minute), while Emma's is limited to 25 kts (the single engine generates the required power of 80 MW while running at 90 rpm). Thus, despite operating at higher propulsion resistance, thanks to operating at a lower engine rate and operating/cruising speed, the Triple E Maersk is expected to be overall more efficient than its closest smaller counterpart Emma Maersk. Figure 4.11 shows the principal differences in design of the two ships.

However, in both cases, utilization of the ships' available capacity (deadweight) is variable mainly due to frequent oversupply on the one hand and fluctuating market conditions on the other. These conditions which will likely become increasingly common in the future.

Propulsion/engines and propellers

Propulsion system/engines

The propulsion system/engines of large advanced container ships are one of their crucial components. Some empirical evidence indicates that their power can be roughly estimated from the modified Admiralty formula as follows:

$$P = c_p * V^{3+k} * \Delta^{2/3} \tag{4.4}$$

where

c_p is the coefficient

V is the ship's designed operating speed (kts); and

Δ is displacement (tons).

The displacement (Δ) is the actual gross weight of the ship consisting of its own empty weight and the weight of its cargo, fuel, fresh water, provisions, and crew. As can be seen, the propulsion power is proportional to the ship's speed (V) to the power of $3 + k$ and to its displacement to the power of 2/3. By proper selection of

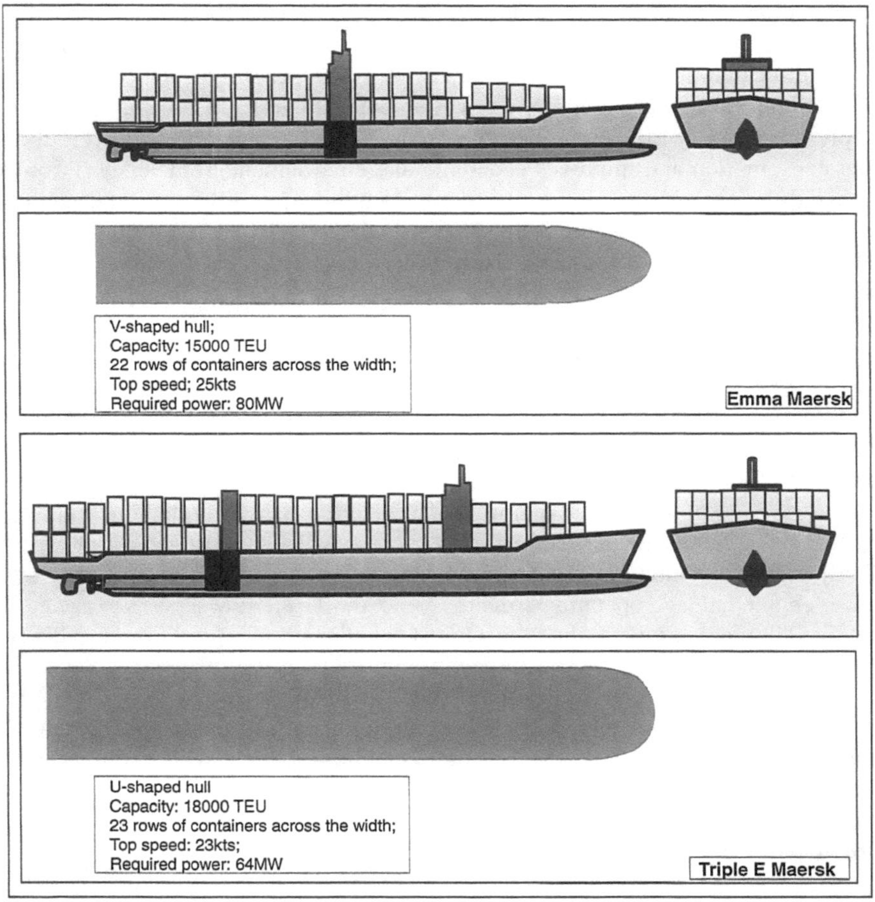

Fig. 4.11 Scheme of large advanced container ships

the coefficients (c_P) and (k), each individual speed–power curve of each individual ship can be expressed. In addition, Eq. 4.4 suggests that more engine power efficiency can be obtained by increasing the ship's size than by increasing the ship's speed (DNV 2012). Figure 4.12 shows the relationship between the size and engine power of large container ships.

In this case, the engine power increases at a decreasing rate as the ship's size increases, thus confirming the above-mentioned trend of designing larger container ships assumed to operate at lower speeds. The exception is the forthcoming Triple E Maersk as the largest container ship in the world with a capacity of 18,000 TEU and two engines delivering about 65 MW of power as compared to its currently largest counterpart (Emma Maersk) with a capacity of 15,000 TEU and a single engine delivering 67.7 MW of power (i.e., 3.61 vs. 4.51 kW/TEU, indicating an improvement of the power efficiency of about 25 %) (Table 4.3).

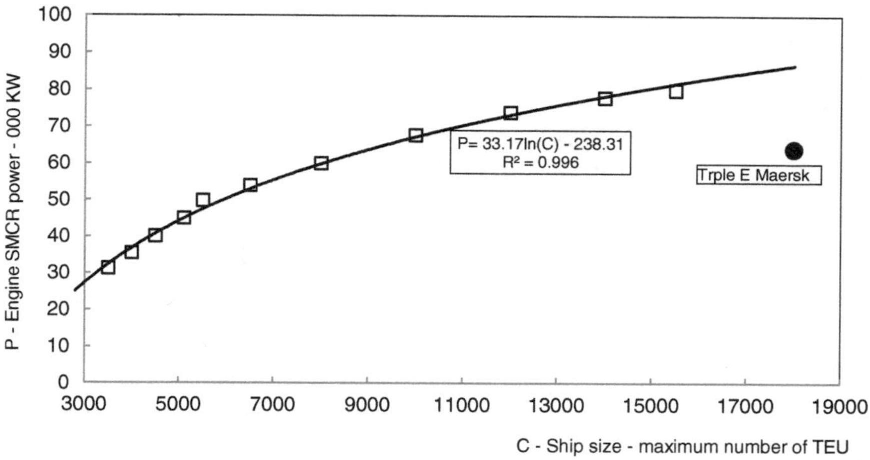

Fig. 4.12 Relationship between engine power and size (carrying capacity) of large advanced container ships (*SCMR* Specific Maximum Continuous Rating) (MAN Diesel 2011)

Propellers

The propulsion system of large advanced container ships is usually placed near the middle of the ship in order to make the best use of the rigidity of the hull and to maximize the carrying capacity. Once the main propulsion system is made available, the crucial element to be designed is the propeller. Propellers of existing container ships are made of nickel aluminum bronze usually with six blades. Their diameter and weight generally increase at a decreasing rate as the engine SMCR (Specific Maximum Continuous Rating) power for the specified speed increases. For example, for an engine of 60 MW SMCR power, the diameter of a propeller rotating at the speed of 94 r/min is about 9.2 m and its weight 95 tons. For an engine of 100 MW SMCR power rotating at the same speed, the diameter is about 10 m weighing 155 tons. For engines of 60 MW SMCR power rotating at the speed of 104 r/min, the diameter is about 8.5 m and weight 90 tons. For an engine of 100 MW power rotating at the same speed, the diameter is about 9.7 m and weight about 140 tons. This indicates that higher rotating speeds enable the design and construction of propellers with smaller diameter and weight (MAN Diesel 2011). For example, the forthcoming Triple E Maersk container ship will be equipped with a twin engine/twin screw propulsion system. Each of the two propellers will have a diameter of 9.8 m and 4 blades as compared to the Emma Maersk ship equipped with a single engine/single screw propulsion system where the propeller has a diameter of 9.6 m and 6 blades. In the twin screw propulsion system, the propellers are lighter, thus reducing any vibration of the hull. In addition, such systems provide greater pushing power and lower water resistance (Tozer 2001).

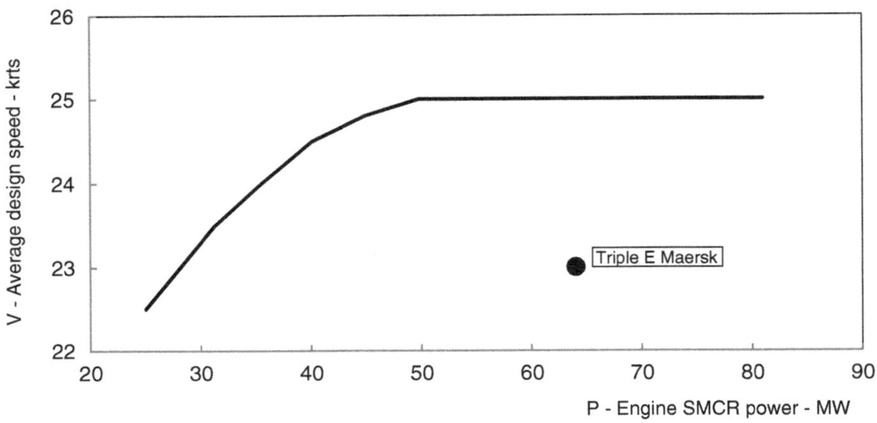

Fig. 4.13 Relationship between the design speed and the engine power of large advanced container ships (*SCMR* Specific Maximum Continuous Rating) (MAN Diesel 2011)

4.3.2.4 Operational Performances

The operational performances of large advanced container ships include speed, maneuverability, turnaround time, technical productivity, and fleet size/capacity.

Speed

Design speed

The design speed of large advanced container ships generally increases as the engine power increases, albeit at a decreasing rate. Since the engine power increases with the size of the ship, the speed also tends to increase in line with the size of the ship. However, it remains constant and independent of the size of the ship and related engine power in ships larger than 5,500 TEU and with an engine power equal or greater to about 50 MW; the latter is shown in Fig. 4.13.

However, the most recent exception from the above-mentioned rule of thumb is the largest Triple E Maersk container ship expected to be launched in 2015, with a design speed of 23 kts and a total twin-engine power of about 645 MW.

Operating/cruising speed

The operating/cruising speed of large advanced container ships differs in practice from their design speed due to the various reasons. One of these reasons is the preference of operators to minimize fuel consumption and related costs by adopting lower speeds. As a result, four speed categories of these ships can be distinguished as follows:

- Nominal speed (20–25 kts; 37.0–46.3 km/h), which represents the optimal cruising speed at which a given container ship and its engine have been designed to operate;
- Slow steaming speed (18–20 kts; 33.3–37.0 km/h), which represents the speed achieved by running the ship's engines below their capacity in order to reduce fuel consumption. (In 2011, more than 50 % of the global container shipping capacity was operating at this speed);
- Super slow steaming speed (15–18 kts; 27.8–33.3 km/h), which is also known as the economic speed aiming at minimizing fuel consumption while still maintaining a competitive commercial service; and
- Minimal cost speed (12–15 kts; 22.2–27.8 km/h), which represents the lowest technically possible speed since even lower speeds do not lead to any significant additional reduction in fuel consumption. (However, since these speeds and the related quality of services are commercially unviable, it is unlikely that maritime shipping companies will adopt them as part of their practice).

The practice of slow steaming emerged during the financial crisis of 2008–2009 when on the one hand, the demand for international trade and containerized shipping was severely affected, and on the other the new capacity ordered during the previous years of economic boom was coming into service (Figs. 4.9, 4.10). In reaction to such an imbalance between the decreased demand and increased capacity, maritime shipping companies adopted slow steaming and even extra slow steaming services on particular routes. Since the lower operating/cruising speeds required longer ship turnaround times, more ships were needed and indeed were available thanks to the new additional capacity.

It seems that slow steaming will remain the operational practice of many shipping companies due to the following reasons: (i) reducing fuel consumption and related costs, particularly if the trend of increasing fuel prices continues; and (ii) reducing emissions of GHG, thus respecting increasingly stricter environmental regulations.

As an innovative operational practice/regime, slow steaming will require adapting engines through their "de-rating," namely involving the timing of fuel injection, adjusting exhaust valves, and exchanging other mechanical components to the new speed and power level of about 70 % instead of the previously regular 80 %.

Maneuverability

One additional important operational advantage of large advanced container ships is their maneuverability. Conventional container ships of all sizes generally satisfy all of the IMO (International Maritime Organization) maneuverability criteria by using conventional steering systems. Some problems emerge around congested ports as the wind loading on the above-water profile of large units can be great. Therefore, for example, on the currently largest Emma Maersk container ship, two bow and two stern thrusters provide port maneuverability, and two pairs of stabilizer fins reduce rolling. When the banking angle is 20°, the bridge sways by about 35 m. In addition, the turning diameter of the ship operating at the speed of 24 kts (44.5 km/h) is about 0.81 nm (i.e., 1.5 km).

Turnaround time

The turnaround time of a large container advanced ship scheduled to operate along a given route is defined as the total round time between the given origin and destination port, and back. This time includes the ship's turnaround time at the origin and destination port, its stop/transit time at the intermediate ports (which usually depends on the pattern and volumes of freight demand to be served in both directions), and the operating/cruising time between the particular ports. Thus, the ship's route between each given origin and destination port can be considered to consist of several segments. If the stops are the same in both directions, the turnaround time of a given ship can be estimated as follows:

$$\tau_{tr} = \tau_0 + 2 * \left[\sum_{i=1}^{N} \tau_i + \sum_{k=1}^{K} \tau_k \right] + \tau_d \tag{4.5a}$$

where

τ_o, τ_d is the ship's turnaround time at the origin and destination port, respectively (days);

τ_i is the ship's operating/cruising time along the (i)-th segment of a given route (days);

τ_k is the ship stop/transit time at the (k)-th intermediate port along a given route (days);

N is the number of segments of a given route; and

K is the number of ports along a given route where the ship stops ($K = N - 1$).

The ship's operating time along the (i)-th segment of the route in Eq. 4.5a can be estimated as follows:

$$\tau_i = s_i / V_i \tag{4.5b}$$

where

s_i is the length of the (i)-th segment of the route (nm); and

V_i is the ship's operating/cruising speed along the (i)-th segment of the route (kts)

In addition, the length of the route in one direction can be estimated as follows:

$$d = \sum_{i=1}^{N} s_i \tag{4.5c}$$

where all symbols are as in the previous equations.

It follows from Eq. 4.5a that reducing the cruising speed increases the ship's turnaround time along a given route, which can be partially compensated by shortening its turnaround time at the origin and destination port, and the stop/

transit time(s) at intermediate ports. These times generally depend on the ship's size, volume of load/cargo, and the rate of ship/container handling, the latter influenced by the available (increasingly automated) loading/unloading facilities and equipment at ports (Tozer 2001).

In order to estimate the performance of the additional loading and unloading facilities and equipment in ports (cranes) needed to compensate the extra travel time due to slow steaming, Eq. 4.5b can be modified as follows:

$$\tau = s * (1/v_s - 1/v_r) \tag{4.5d}$$

where

C is the ship's capacity (TEU);

S is the length of route, i.e., trip distance between origin and destination port (nm);

v_r, v_s is the regular and slow steaming speed, respectively (kts).

Equation 4.5d states that the ship's extra trip time will increase in line with the route length and the difference in the regular and the slow steaming speed. Such trip time extensions generally increase the cost of TEU/goods time while in the chain, which can be estimated as in Eq. 3.1a (Chap. 3) as follows:

$$\Delta c = \beta * C * s * (1/v_s - 1/v_r) \tag{4.5e}$$

where

Δc is the extra cost of freight/goods time ($US/TEU-h);

β is the freight/goods time while in transportation ($US/TEU-h).

The other symbols are analogous to the previous equations. The extra trip and cost of freight/goods can be partially compensated by shortening the ship's loading/unloading time at the origin/destination port, respectively. Modifying Eq. 3.1a (Chap. 3), the time loading units remain in the regular and the slow steaming supply chain can be estimated, respectively, as follows:

$$\tau_r = 2C/n_r * \alpha_r + s/v_r \text{ and } \tau_s = 2C/n_s * \alpha_s + s/v_s \tag{4.5f}$$

where

n_r, n_s is the number of loading/unloading facilities and equipment (cranes) at the origin and destination port serving regular and slow steaming ships, respectively (-); and

α_r, α_s is the service rate of a single facility (crane) in either port serving a ship operating under regular or slow steaming regime (TEU/h).

The other symbols are analogous to those in the previous equations. Consequently, the loading/unloading capacity for a slow steaming ship that could compensate the extra cruising time can be estimated from Eq. 4.5e as follows:

$$n_s\alpha_s = \frac{2C * n_r * \alpha_r}{2C - n_r * \alpha_r * s * (1/v_s - 1/v_r)} > 0 \tag{4.5g}$$

This is reasonable only if the denominator of Eq. 4.5g is positive, i.e., if the reduced speed is not less than:

$$v_r > \frac{n_s * \alpha_s * v_r * s}{v_r * C + n_r * \alpha_r * s} \tag{4.5h}$$

Equation 4.5h states that compensating the ship's extra trip time due to slow steaming by increasing the loading/unloading capacity at ports (i.e., shortening the corresponding time(s)), could be achieved only if the reduced speed is not under a certain threshold. Otherwise, this is not possible, thus forcing users/freight/goods shippers to adapt and accept such changed quality of service.

Technical productivity

As with other vehicles, the technical productivity of a large advanced container ship can be determined as the product of its operational/cruising speed and carrying capacity as follows:

$$TP = V * C(d) \tag{4.5i}$$

where

C(d) is the capacity of a given container ship operating along the route *d* (DWT or TEUs).

The other symbols are as in the previous equations.

The technical productivity of a given route for a given period of time (week, month, year) can be estimated as the product of the technical productivity of a single ship and the service frequency. In many cases, it is assumed that all services are carried out by ships of the same capacity operating at the same speed. The service frequency is directly proportional to the quantity of freight to be transported (TEUs) and inversely proportional to the ship size and its load factor (Chap. 2).

Fleet size/capacity

The fleet size/capacity of large container shipping companies can be expressed as the sum of the capacity of all ships in the fleet. Figure 4.14 shows the relationship between the total fleet capacity expressed in TEUs and the number of ships in the fleet for the 20 largest global container shipping companies.

As can be seen, the total transport capacity expressed in TEUs increases in line (at a constant rate) with the number of ships in the fleet. The average capacity of a container ship in a fleet in the given example is about 4,000 TEU.

The number of ships in the fleet of large advanced container ships operating on the route *(d)* during the time period *(T)* can be determined, based on Eq. 4.5a, as follows:

$$N(d, T) = f(T, d) * \tau_{tr} \tag{4.6}$$

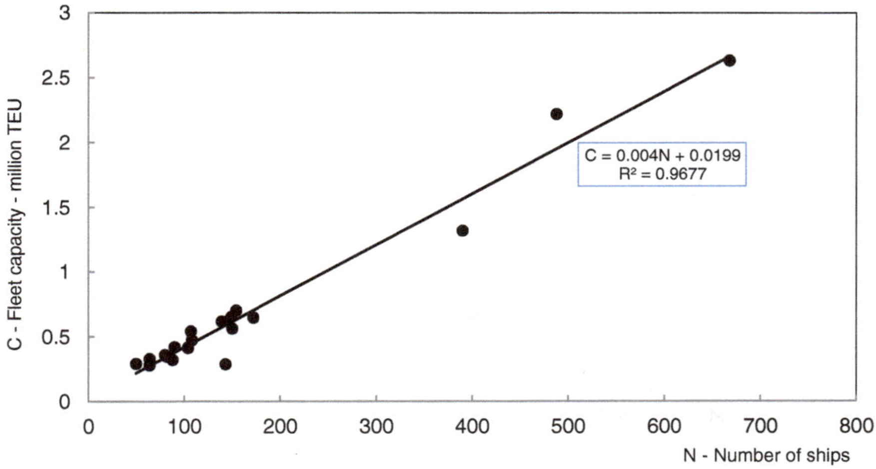

Fig. 4.14 Relationship between the size and capacity of the containership fleet (May 2012) (http://en.wikipedia.org/wiki/Intermodal_freight_transport)

where

f(T, d) is the service frequency on the route *(d)* during time *(T)*.

The other symbols are analogous to those in the previous equations.

Equation 4.6 confirms that increasing the ship's turnaround time due to slow steaming will require a larger fleet of ships of a given size. As the volumes of freight demand increase, so does the number of ships of a given capacity and utilization in the fleet. At the same time, using larger and better utilized ships will require fewer of them.

4.3.2.5 Economic Performances

The economic performances of advanced large container ships include their costs and revenues.

Costs

The costs of container ships, as the costs of other categories of freight/goods transport vehicles, consist of capital and operating cost including the cost of capital, port and terminal call charges, costs of insurance, maintenance and repair, costs of lubes and stores, and fuel, crew, and administration/overhead costs. Experience so far shows that due to economies of scale, the average cost per transported unit of freight/goods (TEU) on a given distance/route decreases more than proportionally as the size of a container ship increases. Figure 4.15 shows an

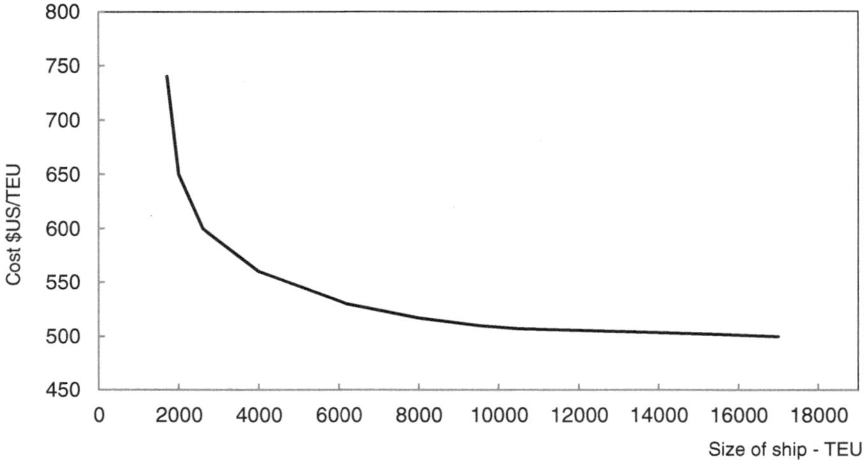

Fig. 4.15 Relationship between the average cost per TEU and the size of container ship (AECOM 2012; Cullinane and Khanna 2000; Nottebon and Rodrigue 2007; Tozer 2001)

example for a Transatlantic route of an average length of about 4,000 nm (AECOM 2012; Nottebon and Rodrigue 2007).

Such diminishing economies of scale pose the question whether large advanced container ships such as the Triple E Maersk will bring more substantial cost-decreasing benefits as compared to their smaller predecessors. In all cases, fuel costs dominate (80–85 %), followed by capital costs (about 8.0 %), Panama Canal tolls (6–10 %), and crew costs (1.2–2.4 %) (AECOM 2012). In addition, for a given ship size, the average unit cost increases in line with the route length and the operating/cruising speed, both approximately at a decreasing rate. The latter is because the increased speed requires greater fuel consumption, thus additionally raising the share of already dominating fuel costs in the total operating costs of the ship.

Revenues

The revenues of container ship operators come from transporting containers between ports.

Usually, their rates are set to cover operating costs while respecting highly competitive market conditions. Since these conditions are all quite different, the rates vary substantially across operators, markets and routes, and prevailing micro- and macroeconomic conditions. In addition, they depend on the transport distance and sometimes also ship size. In addition, generally speaking, the exporting rates when ships are fuller are lower than those of returning trips when ships may transport only empty containers. An example of such high diversity of the average unit rates in container shipping is shown in Fig. 4.16.

In the specified month, the average rate for transporting freight/goods of 1 TEU generally increased in line with the port-to-port distance in all markets, except

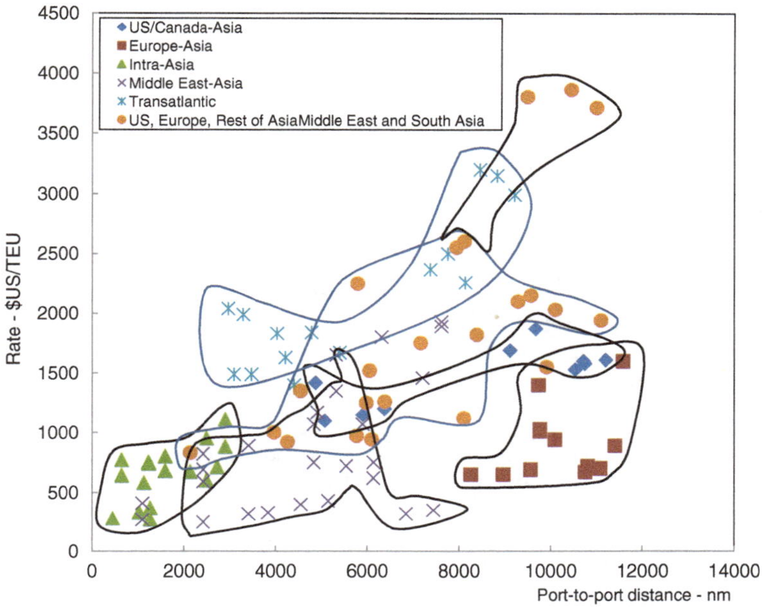

Fig. 4.16 Relationship between the lowest rate and port-to-port distance for transporting containers in particular markets—(March 2010) (DSC 2010)

Europe–Asia, US/Canada–Asia, and partially Middle East-Asia markets. In these markets, the rates remained rather independent of the distance, but very distinctive across the particular routes. All rates were changed the following month, thus indicating their above-mentioned high fluctuation depending on the short-term market and stakeholder-related conditions (DSC 2010).

4.3.2.6 Environmental and Social/Policy Performances

The environmental and social/policy performance factors of large advanced container ships are considered to be the energy/fuel consumption and related emissions of GHG (Green House Gases), land use/take for the container terminals in ports handling such ships (environmental), as well as traffic incidents and accidents (safety) (social).

Fuel consumption and emissions of GHG

Fuel consumption

Large advanced container ships consume MDO (Marine Diesel Oil), sometimes also known as No. 6 Diesel or HFO (Heavy Fuel Oil) or Bunker C fuel adapted to the 2005 standards as MDF (Marine Distillate Fuels) (http://en.wikipedia.org/wiki/

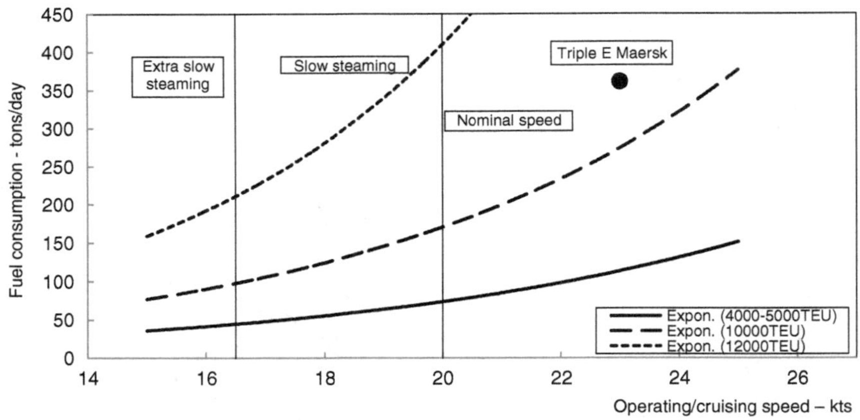

Fig. 4.17 Relationship between the fuel consumption and the operating speed and size of container ships (AECOM/URS 2012; Churchill and Johnson 2012; Notteboom and Carriou 2009)

Heavy_fuel_oil). These are largely unrefined very thick crude oil derivatives, often needed to be heated by steam in order to reduce their viscosity and thus enable them to flow. The fuel consumption of large container ships generally depends on their size and operating speed and usually increases in line with both factors individually and/or simultaneously as shown in Fig. 4.17.

As can be seen, the fuel consumption of a container ship of a given size increases more than proportionally as the operating/cruising speed increases. For example, large container ships of a capacity of 12,000 TEU such as Emma Maersk consume about 400 tons of fuel per day while cruising at a speed of about 20 kts (the length of route is about 15,000 nm) (AECOM/URS 2012). Ships with a capacity of 10,000 TEU consume about 375 and 200 tons of fuel per day while cruising at the (designed) speed of 25 kts and reduced speed of 21 kts, respectively. For container ships of 4,000–5,000 TEU, the corresponding fuel consumption is 150 and 85 tons per day, respectively. The forthcoming largest Triple E Maersk ship of a capacity of 18,000 TEU will consume about 360 tons of fuel per day while cruising at the speed of 25 kts. These figures illustrate the very high sensitivity of fuel consumption to changes in the ship's operating/cruising speed.

In addition, fuel consumption can be expressed in other units. For example, currently the world's largest single diesel Wärtsilä-Sulzer 14RTFLEX96-C engine powering the Emma Maersk largest container ship delivers the maximum power of 80–81 MW for the designed cruising speed of 25 kts. Under such regime it consumes about 19,000 l or 16.7 tons of HFO/h or 198 g/KWh. The forthcoming Triple E Maersk container ship with two MAN diesel engines delivering total power of 64 MW enabling an operating/cruising speed of 23 kts will consume 15.04 tons of HFO/h, or about 231 g of HFO/kWh. At the designed operating/

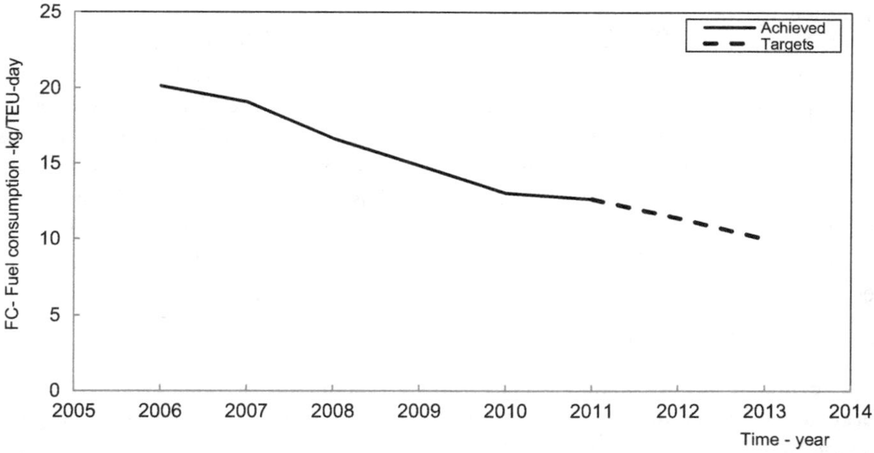

Fig. 4.18 Changes in the average unit fuel consumption of a large shipping company over time—Maersk Line (Maersk Line 2011)

cruising speeds, these give 1.39 kg of HFO/TEU/h for Emma Maersk and 0.84 kg of HFO/TEU/h for Triple E Maersk ship, which is a reduction of about 40 % (MAN Diesel 2011; Tozer 2001).

Endeavors to reduce fuel consumption

Modern large advanced container ships use HFO (Heavy Fuel Oil), the burning of which produces GHG such as SO_x (sulfur oxides), NO_x, (nitrogen oxides), PM (particulate matter), and CO_2 (carbon dioxide). Therefore, due to the permanent increase in the total emissions of GHG, the maritime industry and its national and international organizations have made efforts to at least control such emissions. For example, the WSC (World Shipping Council) and its members have been engaged through the IMO (International Maritime Organization) in numerous efforts to improve the energy efficiency of the maritime sector through reducing the ships' fuel consumption and related emissions of GHG. At the level of individual shipping companies, this has been carried out through medium- to long-term sustainability plans. Figure 4.18 shows the achievements of a large shipping company—Maersk Line—over the 2006–2011 period.

As can be seen, the company has certainly followed a downward path toward the established target of an average fuel consumption of about 10 kg/TEU/day to be achieved by 2013. This also implies corresponding savings in the emissions of GHG.

Other efforts have been institutionalized through Annex VI of MARPOL, an international treaty developed through the IMO, which has established legally binding international standards for regulating the energy efficiency of existing and future ships. Consequently, the main environmental and social/policy performances of large advanced container ships are contained in these standards specified by the MEPC (The Marine Environment Protection Committee) of the IMO (International Maritime Organization) aimed at reducing the energy consumption and related emissions of GHG (Green House Gases) over the forthcoming 2013/14–2025 period and beyond (the standards are also specified for all other types/categories of ships). The main quantitative attributes of these standards for container ships with a deadweight of over 15000 tons imply the following targets for reducing GHG (CO_2) emissions: Phase 0–0 % over 2013–2014; Phase 1–10 % over 2015–2019; Phase 2–20 % over 2020–2024; and Phase 3–30 % beyond 2025 (MEPC 2012). This is expected to be achieved by: (i) technical/technological measures; (ii) operational measures; and (iii) economic measures.

- **Technical/technological measures**

The technical/technological measures aim at enhancing the energy efficiency of large advanced container ships through improving their technical/technological performances. For existing ships, this can be carried out through different modifications. For new ships, this is to be carried out through their design. Some of these measures include:

- Reducing propulsion resistance by modifying the hull form;
- Ensuring enhanced propulsion efficiency by modified propeller(s);
- Increasing the hull size in order to increase the deadweight (capacity);
- Using energy from exhaust heat recovery; and
- Using renewable energy (wind, solar power, etc.).

EEDI (Energy Efficiency Design Index): This index is proposed by the IMO in order to evaluate the effects of the particular above-mentioned technical/technological measures for existing and new ships. In general, the rather complex expression for the original EEDI can be simplified as follows (IMO 2011; LR 2011):

$$\mathrm{EEDI}_{\mathrm{ref}} = \frac{P * \mathrm{SFC} * C_f}{\mathrm{DWT} * V} (\mathrm{gCO_2/ton - mile}) \tag{4.7a}$$

where

P is the engine power including the main engine and auxiliary engines (kW);
SFC is the specific fuel consumption (the recommended value is 190 g/kWh)
C_f is the carbon emissions factor (3.1144 gCO_2/g of fuel for HFO);
DWT is the ship's deadweight (tons); and
V is the speed that can be achieved at 75 % of P of the main engine.

As indicated in Eq. 4.7a, EEDI decreases more than proportionally as the ship's deadweight *DWT* and operating/cruising speed *V* increase, and increases in proportion with the engine power *P* and fuel efficiency *SFC*.

Using the data for container ships built over the period 1999–2008, the average EEDI is calculated and regressed with the deadweight DWT. The average regression line is obtained as follows (Flikkema et al. 2012; IMO 2011):

$$EEDI_{ref} = 174.22 * DWT^{-0.201} \tag{4.7b}$$

where all symbols are as in the previous equations.

This Reference Line is used as the terms of reference for existing and especially new-build ships. If the attained EEDI value of a given ship is above the Reference Line, the ship is considered energy inefficient, and vice versa. Consequently, the required EEDI can be defined as the allowable maximum attained EEDI for a given container ship, which is below and/or at most at the Reference Line. Regarding the above-mentioned policy targets for improving the energy efficiency of container ships over the forthcoming period (i.e., by 2025 and beyond), the required EEDI can be estimated as follows (IMO 2011):

$$EEDI_{req} = (1 - X/100) * EEDI_{ref} \tag{4.7c}$$

where

X is the target for improving the energy efficiency of container ships during the specified period of time (%).

As an example, using Eq. 4.7a, the attained EEDI is calculated for existing Post-Panamax large container ships which entered service over 2003–2008. Capacity utilization (DWT) is assumed to be 70 %, the power 75 % of the maximum engine power, and the speed 1 kts below the maximum designed speed (MAN Diesel 2011). The required EEDIs respecting the above-mentioned energy efficiency improvement targets can be calculated using Eqs. 4.7b, c. Figure 4.19 shows the results. As can be seen, all considered container ships fulfill the required 2013–2014 EEDI. Ships larger than 85,000 DWT will be able to satisfy the required 2015–2019 EEDI. None of these ships will be able to satisfy the required 2020–2024 EEDI, or those set for 2025 and beyond.

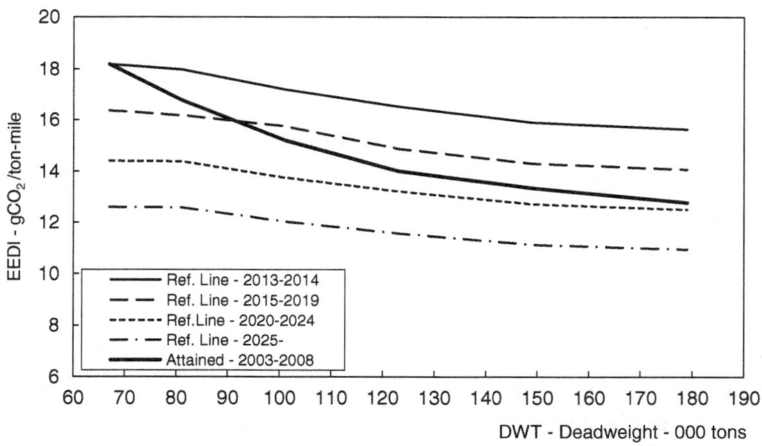

Fig. 4.19 Environmental performances of the large advanced container ships—relationship between the existing and required EEDI, and the capacity (IMO 2011; MAN Diesel 2011)

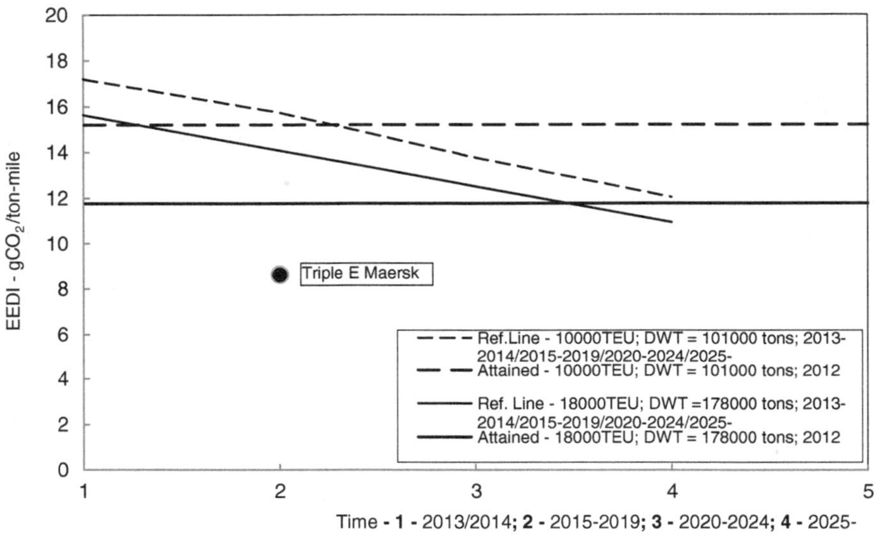

Fig. 4.20 Environmental performances for the selected large advanced container ships—attained and required EEDI specified by the MEPC policy for the 2013–2025 period and beyond

In addition, Fig. 4.20 shows that container ships larger than 85,000 DWT including the largest forthcoming Triple E Maersk to be launched in 2015 will be able to satisfy the required EEDI from 2012 until 2020 and slightly beyond, but not later. If, however, the Triple E Maersk ship is to be designed as expected (SFC = 168 g/kWh and engine power $P = 65$ MW), its attained EEDI will be comfortably below the required EEDI over the entire period (2015–2025 and beyond).

- **Operational measures**

Operational measures aim at improving the energy efficiency of large advanced container ships through innovative operations. They can be applied to existing ships by shipping companies in the scope of their efforts to improve energy and consequently economic efficiency. Some of these measures include:

- Optimizing operations of individual ships and fleets;
- Operating/cruising at reduced speed, i.e., slow steaming;
- Entering and leaving ports on time;
- Maintaining the hull clean in order to reduce propulsion resistance; and
- Ensuring regular maintenance of the ship's overall machinery.

In order to promote, stimulate, and implement some and/or all above-mentioned operational measures applicable to existing large advanced container and all other ships, the IMO has also proposed two indicators/tools: EEOI (Energy Efficiency Operational Indicator) and SEEMP (Ship Energy Efficiency Management Plan).

EEOI (Energy Efficiency Operational Indicator): This indicator was introduced on a voluntary basis in 2005 and is expected to be used by the owners and operators of large advanced container and other ships as an indicator expressing the energy efficiency of a given ship in operation. It can be estimated as follows (IMO 2012):

$$\mathrm{EEOI} = \frac{\mathrm{FC} * C_f}{W_c * d}(\mathrm{gCO_2/ton - mile}) \tag{4.7d}$$

where
FC is the fuel consumption during a trip (tons);
W_c is the actual weight of freight/goods (tons); and
d is the length of route, i.e., actual trip distance (nm).

The other symbols are analogous to those in the previous equations.

Equation 4.7d indicates that EEOI is proportional to the fuel consumed during a given trip and is inversely proportional to the actual weight of freight/goods on-board and length of route. Thus, the EEOI can be improved by decreasing the fuel consumption, as mentioned above, through reducing the operating/cruising speed, i.e., slow steaming, while transporting larger quantities of freight on longer distances.

Despite being expressed in the same units as EEDI, the EEOI is estimated from the values of particular variables measured during or just after a given trip. Therefore, it can be used for measuring changes in the energy efficiency of the same ship operating along different routes/markets under different conditions. Due to such inherent diversity of the independent variables already used for the same ship, the EEOI appears inappropriate for the comparison of different ships.

SEEMP (Energy Efficiency Management Plan): This management plan is used for implementing improvements in the ship's and fleet's energy efficiency through operational measures. These include planning the trip in terms of weather routing, arrivals and departures from ports on time, optimization of speed, etc., optimizing the ship's handling and maintenance of the hull, use of engines and waste heat recovery, as well as energy management and reporting. The implementation of SEEMP voluntarily by ship and fleet owners and/or operators can be carried out through five procedures comprising the energy efficiency improvement cycle as follows (Sala 2010):

- *Planning* identifies measures for improving energy efficiency, sets up the targets, defines the activities and persons in charge, and establishes all these as a system with a roadmap over a 3–5 year period;
- *Implementation* implies realization of the energy efficiency improvements according to the plan; these include zero or low cost simple improvements made during daily operation and maintenance, less than 2 year pay-back improvements of systems by minor conversions during regular operations, and improvements of the systems and haul requiring the ship to dock;

- *Monitoring* implies developing the method and performing energy efficiency monitoring continuously by collecting and processing quantitative information/ data during and/or at the end of the improvement cycle; and
- *Self-evaluation* implies assessing the effects of the implemented measures by using the monitored results and feeding them back to the next energy efficiency improvement cycle; and
- *Publication* of the results voluntarily implies presenting the results to the professional public and enabling third party evaluation.

Due to the need for collecting a relatively large quantity of information even for a single trip, different support systems are being developed for calculating, analyzing, and preparing reports on energy efficiency improvement cycle(s).

- **Economic measures**

Economic measures aim at promoting and implementing the above-mentioned technical/technological and operational measures. IMO has proposed several measures classified into two broad categories as follows:

- The fuel pricing system (proposed by Denmark); and
- The emissions trading system (proposed by Norway, Germany, and France).

The former measure implies automatically charging an amount on the purchase of fuel. The collected funds would be used for different projects aiming at reducing emissions of GHG, particularly those in developing countries. In addition, a part of the amount collected would be awarded back to ships achieving substantial improvements in energy efficiency. The latter implies that the total amount of GHG generated by the shipping industry would be regulated by the emission trading system. In such a case, each ship would be assigned a credit in terms of the annual allowable emissions of GHG (CO_2). Subsequently, the differences between the actual and credited (assigned) emissions would be traded with other ships and/ or the rest of the transport and other non-transport sectors.

Potential impact on global emissions of GHG

The presented EEDI and SEEMP measures are expected to significantly reduce emissions of GHG from the world's freight ships over the long-term 2010–2050 period. Figure 4.21 shows one such scenario.

As can be seen, both EEDI and SEEMP will contribute to reducing the total emissions of GHG (CO_2) as compared to the emission levels in 2010. For example, the reduction will be between 13–23 % over the 2020–2030 period. However, these measures will not be able to prevent a further increase in the total emissions of GHG according to an upward trend although with some reduced rates as compared to the BAU (Business As Usual) scenario, mainly driven by the expected growth in global trade.

The contribution of EEDI and SEEMP will most likely be proportional to the product of their share in the total global freight ship fleet and the above-mentioned

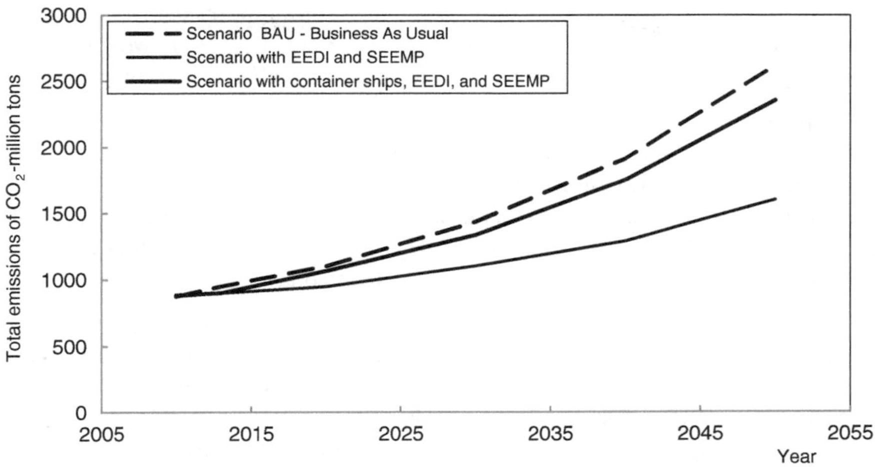

Fig. 4.21 Scenarios of development of emissions of GHG (CO_2) by the global freight ship fleet over time (IMO 2011; MEPC 2012)

required EEDI targets. In addition to the required EEDI targets, let the "what-if" scenario assume an increase in the share of large advanced container ships in the global fleet at an average rate of 5 %/10 years by 2050. This will amount to shares in the corresponding totals of 5, 7, 17, 23, 28, and 33 % in 2010, 2013, 2020, 2030, 2040, and 2050, respectively. As applied to the total amount of emissions of GHG, this again produces an upward trend as shown in Fig. 4.21. As can be seen, the contribution to decreasing of the total emissions of GHG (CO_2) would be about 3–10 % over the observed period (2010–2050).

Future measures and technologies for mitigating emissions of GHG

The main drivers of design of future large advanced container ships will be conditioned by the strategic plans of shipping companies and environmental constraints aiming at:

- Improving economics by reducing the staff and increasing productivity;
- Increasing flexibility of services by modifying routes and networks, and ship deployment;
- Optimizing utilization of containers, i.e., securing return freight/goods volumes;
- Minimizing delays at ports; and
- Minimizing fuel consumption and related emissions of GHG (Green House Gases) by meeting the current and prospective energy efficiency regulatory requirements, i.e., required EEDI.

In particular, the options for minimizing fuel consumption and related emissions of GHG through ship design include: (i) Reduction of power; (ii) New technology for power generation; and (iii) Renewable fuel/energy primary sources.

- Reduction of power can generally be achieved by designing/developing the hull form, reducing the weight and power for the ship's own use, frictional and wind resistance, and improving the engine efficiency;
- New technologies for power generation include use of alternative fuels such as biofuels, LNG (liquefied natural gas), LH_2 (liquid hydrogen) and fuel cells; and
- Renewable fuel/energy primary sources include solar and wind energy.

In addition, an option for minimizing the fuel consumption and related emissions of GHG includes forthcoming trip support systems, one of which is "Sea-Navi." These are designed to support online optimization of the ship's routing respecting the shortest distance, weather, characteristics of the hull, and regime of engine operation, thus contributing to improving EEOI and SEEMP.

Reduction of power

Some achievements have already been made by reducing power. An example of using existing technologies is the Triple E Maersk container ship, which will consume about 35 % less fuel (HFO) and emit about 20 % less CO_2 per TEU as compared to today's most energy efficient container ships, and about 50 % less of both as compared to the industry average for container ships operating in the Asia–Europe market. Such energy efficiency of about 9 % will be partially achieved also thanks to an advanced energy efficient waste heat recovery system, the purpose of which is to reduce the engine's need for fuel and consequent emissions of CO_2. The system operates by capturing the heat and pressure contained in the exhaust gases and then using them to move turbines creating mechanical energy for operating an electrical generator.

In addition, a future container ship concept has been proposed by Odense Steel Shipyard Ltd. The reference case is the existing A-class Post-Panamax container ship of a capacity of 8,500 TEU, 109000 tons DWT, length 352 m, draught 15 m, the power of main engine 63 MV at 100 rpm, and design speed of 26.5 kts. In parallel with improving the main engine and propeller design, an innovative WIF (Water in Fuel), WHRS (Waste Heat Recovery System), and EGR (Exhaust Gas Recirculation) system aiming at improving the ship's energy efficiency would be introduced in four of five designs as shown in Table 4.4 (OSSL 2009).

The given cases indicate that de-rated engine power, modified propeller design and slow steaming, in combination with other systems (WHRS, WIF, and EGR), could improve the ship's energy efficiency in terms of energy consumption and related emissions of GHG (CO_2) by about 30 %.

Furthermore, some other advanced container ship designs with existing HFO may be promising. One of these is the new method of powering large freight ships developed by Gamma Light and Heavy Industries Ltd. The method implies that instead of placing diesel engines at the rear of the ship as is common, the sets of

Table 4.4 Improving performances through the design of advanced future container ships (OSSL 2009)

Performance	1	2	3	4	5
WHRS	No	Yes	Yes	Yes	Yes
WIF and EGR	No	Yes	Yes	Yes	Yes
Capacity (TEU)	8,500	8,500	8,500	8,500	8,500
Design speed (kts)	26.5	26.5	26.5	24.1	22.08
Engine (de-rated) power (MW)	63.0	62.6	58.3	40.9	31.4
RPM[1)]	100	94	94	78	76
Propeller					
Diameter (m)	8.9	9.2	9.2	9.2	8.8
Number of blades	6	6	6	6	6
Fuel consumption (HFO) (tons/day)	278	246	227	160	121
Emissions of GHG (CO_2) (tons/day)	866	766	707	498	377

RPM Rotations per minute; (*1*) A-class as built (reference design); (*2*) New engine, larger propeller diameter + WHRS, + WIF&EGR; (*3*) 2 + New propeller blade design, hull coating, and advanced rudder; (*4*) 3 + Lower steaming I; (*5*) 4 + New Engine, Smaller propeller diameter, new propeller blade design, hull coating, advanced rudder + Lower steaming II

diesel electric units are placed down the side and along the entire length of the ship, enabling much higher efficiency due to much reduced use of the available power and proportionally less fuel consumption while maneuvering. Such a constellation is expected to improve the nominal (present) energy efficiency of ship(s) by about 75 %.

New technologies

Biofuels are considered as alternative fuels. They include biodiesel/vegetable oils and biogas. The former is applied to FAME (Fatty Acid Methyl Ester), which can be used exclusively or as an ingredient of conventional HFO. Generally, it can be produced from oleaginous crops such as rapeseed, sunflowers, soy beans, palm oils, etc. The latter, also known as SNG (Substitute Natural Gas), can be produced with similar characteristics as LNG (Liquefied Natural Gas) allowing use in LNG engines. The primary source of biogas is organic waste and energy crops. It can be used by LNG engines and also fuel cells (UniCredit 2009).

LNG is also an alternative fuel, the main component of which is methane. It has come under focus mainly due to its CO_2 content which is about 20–25 % lower than that of HFO. Also, it can reduce emissions of SO_x (Sulfur Oxide) by about 90–95 % and NO_x to the level complying with IMO Tier III limits to be in effect from 2016. In addition, the price of LNG would be comparable to that of HFO. However, the tanks for storing LNG are much larger than those for storing HFO, thus requiring more space, which can compromise the ship's loading capacity. Nevertheless, this could be compensated by improved energy efficiency. The engines powering the ship would be the dual-fuel hybrid constructions enabling operation in both HFO and LNG mode. Some designs such as Quantum

(DNV—Det Norske Veritas) indicate that the EEDIs of large container ships using a mixture of LNG/HFO could be significantly lower than that required beyond 2025 (by about 30 %). In addition, particular attention needs to be devoted to the safety and reliability of LNG bunkering systems by excluding any spillage (GL 2012).

In principle, fuel cells convert chemical energy (for example contained in hydrogen H_2 and oxygen O_2) directly into electricity used for powering the ship's electromotor(s). Such direct conversion makes hydrogen fuel cells highly efficient. On container ships they are located inside the container units and they can be of the PEM (Polymer Electrolyte Membrane) type characterized by a high power density and flexible behavior in operation. As such, they enable optimization of power use on a case by case (trip) basis depending on the prevailing conditions. The NYK Super Eco Ship 2030 is an example of a future container ship that could alternatively be powered by fuel cells (UniCredit 2009).

Renewable fuel/energy primary sources

Wind energy was commonly used in maritime operations in the middle ages. One modern example is the MS Beluga SkySails ship developed by two companies—KiteShip and SkySails—and launched in 2007 (EC 2010b). The wind assisting system includes larger sails attached, for example, to the container ship, which pull the ship through the water by using high-altitude wind(s). Depending on the ship's size, the sails can have up to about 5000 m^2 of surface area. They are divided into compartments with compressed air keeping them rigid. Such sails are controlled by computer in addition to an autopilot system used to determine the optimal shipping route(s) depending on the weather (prevailing wind) conditions. In the given case, the wind energy is partially used as a means of assistance, since a diesel HFO engine still remains in place. The potential improvement in the ship's energy efficiency is estimated to be up to about 10–35 %.

In addition, solar and wind energy will be used exclusively and/or in combination with LNG for powering future large container ships. In such cases, both solar and wind energy will be harnessed by solar panels (or solar cells on foils) and sails, respectively, and then either directly converted into electricity and consumed or stored for later use. Table 4.5 gives a comparison of the existing MV NYK VEGA and future NYK Super Eco Ship using different power sources.

The future NYK Super Eco ship will be equipped with an LNG engine with about 30 % less power, but supported by about 5–13 % of renewable (solar and wind) energy. Both will enable improvement of the ship's energy efficiency by about 70 %.

Another example is the Aquarius Eco Ship designed by EMP (Eco Marine Power) from Fukuoka (Japan). The central component of the ship is the Aquarius MRE System based on EnergySail technology. This is a renewable energy platform also designed by EMP fitted with different renewable energy technologies incorporating solar panels and wind power devices, energy storage modules, and a positioning system. The first enable tapping wind and sun power while at sea or

Table 4.5 Main characteristics of existing and future advanced large container ships (NYK Line/MTI 2010)

Basic characteristics: 9000 TEU/25 kts	MV NYK VEGA (2006)	NYK Super Eco Ship (2030)
Length (m)	338	353
Weight (m)	45.8	54.6
Draught (m)	13.0	11.5
Engine type	Diesel engine (HFO)	Fuel cell (NLG)
Required power (MW)	64	40
Renewable energy (MW)	0	Solar: 1–2/Wind: 1–3
Emissions of CO_2 (g/TEU-mile)	195	62

even in the port. The second enable energy storing for eventual future use. The latter contributes to sailing optimization. The solar panels and wind devices are located in an array of rigid sails made of composites whose number and area depend on the required power compensated by the assisting LNG engine. Such rigid sails would be automatically positioned to best suit the prevailing weather conditions including being lowered and stored during inconvenient weather. In addition to the Aquarius MRE System, the future Aquarius Eco Ship would be equipped with other energy efficiency improving components such as an advanced electrical propulsion system and waste heat recovery technologies. A computer system will monitor and control operation of all above-mentioned components. Such combination of technologies could lead to improving the energy efficiency of Aquarius Eco Ship(s) by about 40 % (http://www.ecomarinepower.com).

Land use

Large advanced container ships are handled at the port's container terminals. Each terminal consists of water-side berths for ship docking, a large coastal area of land for the storage of containers, specialized berth cranes and yard cranes for container loading/unloading to/from the ship and within the storage area, tractors and other equipment for handling containers from the ships to the storage area, gates for inland road trucks, in many cases yards, and barges and various maintenance and administrative buildings. The size of this used land generally increases proportionally or more than proportionally as the number and size of berths increases, as shown in Fig. 4.22 (an example for seven U.S. port container terminals). In addition, most ports set aside 'land banks' for future expansion, which can range from a few hundreds to a few thousand percent of the land occupied by existing terminals (CGI 2007).

Safety

Traffic incidents and accidents have happened to all container ships, including large advanced ones. Incidents and accidents usually refer to fatalities during particular stages of operations such as loading and unloading at the port terminals, operations in

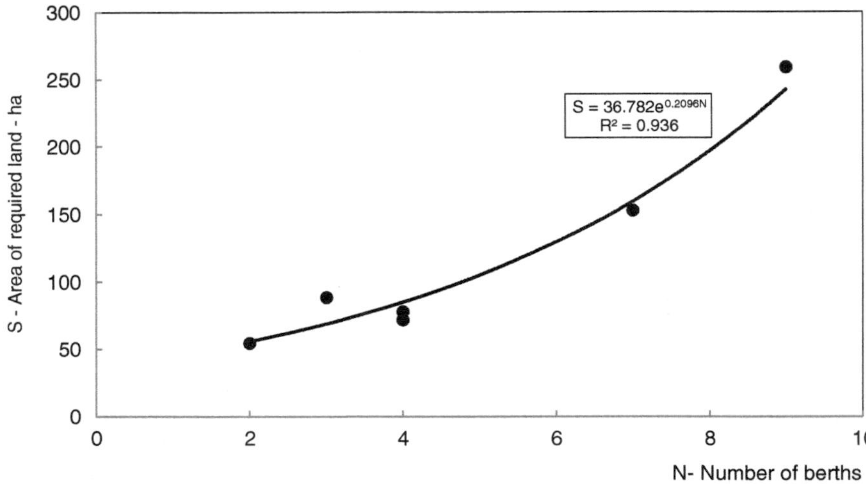

Fig. 4.22 Relationship between the area of land occupied by U.S. port container terminals and the number of berths (CGI 2007)

ports, restricted and coastal waters, and open sea transit. Similarly as in other transport modes, the risk of loss of life is expressed by the number per unit of output. In addition, the impacts of incidents and accidents on the environment are taken into account. They are usually expressed in absolute terms. For example, some figures indicate that the average number of fatalities was 3.52×10^{-3} fatalities/ship-year, the number of environment-pollution events 4.36/year, and the average number of lost containers 182/year, all over the 1993–2004 period (IMO 2007).

In order to assess the risk of potential fatalities in container ship crew members, the main factors of risk include collisions, contact, grounding, fire/explosion, and heavy weather. Respecting this classification, the risk of potential fatalities among ship crew members has been assessed as 9.00×10^{-3}, and that for an individual crew member as 2.25×10^{-4}. The latter is lower than the maximum perceivable risk for a crew member of 10^{-3}, but higher than the "negligible" risk of 10^{-6}. At the same time, the perceived environmental risk causing release of substantial quantities of dangerous substances and fuel has been estimated to be 1.01 (IMO 2007). This implies that large advanced container ships are at least as safe as their smaller counterparts.

4.3.3 Evaluation

Large advanced container ships possess both advantages (Strengths and Opportunities) and disadvantages (Weaknesses and Threats) as compared to their smaller counterparts.

Advantages

- Strongly supporting international trade, thus contributing to further globalization of the national and global economies;
- Supplying relatively substantial transport capacity per service, which in turn can compensate the diminishing technical productivity during slow steaming;
- Slow steaming can improve the economic performances by reducing the operating costs and the environmental performance by improving energy efficiency, i.e., reducing fuel consumption and related emissions of GHG (CO_2); in addition, slow steaming can contribute to increasing the number of transport units/ships in operation serving given volumes of demand and consequently partially compensate short-term capacity oversupply caused by market volatility; and
- Large advanced container ships are convenient in terms of technical/technological feasibility and testing and implementing different alternative technical/technological and operational innovations for improving technical/technological, operational, economic, and environmental performances (particularly energy efficiency).

Disadvantages

- Calling or serving, i.e., being able to access, a limited number of ports due to limitations of the maximum draught, and efficient and effective maneuverability;
- Requiring substantial investments in some port/terminal infrastructure, facilities, and equipment (berths, cranes, store areas for containers, etc.) for efficient and effective handling;
- Requiring increasing use of coastal land for building larger berths, terminals, and inland transport infrastructure;
- Vulnerability to market volatility easily creating imbalances between demand and capacity, thus compromising the overall economic feasibility; adapting to such conditions requires modification of the service network(s) into stronger hub-and-spoke configuration(s) with a smaller number of ports serving mature markets with relatively stable freight demand in both directions;
- Economies of scale disappear rapidly beyond a certain size, i.e., over 10,000 TEU;
- Contributing to a decrease in the overall speed of supply chains and prolonging the period freight/goods remain within the chains, thus causing raising inventory costs due to slow steaming (Chap. 3);
- Remaining concerns relating to energy efficiency in terms of the current and prospective contribution to total fuel consumption and related emissions of GHG (CO_2) despite forthcoming technical/technological (EEDI) and operational (SEEMP) improvements, at least until HFO is mainly used; and
- Some reservations including criticism as to whether it is correct to use EEDI for assessing energy efficiency, implying the need for further modifications and improvements.

Finally, regardless of the above-mentioned advantages and disadvantages, the fleet of large advanced container ships will likely continue to grow, while being continuously modernized by simultaneously improving their technical/technological, operational, economic, and environmental performances. In combination with their size, these have already and will continue to make such ships advanced.

4.4 Liquid Hydrogen-Fuelled Commercial Air Transportation

1957	The first successful test of the bomber aircraft B57 modified to use LH_2 (Liquid Hydrogen) under military auspices (U.S.)
1988	The first flight of the TY155 aircraft with modified Engine No.3 by *Kuznetsov* to use LH_2 or NG (Natural Gas) (USSR)
1990	The standard reference book "*Hydrogen Aircraft Technology*" is published by D. Brewer (U.S.)
1990s	The European–Canadian "*Euro-Quebec Hydro-Hydrogen Pilot Project*" covers many aspects of hydrogen use (Europe, Canada)
2003	The study carried out in the scope of the 5th EC FMP (Framework Program) covers different aspects of making the transition from conventional to LH_2 fuel such as: aircraft configuration, systems and components, propulsion, safety, environmental compatibility, fuel sources and infrastructure, and transition processes, from both global and regional perspective (Europe)

4.4.1 Definition, Development, and Use

Mitigating the medium- to long-term impacts of the APT (Air Passenger Transport) system on the environment in terms of energy consumption and related emissions of GHG (Green House Gases) and society can be achieved, among other endeavors, also by further development of aircraft propulsion systems (engines) as follows:

- Improving existing turbofans in combination with using fuels synthesized from alternative sources such as coal and natural gas, and biomass from plants and algae; these engines are expected to be generally more fuel efficient by about 15 % and quitter by about 25 dB; and
- Improving existing turbofans by using advanced materials enabling adapting combustion thanks to higher combustion temperatures, including developing advanced concepts such as the following:
  - Ultrahigh by-pass ratio engines (Geared turbofan (GTF)) developed by Pratt and Whitney and NASA (National Aeronautics and Space Administration) in the U.S., which would be about 15 % more fuel efficient and 30 dB quitter as compared to existing turbofans;

- Open-rotor engine (GTF) or Unducted Fan (UDF) developed by GE (General Electric) and NASA, which would be about 25 % more fuel efficient and 15–20 dB quitter then their existing counterparts;
- Hybrid and electric engines such as Boeing/NASA SUGAR Volt-hybrid and Voltair (all-electric aircraft concept), which would require major advances in the battery energy density such as the Lit-Air (Lithium Air) battery concept; the theoretical energy density of this concept is about 11.5 kWh/kg, which is close to that of gasoline/kerosene of about 13 kWh/kg; and
- Improvements and modification of existing turbofan engines enabling them to use LH_2 (Liquid Hydrogen) as fuel.

The latest concept as an alternative for mitigating the global emissions of GHG by APT over the future long-term period implies switching from conventional jet fuels (kerosene) to alternative fuels, one of which is LH_2 (Liquid Hydrogen) (Janic 2008, 2010).

4.4.2 Analysis and Modeling Performance

4.4.2.1 Liquid Hydrogen (LH_2) as Fuel

Manufacturing, logistics, and economics

Some of the methods for manufacturing hydrogen as a fuel are already commercially available, but the produced quantities are used only in small niche markets as a chemical substance and not as an energy commodity. For commercial use in general, hydrogen can be produced from chemically reformed natural gas, fossil fuels, and/or biomass feedstock using conventional chemical processes. In addition, it can also be produced by using electricity or heat, sunlight, and/or specialized microorganisms for dissociating water. In cases of producing relatively large quantities, the process will be mainly driven by economic reasons including the full logistics costs (Chevron 2006; IEA 2006).

The logistics of hydrogen includes its transport and storage. Hydrogen can be transported and stored after being converted into a highly concentrated form either by increasing the pressure or by lowering the temperature. In general, over shorter distances, it is transported as a compressed gas by dedicated vehicles and/or pipeline systems. Over longer distances, it is exclusively transported as a liquid by dedicated vehicles operated by all transport modes. It is stored in high-pressure cylindrical tanks and containers (IEA 2006).

The economics of hydrogen implies the amount of energy consumed for its production, packaging, transport, and storage, all depending on whether it is a liquid or gas. For example, the energy input could be about 2.12 and 1.65 times higher than the energy content of the delivered liquid and gas hydrogen, respectively (i.e., loss factor). In both cases, the loss factor is considerably higher than that of conventional jet fuels (about 1.12) (Bossel and Eliasson 2003). In addition, the most important issue in supplying hydrogen as an energy commodity is a competitive price. This depends greatly on the primary sources and the related manufacturing processes on the one hand, and some market mechanisms such as, for example, taxes on the emissions of CO_2 (Carbon Dioxide), on the other. The price of hydrogen should in general be comparable to that of conventional jet fuel. In this context, in the long term the prices of conventional jet fuel are expected to increase and that of hydrogen to decrease, which makes the expectation of comparable prices for both fuels more realistic. Some estimates indicate that in 2035, the production costs of hydrogen will range between 0.8 and 3.5 $US/kg H_2 (IEA 2006).

Operational and environmental performances

Hydrogen as a fuel for APT is going to be in the liquid aggregate state (i.e., LH_2—Liquid Hydrogen). Its main operational characteristics include its specific energy (120 MJ/kg), specific density (0.071 kg/m^3 at 15 °C), energy density (8.4 MJ/l), and boiling point (−253 °C) (Chevron 2006; Daggett et al. 2006). As compared to conventional jet fuel, LH_2 has the following advantages in terms of emissions of particular GHG: 0 versus 0.50 g CO; 0 versus 0.75 kg CO_2; 0.78 versus 0.30 kg H_2O; 0.02–0.102 versus 0.41 g NO_x; and 0 versus 0.20 g UHC (this is based on 10 MJ of energy content obtained from 0.5 l of LH_2 and 0.3 l of Jet A) (Daggett et al. 2006; EEC 2005). Thus, its burning does not produce CO_2 or SO_x. Increased emissions of H_2O and NO_x (the latter under specific conditions) remain the only matter of concern. Figure 4.23a, b and c shows the net relative impact of GHG from conventional jet fuel/kerosene (K) and LH_2 (H) on global warming.

In general, the net impact of GHG such as H_2O and NO_x on global warming increases more than proportionally, and that of CO_2 decreases less than proportionally as the aircraft flying altitude increases. In the case of kerosene, increasing the flying/cruising altitude from 9 to 11 km increases the net impact of NO_x by about a half and that of H_2O by about 75 %, the latter due to the formation of contrails since the aircraft fly in the troposphere. At the same time, the impact of CO_2 decreases by about 30 %. In the case of LH_2, increasing the flying/cruising altitude from 9 to 11 km increases the impact of NO_x up to about 10 % and that of H_2O by about 50 %. There is no CO_2 impact.

Social/policy performances

The main social/policy performances of LH_2 are considered to be safety, i.e., not causing incidents/accidents resulting in injuries, loss of life, or damaging properties due to known reasons. In general, LH_2 can be a safe fuel. Nevertheless, its main potential disadvantages are its explosive rate of 13–79 % concentration in the air and its very low ignition energy (about only 0.02 mj). LH_2 also mixes faster with air than jet fuel vapor, and disperses rapidly through the air in contrast to jet fuel, which pools on the ground. It burns with a nearly invisible, colorless, and odorless flame, which is also an important safety concern (IEA 2006).

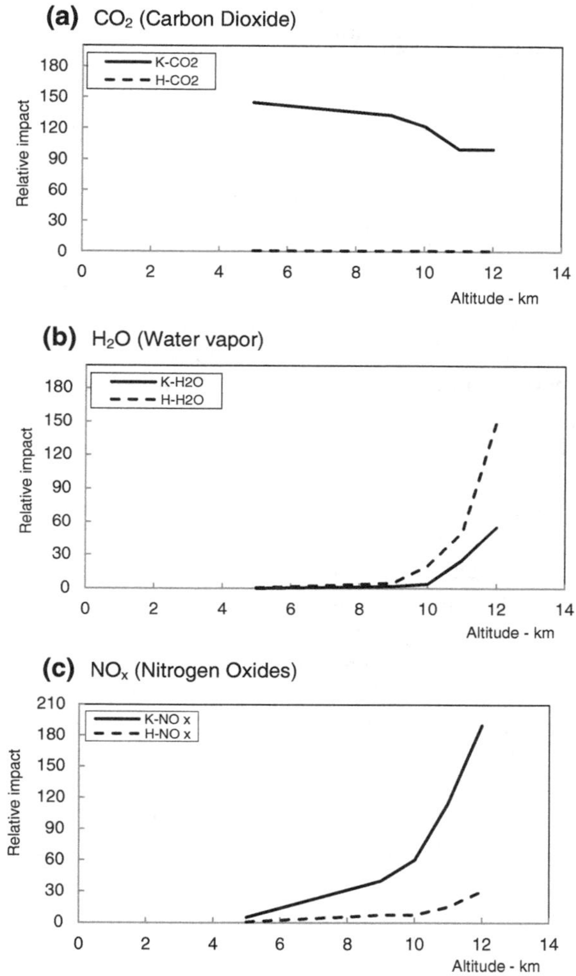

Fig. 4.23 Impacts of GHG (Green House Gases) from conventional jet fuel (kerosene) and LH_2 (Liquid Hydrogen) on global warming depending on the aircraft flying altitude (Penner 1999)

4.4.2.2 The LH_2-Fuelled Commercial Aircraft

Development

Research on using hydrogen for commercial aircraft has been carried out in Europe, the U.S., and the Russian Federation for a considerable length of time. However, the first ideas and related experiments emerged more than 70 years ago and continue until nowadays. Recently, different projects have provided a vision of the prospective technical/technological, operational, economic, environmental, and social/safety performances of cryogenic aircraft expected to be fully developed by around 2020 and consequently commercialized by around 2040 (EC 2003; EEC 2005).

Technical/technological and operational performances

In light of the characteristics of LH_2 as compared to conventional jet fuel (2.8 higher specific energy and about 11 times less specific density), LH_2 fuelled or cryogenic aircraft will require about 4.3 times more fuel volume for equivalent energy output than conventional aircraft. Therefore, their main design characteristic will be a relatively large volume of well-insulated cylindrical fuel tanks. They can have different positions within the aircraft configuration: above the payload (passengers and freight), above and aft of the payload, and fore and aft of the payload section. The wings could, with no fuel storage space, be smaller. The results will be increased aerodynamic resistance and aircraft empty weight as compared to conventional aircraft. However, the much lower weight of LH_2 is expected to compensate for such an increase in the empty weight and consequently contribute to reducing the maximum take-off weight (Brewer 1991; EC 2003).

Cryogenic jet engines will retain the basic structure of conventional jet engines but with some necessary modifications, such as fuel pumps, fuel control unit, and combustion chambers. Experiments so far have shown that these engines will have about 64 % lower Specific Fuel Consumption (SFC) than conventional jet engines (0.0976 vs. 0.2710 (kg/h)/kg for cruising and 0.0512 vs. 0.1420 (kg/h)/kg for the take-off phase of flight). In addition, they are expected to be 1–5 % more efficient in generating thrust from the given energy content. For Supersonic Transport Aircraft (STA), the specific consumption of LH_2 and Jet A fuel during cruising is expected to be about 0.260 (kg/h)/kg and 0.680 (kg/h)/kg, respectively, (the ratio for Jet A/LH_2 is 2.61). Last but not least, the hydrogen engines for either aircraft category are expected to operate with a slightly lower turbine entry temperature, which in turn will extend their life and reduce maintenance costs (Brewer 1991; Corchero and Montanes 2005; EC 2003; Guynn and Olson 2002; Svensson et al. 2004; http://www.tupolev.ru).

Economic performances

Over the 2003–2007 period, the share of fuel cost in the total operating costs of commercial airlines stood at about 30 % (EC 2003). Respecting the unit price of LH_2, the latest price of conventional jet fuel, and the lower Specific Fuel Consumption of cryogenic engines of about 64 % (i.e., 1 kg Jet A is equivalent to 0.36 kg LH_2), estimates show that the share of fuel costs in the total operating costs of cryogenic aircraft could vary between 45 % (1$US/kg LH_2) and 78 % (1.73$US/kg LH_2), if the prices of other inputs are assumed constant. Equalizing the prices of both fuels to 1$US/kg, the shares of corresponding costs would amount to about 60 and 35 %, respectively, mainly due to the lower Specific Fuel

Consumption of cryogenic aircraft. This scenario appears realistic since the prices of conventional jet fuel are expected to continue to rise, while those of LH_2 are assumed to decrease as both the efficiency of production and the overall logistics improve.

Environmental performances

Cryogenic aircraft powered by LH_2 do not emit CO_2. However, water vapor (H_2O) emitted in quantities about 2.6 times higher than by conventional aircraft at and above cruising altitudes of 31,000 feet (FL310; FL-Flight Level) will be the main GHG. However, its impact as compared to that of conventional aircraft appear to be much lower (Marquart et al. 2005). Reducing the cruising altitude is an option for eliminating these impacts. However, this could compromise other performances. In addition, cryogenic aircraft are assumed to emit about 5–25 % less NO_x than their conventional counterparts, which is expected to be achieved through the design of the combustion chamber of cryogenic engines (EC, 2003). Table 4.6 gives the environmental performances of typical long-range conventional and cryogenic aircraft in relative terms, mainly for comparative purposes.

Social/policy performances

As with fuel, the main social performance of cryogenic aircraft should be safety. As applies to fuel in general, cryogenic aircraft should be at least as safe as their conventionally fueled counterparts. In the event of an accident, LH_2 burns much faster (15–22 s) with low heat radiation, thus mitigating the fire impact in cases of collapsing fuselage. This contrasts to the impact of fire from conventional jet fuel. In addition, burning LH_2 covers a much smaller surface area (EC 2003). The overall safety figure also includes the appropriate design and operation of the airport fuel supply system. It seems likely that the manufacturing of LH_2 will take place at the airport fuel storage area that reserves will be stored in the large storage tanks, and that fuel will be usually delivered to the aircraft at the airport parking stands through a dedicated underground pipeline system.

Table 4.6 Environmental performances of typical long-range conventional and cryogenic aircraft—ratio (EC 2003; Janic 2008)

Characteristic	Conventional aircraft (Jet A)	Cryogenic aircraft (LH_2)
Fuel energy content	1	0.36
Volume of fuel	1	11
Volume of fuel tanks	1	4.3
MTOW	1	0.85–1.05
Aerodynamic resistance	1	1.1
Pollutants CO, CO_2, SO_x, HC	1	0
H_2O	1	2.6
NO_x	1	0.05–0.25

4.4.2.3 Modeling Performances of a LH_2-Fueled Air Transport System

Modeling the performances of a LH2-fuelled commercial air transportation includes the structure and application of the methodology. The former consists of the models for a single and two aircraft fuel technologies. The later embraces deriving the input data and analysis of results.

Structure of the methodology

LH_2-fueled APT is expected to be a "carbon-neutral" system. This implies that despite the continued growth of commercial air passenger transport demand over the future period of time, the energy/fuel consumption and related emissions of GHG of the APT system will remain constant or even decrease. The methodology for assessing such developments consists of the models for estimating the annual quantities of emissions of GHG by the APT system using: (i) one single (conventional), and (ii) two (conventional and cryogenic) aircraft fuel technologies, both on the global scale (Janic 2008).

The model for a single aircraft fuel technology

The global emissions of GHG generated by commercial air transportation in the given year of the observed period can be estimated as follows:

$$E_n = V_0 * (1 + i_v)^n * F_{C0} * \left(1 - i_f\right)^n * \sum_{l=1}^{L} e_l \tag{4.8a}$$

where

E_n is the total emission of GHG in year (n) counted from the beginning of a given period of N years, i.e., the base year "0" (tons);

V_0 is the volume of air traffic demand in the base year (0) of a given period (RPK—Equivalent Revenue Passenger Kilometers)[2];

F_{C0} is the average consumption of conventional jet fuel in the base year (0) of a given period (g/RPK);

i_v is the average annual rate of growth of traffic demand in terms of equivalent RPKs over a given period of time (%);

i_f is the average annual rate of improvement of the average unit fuel consumption over a given period of time (%); and

e_l is the emission rate of the l-th green house gas (g/g of Jet A fuel).

According to Eq. 4.8a, the total emissions E_n can be affected through the influencing variables in the given (target) year (n) as follows:

[2] Equivalent RPKs are regarded as the sum of RPKs and RTKs (Revenue Ton Kilometers) (1 RTK = 10 RPK).

- Achieving a rate of improvement of the average unit fuel consumption compared to the rates of air traffic growth, i.e., $i_f \geq i_v/(1 + i_v)$;
- Slowing air traffic growth according to the rate of improvement in the unit fuel consumption, i.e., $i_v \leq i_f/\left(1 - i_f\right)$;
- Constraining air traffic growth by imposing a cap on the total emissions of green house gases, i.e., $i_v = \left[E_n^*/[V_0 * F_{CO} * \left(1 - i_f\right)^n * \sum_{l=1}^{L} e_l]\right]^{1/n} - 1$, where E_n^* is the "cap" on the total emissions of green house gases in the target year (n); and
- Affecting the air traffic growth rate by weakening its relationship with the main internal and external demand-driving forces.

The first three above-mentioned conditions are not likely to be achieved before 2025/26 and beyond, mainly because of the relatively wide differences between the current and predicted average annual air traffic growth rates (3.1 %, IPCC 1999; 5.4 %, Airbus 2006; Boeing 2007) and the rates of improvement in fuel efficiency (1.2–2.2 %; EEC 2005; IPCC 1999; Learmount 2007; Lee et al. 2004). For example, in the first case i_f should be not less than 4.3–4.8 %, respectively, which is almost twice as much as the current very optimistic 2.2 %. In the second case, the air traffic growth rate i_v should not be greater than the expected rate of improvement in fuel efficiency, i.e., about 1.2–2.2 %. In the third case, the main problem appears to be criteria for setting up the annual cap E_n^* and its monitoring and control (IPCC 1999, 2001). The last case seems highly uncertain.

Consequently, the above-mentioned expected reductions in fuel consumption and related emissions of GHG by technological and operational improvements appears to be the only realistic but certainly insufficient alternative. This indicates that achieving a "carbon-neutral" air transport system will be extremely difficult if not impossible with conventional aircraft jet fuels.

The model for two aircraft fuel technologies

The introduction of cryogenic aircraft powered by LH_2 is expected to be a process of gradually replacing part of the conventional aircraft fleet. This process will be able to start if and after the following conditions are fulfilled:

- A pallet of different categories of cryogenic aircraft are fully developed regarding the size–range (small-short, medium–medium, large–long);
- The sufficient manufacturing capacities of cryogenic aircraft and LH_2 are available to satisfy the given rate of replacement;
- The airport infrastructure for supplying LH_2 is fully operational;
- The market prices of LH_2 are competitive to the prices of conventional jet fuel; and
- The emissions of GHG during the manufacture of LH_2 are captured and stored.

The gradual replacement process will take place over a "transitional" period during which both conventional and cryogenic aircraft will be used. The contribution of such a "hybrid" fleet to the total emissions of GHG in the year (k) of the "transitional" period of K years can be estimated, based on Eq. 4.8a, as follows (Janic 2008):

$$E_k = V_0 * (1 + i_v)^k * \left[F_{C01} * (1 - i_f)^k * (1 - ki_h) * \sum_{l=1}^{L} e_l + F_{C02} * (ki_h) * \sum_{m=1}^{M} e_m \right] \quad (4.8b)$$

where

i_h is the average share of the total volume of traffic (RPKs) carried out by cryogenic aircraft in each year of the observed period ($0 \leq ki_h \leq 1$; $k = 1,2, \ldots, K$);

F_{CO1}, F_{CO1} is the average unit fuel consumption of conventional (Jet A) and cryogen (LH_2) fuel, respectively, in the base year (0) of the given "transitional" period (g/RPK); and

e_m is the emission rate of the m-th GHG from cryogen fuel (LH_2) (g/g of JetA fuel).

The other symbols are analogous to those in Eq. 4.8a. The parameter F_{CO1} in Eq. 4.8b is assumed to be at the level achieved when the process of introducing cryogenic aircraft starts, i.e., at the beginning of the "transitional" period, and will continue to improve over the said period. The parameter E_{CO2} will be lower than E_{CO1} approximately proportionally to the ratio between the specific energy of conventional jet fuel and LH_2, i.e., 43.2/120 = 0.36. This ratio is assumed to remain constant over the "transitional" period. The cryogenic aircraft replacing conventional aircraft will be introduced each year in constant proportions, thus implying their constantly increasing share in satisfying air traffic demand (RPKs). In this case, the eventual stabilization and/or even reduction in emissions of GHG in the given (target) year could be achieved by the same alternatives as in Eq. 4.8a. In addition, one additional alternative could consist of adjusting the rate of introducing cryogenic aircraft in Eq. 4.8b as follows:

$$i_h = \left[i_v * F_{C01} * (1 - i_f) * \sum_{l=1}^{L} e_l \right] / \left\{ \left[F_{C01} * (1 - i_f) * \sum_{l=1}^{L} e_l - F_{C02} * \sum_{m=1}^{M} e_m \right] * [1 + i_v(k + 1)] \right\} \quad (4.8c)$$

where all symbols are as in previous equations.

Application of the methodology

Input

The methodology is applied to the long–term development of the APT system, related fuel consumption, and emissions of GHG. The time horizon is divided into three subperiods: 2006–2025/26, 2025/26–2040, and 2040–2065. The first sub-period is specified by the two leading aircraft manufacturers (Airbus 2006; Boeing 2007). The second sub-period is specified as the period until the start of the "en-masse" introduction of cryogenic aircraft (2040). The last period represents the "transitional" period of gradually replacing a certain proportion of conventional aircraft with cryogenic aircraft. This implies that at the end of the final period, a "hybrid" aircraft fleet consisting of both aircraft categories will operate. The potential "what-if" scenarios of prospective development of APT demand and cryogenic aircraft over the specified periods of time are used as inputs for the methodology and are given in Table 4.7.

The growth rates of air traffic demand are assumed to be constant during each sub-period and to decrease when looking further into the future.[3] This reflects the increasing maturity of the air transport market combined with the weakening dependency of air transport demand and its main driving forces. The fuel efficiency of conventional aircraft is assumed to permanently improve over time, albeit at a decreasing rate. Aircraft utilization is assumed to generally increase over time at a decreasing rate, which implies the number of aircraft increasing at a decreasing rate. The rate of introduction of cryogenic aircraft is assumed to be constant in each year of the "transitional" period, thus providing the share of cryogenic aircraft in the total of RPKs of 22 and 50 % by the end of the year 2065.

Table 4.7 Scenarios of the future development of the APT demand and aircraft fleet (Janic 2008)

Input variable	Period		
	2006–2005	2026–2040	2040–2065
Basic annual traffic volume: V_O (trillion Equivalent RPKs)	6.26[a]	13.78	22.61
Average traffic growth rate: i_v (%)	5.4[a]	3.5	2.0
The number of aircraft at the beginning of the period	18230[a]	36420	48823
Average aircraft utilization at the beginning of the period (trillion RPK/year)	0.3615	0.3784	0.4632
Rate of improvement of aircraft utilization: (%/year)	1.50	1.25	1.00
Average unit fuel consumption of conventional aircraft: E_{CO1} (g/RPK)	27.7	19.66	16.28
Rate of improvement in E_{CO1}: i_f^- (%/yr)	1.70	1.25	1.00
Average unit fuel consumption of cryogenic aircraft: E_{CO2} (g/RPK)	N/A	N/A	5.86
Average share of the total traffic carried out by cryogenic aircraft: i_h (%/yr)	0.00	0.00	1.00/2.00

[a] Airbus 2006; Boeing 2007

[b] $E_{CO2} = 0.36\ E_{CO1}$; *N/A* Not Applicable

[3] The average growth rate of APT demand over the entire time horizon is about 3.2 %, which is similar to the growth rate of 3.1 % over the 1990–2050 period in one of the scenarios of the air traffic growth developed by IPCCs. This rate produces a total of about 16.5 trillion RPKs in 2050 and 26.02 trillion RTKs in 2065 (IPCC 1999).

The eventual improvements in the unit fuel consumption of cryogenic aircraft are not considered due to the lack of realistic data.

Results

The results from application of the methodology using the above-mentioned inputs in Table 4.7 are shown in Fig. 4.24a, b, c. It shows the development of APT demand and related emissions of GHG (CO_2, H_2O, and NO_x, respectively) over time in relative terms (Index).

Figure 4.24a shows that if only conventional aircraft continue to be used, the future emissions of CO_2 will continue to increase driven by increasing volumes of air traffic. However, the emissions of CO_2 will rise more slowly than the traffic, mainly due to permanent improvements in aircraft fuel efficiency on the one hand, and aircraft utilization on the other. For example, at the end of the period (2065), air traffic will have increased six fold and the related emissions of CO_2 by 3.5 times as compared to those in the base year (2006). This is lower than in the IPCC's Reference Scenario where the CO_2 emissions in 2050 are predicted to be about 3.9 times greater than in 2006 (IPCC 1999). Consequently, it becomes evident that independently of the rate of improvement of conventional aircraft, stabilization of the annual global emissions of CO_2 will not be possible under conditions of unconstrained growth of air traffic demand implying that achieving a "carbon neutral" system will not be possible. However, from the time of introducing cryogenic aircraft even at a modest proportion of only about 1 % per year, the emissions of CO_2 will start to gradually slow down, stagnate, and finally stabilize by 2065 at a level about 2.8 times higher than in 2006, despite continuous traffic growth. If the rate of introduction of cryogenic aircraft is about 2 % per year, the rate of CO_2 will immediately start to decrease and be about 1.8 times higher in 2065 as compared to the base year (2006). This indicates that cryogenic aircraft may enable the decoupling of growth of air traffic and related emissions of CO_2 and thus contribute to achieving a "carbon neutral" APT system.

Figure 4.24b shows that emissions of H_2O will continue to increase in line with air traffic demand independently of the aircraft technology. If only conventional aircraft are used, the level of H_2O in 2065 will be about 3.3 times greater than in the base year (2006). At the same time, air traffic demand will be about 7 times higher. This indicates that, as in the case of CO_2, improving the aircraft fuel efficiency and daily utilization will slow down the increase in H_2O emissions. Introducing a relatively low proportion (1 %) of cryogenic aircraft will slightly (negligibly) increase this level during the period of replacement (2040–2065). However, if the proportion of introduced cryogenic aircraft is 2 %, the level of H_2O in 2065 will be about 4.2 times higher than in the base year (2006). These figures confirm the present concern that cryogenic aircraft will not stabilize emissions of H_2O, but, to the contrary, contribute to their substantial rise and thus the increased risk of more intensive formation of contrails.

Figure 4.24c shows the prospective long-term emissions of NO_x. As can be seen, similarly as in the case of the other two green house gases, when conventional

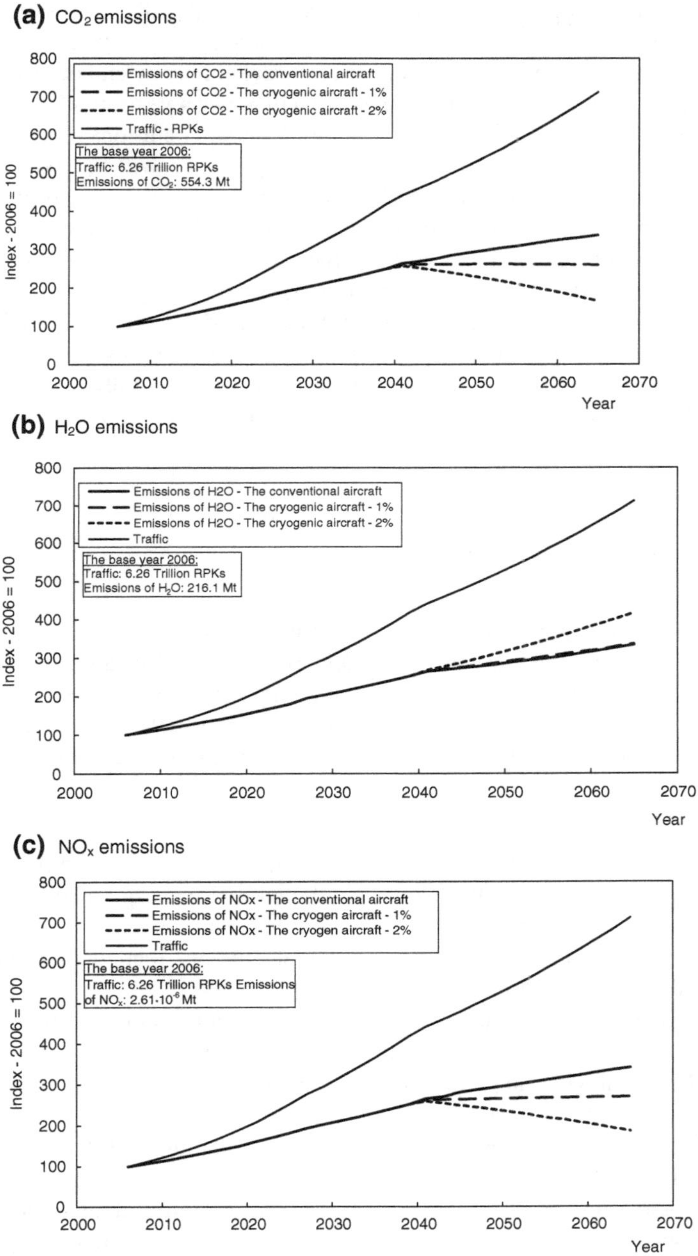

Fig. 4.24 Influence of cryogenic aircraft on the long-term global emissions of GHG (Janic 2008)

aircraft are exclusively used over the entire period (2006–2065), emissions of NO_x will continue to rise driven by the growth of air traffic demand, but again at a slower rate, mainly thanks to improvements in aircraft fuel efficiency and daily utilization. This again indicates that conventional aircraft will not be able to stabilize the level of NO_x and thus make the system "carbon-neutral" under conditions of unconstrained air traffic growth. For example, the level of NO_x in 2065 will be about 3.5 times greater than that in the base year (2006), driven by an increase in air traffic demand by about 7 times. If cryogenic aircraft really achieve NO_x emission rates of about 5–25 % of that of conventional aircraft, their gradual introduction will certainly stabilize and even decrease the total emissions of NO_x despite growing air traffic demand. For example, if the rate of introduction of cryogenic aircraft is 1 %, emissions of NO_x in 2065 will stabilize at a level about 2.8 times higher than that in the base year (2006). If the rate of introducing cryogenic aircraft is 2 %, the emissions of NO_x will decrease by 2065 to a level of about 2 times higher than in the base year (2006).

4.4.3 Evaluation

The LH_2 (Liquid Hydrogen)-fueled APT (Air Passenger Transport) system possesses both advantages (Strengths and Opportunities) and disadvantages (Weaknesses and Threats) as compared to its conventional crude oil-based jet fuel/kerosene counterpart. In general, the advantages and disadvantages relate to LH_2 as a fuel and the aircraft/engine technology, as well as general environmental effects/impact.

Advantages

- The production cost of LH_2 can be comparable to those of conventional fuels by using primary sources such as solar, wind, and hydro energy, and SRM (Steam Methane Reforming) biofuels and biomass fuels;
- LH_2 aircraft engines are slightly more fuel efficient than their conventional crude oil-based counterparts;
- The life and related maintenance costs of LH_2 aircraft engines can be extended thanks to LH_2 fuel burning at slightly lover turbine temperatures;
- Using LH_2 in a wide range of margins enables reduction of NO_x emissions due to its burning characteristics;
- Using LH_2 definitely has a potential for mitigating, stabilizing, and even decreasing the cumulative emissions of GHG (Green House Gases) except H_2O (water vapor) in the future medium- to long-term period of time, despite continuous air traffic growth; this could be achieved by the gradual replacement of conventional aircraft by cryogenic (LH_2-fuelled) aircraft over the long-term future;
- The lower take-off weight and smaller engines of cryogenic aircraft make them less noisy than their conventional counterparts; and

- Increased use of LH_2 ensures national independence of fuel supply since LH_2 can, in contrast to crude oil, be produced in any country.

Disadvantages

- Increased emissions of H_2O remain a matter of concern;
- The benefits from savings in the emissions of GHG by using LH_2 only arise if the primary sources for its production are wind, solar, and hydro energy;
- LH_2 aircraft engines are just slightly more efficient than their conventional crude oil-based counterparts;
- A relatively substantial commercialization of LH_2-fuelled aircraft of a given category (long range in this case) is needed in order to produce the desired environmental effects;
- These long-range LH_2-fuelled aircraft would produce less relative savings in energy consumption compared to smaller short- and medium-range aircraft;
- LH_2-fuelled aircraft, related manufacturing plants for both vehicles and fuel, and the fuel supply infrastructure at airports do not exist yet and still need to be built; and
- Switching from conventional jet fuel to LH_2 seems, at least at present, to be technologically, economically, and environmentally rather risky; the latter in particular because the share of emissions of GHG by commercial air transportation in the total man-made emissions of GHG is expected to range between 3 and 5 % over the forthcoming medium- to long-term period of time.

Finally, it can be said that LH_2 (Liquid Hydrogen) and the related adaptation of vehicles/aircraft and logistics of fuel manufacturing, supply, and distribution at airports are the main characteristics that will enable further advances of the already advanced commercial air transport system.

References

AECOM. (2012). *NC maritime strategy: Vessel size vs. cost*. Prepared for the North Carolina Department of Transportation, Architecture, Engineering, Consulting, Operations and Maintenance Los Angeles, California, USA.

AIRBUS. (2006). *Airbus global market forecast*. Toulouse: Airbus Industrie.

An, P., Sauer, A. (2004). *Comparison of passenger vehicle fuel economy and GHG emission standards around the world*. Arlington: World Resource Institute, New Center for Global Climate Change.

Archer, D. (2008). *The long thaw: How humans are changing the next 100,000 years of earth's climate*. Princeton: Princeton University.

Bodek, K., & Heywood, J. (2008). *Europe's evolving passenger vehicle fleet: Fuel use and GHG emissions scenarios through 2035*. Cambridge: Laboratory for Energy and Environment, Massachusetts Institute of Technology (MIT).

Boeing. (2007). *Current market outlook 2007: How will you travel through life?* Seattle: Boeing Commercial Airplanes.

Bossel, U., Eliasson, B. (2003). *Energy and the hydrogen economy*. Baden-Dattwil: Report ABB Switzerland Ltd. Corporate Research.

Brewer, G. D. (1991). *Hydrogen aircraft technology*. Boca Raton: CRC Press.
Cambell, C. J. (2002). *Peak oil: An outlook on crude oil depletion*. Retrieved from http://www.greatchange.org/ov-campbell,outlook.html
Chen, C., & Ren, Y. (2010). Exploring the relationship between vehicle safety and fuel efficiency in automotive design. *Transportation Research D, 15*(2), 112–116.
Chevron, (2006), *Alternative jet fuels, addendum 1 to aviation fuels technical reviews (FTR-3/A1)*, Chevron Corporation, USA.
CGI. (2007). *Container terminal parameters: A white paper*. Prepared for Marine Department of Transportation, The Cornell Group, Inc., Fairfax.
Chi, G., Stone, B. Jr. (2005) Sustainable transport planning: Estimating the ecological footprint of vehicle travel in future years. *ASCE Journal of Urban Planning and Development, September*, 170–180.
Churchill, J., Johnson, B. (2012). Saving billions on bunkers. *Maersk Post*. May 2012, pp. 9–12.
Coelho, M. C., & Luzia, M. B. (2010). Evaluating the energy performance of a SUV hybrid electric vehicle. *Transportation Research D, 15*(8), 443–450.
Corchero, G., Montanes, J. l. (2005). An approach to use hydrogen for commercial aircraft engines. *Journal of Aerospace Engineering, 219G*, 35–44.
Cullinane, K., & Khanna, M. (2000). Economies of scale in large containerships: Optimal size and geographical implications. *Journal of Transport Geography, 8*(3), 181–195.
Daggett, L. D., Hebdicks, L. C., Walhter, R., Corporan, E. (2006). *Alternate fuels and their potential impact on aviation, NASA/TM-2006-214365*. National Aeronautics and Space Administration, Glenn Research Centre, Ohio, USA.
DeLuchi, M. A. (1989). Hydrogen vehicles: An evaluation of fuel storage, performance, safety, Environmental impact, and costs. *International Journal of Hydrogen Energy, 14*(2), 81–130.
DNV. (2012). What is the optimum speed for a 20kTEU container ship? *Container Ship Update, 1*, 15–16.
DSC. (2010). *Container freight rate insight: Bi-monthly pricing benchmarks on the CONTAINER market*. London: Drewry Shipping Consultants Ltd.
Eberhard, M., & Tarpening, M. (2006). *The 21st century electric car*. Paolo Alto: Tesla Motors Inc., Tesla North America.
EC. (2003). *Liquid hydrogen fuelled aircraft- system analysis (CRYOPLANE)*. European Commission, 5th R&D Framework Program (Growth 1998–2002), Brussels, Belgium.
EC. (2010a). *Energy, transport, and environment indicators*, EUROSTAT, Statistical books, European Commission, Luxembourg.
EC. (2010b). *EU transport GHG: Routes to 2050? Review of potential radical Future transport technologies and concepts*. Report VI, The European Commission's Directorate General Environment, European Commission, Brussels, Belgium.
EEA. (2008). *EU27 electricity production by fuel*. Copenhagen: European Environmental Agency.
EEA. (2010). *European union emission inventory report 1990–2008 under the INECE convention on long range trans boundary air pollution (LRTAP)*. Copenhagen: European Environmental Agency.
EEA. (2012). *The contribution of transport to AIR quality Term 2012: Transport indicators tracking progress towards environmental targets in Europe*. Copenhagen: European Environment Agency.
EEC. (2005). *GAES-Future Engine Technology Environmental Impact*. Report No. EEC/SEE/2005/002, EUROCONTROL Experimental Centre, Bretigny Sur Orge, France.
Ewing, G. O., Sarigöllü, E., & Gordon, O. (1998). Car fuel-type choice under travel demand management and economic incentives. *Transportation Research D, 3*(6), 429–444.
Flikkema M., Nieuwenhuis, J. J., Duursema, W. (2012). *MIP Project "EEDI, Ontwikkeling van alternatieve CO2 index voor kleine handelsvaart schepen en short sea schepen: Current State of Regulations and Literature Survey*. Report 1, MARIN Conoship, Groningen, The Netherlands.
Funk, K., & Rabl, A. (1999). Electric versus conventional vehicles: Social costs and benefits in France. *Transportation Research D, 4*, 397–411.

Georgakellos, D. A. (2008). A polygon-based environmental appraisal of new vehicle technologies combined with renewable energy sources. *Transportation Research D, 13*(4), 283–288.

GL. (2012). *In focus: Containerships: Boosting performance with efficient solutions*. Germanischer Lloyd, Hamburg.

Greene, D. L., & Hopson, J. L. (2003). *Running out of and into oil: Analyzing global depletion and transition through 2050 ORNL/TM-2003/259*. Oak Ridge: Oak Ridge National Laboratory.

Guynn, M. D., Olson, E. D. (2002). *Evaluation of an aircraft concept with over-wing hydrogen-fuelled engines for reduced noise and emissions*. Technical Memorandum, NASA/TM = 2002-211926, National Aeronautics and Space Administration, Langley Research Centre, Hampton, Virginia, USA.

Haller, M., Welch, E., Lin, J., & Fulla, S. (2008). Economic costs and environmental impacts of alternative fuel vehicle fleets in local government: An interim assessment of a voluntary ten-year fleet conversion plan. *Transportation Research D, 12*(3), 219–230.

Hamilton, W. (1980). Energy use of electric vehicles. *Transportation Research A, 14A*, 415–421.

Heffner, R. R., Kurani, K. S., & Turrentine, T. S. (2007). Symbolism in California's early market for hybrid electric vehicles. *Transportation Research D, 2*(6), 396–413.

Higgins, C. J. H., Matthews, S., Hendrickson, C. T., & Small, M. J. (2007). Lead demand of future vehicle technologies. *Transportation Research D, 12*(2), 103–114.

Hörmandinger, G., & Lucas, N. J. D. (1996). Is clean enough? The influence of environmental externalities on markets for fuel cells in transport. *Transportation Research D, 1*(1), 63–78.

ICAO. (2008). *ICAO environmental report 2007*. Montreal, Canada: International Civil Aviation Organization.

ICG. (2010). *Electric car markets in Europe: 10 countries analysis and strategic review*. Edinburgh: International Consultancy Group Ltd.

IEA. (2006). *Hydrogen production and storage: Research & development priorities and gaps*. Paris: International Energy Agency.

IEA. (2009). *Hybrid & Electric Vehicles: Implementing Agreement*. IA-HEV Outlook 2009, International Energy Agency, Paris, France

IMO. (2007). *FSA (Formal Safety Assessment)—Container vessels: Details on the formal safety assessment, MSC83/INF 8*. London: International Maritime Organization.

IMO. (2011). *Main Events in IMO's Work on Limitation and Reduction of Green house Gas Emissions from International Shipping*. International Maritime Organization, London, UK.

IPCC. (1999). *Aviation and the global atmosphere, intergovernmental panel of climate change*. Cambridge: Cambridge University Press.

IPCC. (2001). *Climate change 2001: Synthesis report*. Contribution of Working Groups I, II, and II and III to the Third Assessment Report of IPCC, Intergovernmental Panel of Climate Change, Cambridge University Press, Cambridge, UK

IPTS. (2003). *Dynamics of introducing of new passenger car technologies: The IPTS transport technologies model*. Seville: Institute for Progressive Technological Studies, p. 76.

IPTS. (2008). *Environmental Improvement of Passenger Cars (IMPRO-car)*. Seville: Institute for Progressive Technological Studies.

Janic, M. (2008). The potential of liquid hydrogen for the future 'Carbon Neutral' air transport system. *Transportation Research D, 13*(7), 428–435.

Janic, M. (2010). Is liquid hydrogen a solution for mitigating air pollution by airports? *International Journal of Hydrogen Energy, 35*(5), 2190–2202.

Johansson, B. (1999). The economy of alternative fuels when including the cost of air pollution. *Transportation Research D, 4*(2), 91–108.

Johansson, B., & Åhman, M. (2002). A comparison of technologies for carbon-neutral passenger transport. *Transportation Research D, 7*(3), 175–196.

Kang, J. E., Recker, W. W. (2009). An activity-based assessment of the potential impacts of plug-in hybrid electric vehicles on energy and emissions using 1-day travel data. *Transportation Research D, 14*(8), 541–556.

Kempton, W., & Letendre, S. E. (1997). Electric vehicles as a new power source for electric utilities. *Transportation Research D, 2*(3), 157–175.

Koyanagi, F., & Uriu, J. (1997). Modeling power consumption of electric vehicles and its impacts on power demand. *Electrical Engineering in Japan, 120*(4), 41–46.

Kurani, K. S., Turrentine, T. S., & Sperling, D. (1996). Testing electric vehicle demand in 'HYBRID Households' using a reflexive survey. *Transportation Research Part D, 1*(2), 131–150.

Lave, L. B., & MacLean, H. L. (2002). An environmental-economic evaluation of hybrid electric vehicles: Toyota's Prius vs. its conventional internal combustion engine corolla. *Transportation Research D, 7*(2), 155–162.

Learmount, D. (2007). *New-technology aircraft to reduce average fuel consumption.* Retrieved from http://www.flightglobal.com

Lee, J. J., Lukachko, S., Waitz, I. A. (2004). Aircraft and energy use. *Encyclopedia of Energy* (Vol. 1, pp. 1–11). Philadelphia: Elsevier Science Publisher.

LR. (2011). *Assessment of IMO mandated energy efficiency measures for international shipping: Estimated CO_2 emissions reduction from introduction of mandatory Technical and operational energy efficiency measures for ships.* MEPC 63/INF.2 Annex, Lloyd's Register, London, UK.

Mabit, L. S., & Fosgerau, M. (2011). Demand for alternative-fuel vehicles when registration taxes are high. *Transportation Research D, 16*, 225–231.

Diesel, M. A. N. (2011). *Propulsion trends in container vessels.* Copenhagen: MAN Diesel-Powering the World.

Marquart, S., Ponater, M., Strom, L., & Gierens, K. (2005). An upgraded estimate of the relative forcing of cyroplane contrails. *Meteorologische Zeitschrift, Gebruder Bontraeger, 14*, 573–582.

MEPC. (2012). **2012** *Guidelines on the Method of Calculation of the Attained Energy Efficiency Design Index (EEDI) for New Ships* (Vol. 212, no. 63). [Annex 8, Resolution MEPC]. London, UK: The Marine Environment Protection Committee.

Line, Maersk. (2011). *Sustainability progress report 2011-Route 2.* Copenhagen: Maersk Line.

Nakata, T. (2000). Analysis of the impact of hybrid vehicles on energy systems in Japan. *Transportation Research D, 5*(5), 373–383.

Nottebon, T., Rodrigue J. P. (2007). The next fifty years of containerization: container vessels, linear shipping, and port terminal. In *Proceedings of the 52nd Annual Meeting of Association of American Geographers*, San Francisco, California, USA.

Notteboom, T. & Carriou, P. (2009). Fuel surcharge practices of container shipping lines: is it about cost recovery or revenue making?. In *Proceedings of the 2009 International Association of Maritime Economists (IAME) Conference, June 2009*, Copenhagen, Denmark.

NYK Line/MTI. (2010). *NYK Super Eco Ship 2030 – Our Concept Ship in the Future.* Presentation, Nippon Yusen Kabushiki Kaisha, Tokyo, Japan.

Ogden, M. J. (1997). *Infrastructure for hydrogen cell vehicles: A Southern California case study.* '97 World car conference, Riverside, California, USA.

OI. (2011). *The vision scenario for the European union: 2011 update for the EU-27.* Oko-Institute, Institute for Applied Ecology, Berlin, Germany.

OSSL. (2009). *Green ship of the future: 8500 TEU container ship concept study.* Odense: Odense Steel Ship Yard Ltd.

Penner, J. E. (1999). *Aviation and the global atmosphere* (p. 257). Cambridge: Cambridge University Press.

Pfeiffer, A. D. (2004). *The end of oil age.* Raleigh: Lulu.com.

Rienstra, S. A., & Nijkamp, P. (1998). The role of electric cars in Amsterdam's transport system in the year 2015: A scenario approach. *Transportation Research D, 3*(1), 29–40.

Rudolf, C. D, I. I. I. (2007). Ship-to-shore productivity: Can it keep up with mega-ship size increases? Part 1. *Port Technology International, 34*(3), 1–5.

Sala, S. (2010). Energy efficiency and the shipping industry. DELTAMARIN, www.deltamarin.com

Schock, R. N., Berry, C. D., Smith, R., Rambach, G. D. (1995). *Hydrogen as a new transportation fuel*. Lawrence Livermore National Laboratory, University of Melbourne, Australia

Schwoon, M. (2007). A tool to optimize the initial distribution of hydrogen filling stations. *Transportation Research D, 12*(2), 70–82.

Spiegel, R. J. (2004). Platinum and fuel cells. *Transportation Research D, 9*(5), 357–371.

Svensson, E., Hasselrot, A., & Moldanova, J. (2004). Reduced environmental impact by lowered cruise altitude for liquid hydrogen-fuelled aircraft. *Aerospace Science and Technology, 8*, 307–320.

Tozer, D. R. (2001). Ultra-Large Container Ships (ULCS). In *Proceedings of Lloyd's Register Technical Association*, London, UK.

UNCDAT. (2012). *Review of maritime transport 2012, Report of the UNCTAD Secretariat*. United Nations conference on trade and development, United Nations, New York, USA.

UniCredit. (2009). *Study: Green shipping*. HypoVereinsbank, Global shipping division, UniCredit Corporate Banking, Hamburg, Germany.

Wang, Q., DeLucchi, M. A. (1991). *Impact of electric vehicles on primary energy consumption and petroleum displacement*. Working Paper UCTC No. 6, University of California Transport Center, University of California Berkeley, Berkeley, USA.

Wang, G., Ogden, J. M., & Sperling, D. (2008). Comparing air quality impacts of hydrogen and gasoline. *Transportation Research D, 13*(7), 436–448.

Chapter 5
Advanced Transport Systems: Infrastructure, Technologies, Operations, Economics, Environment, and Society/ Policy

5.1 Introduction

This chapter describes selected HS (High Speed) passenger transport systems and their multicriteria ranking based on their infrastructural, technical/technological, operational, economic, environmental, and social/policy performances. These include the constantly growing APT (Air Passenger Transport) and HSR (High Speed Rail), as well as forthcoming TRM (TransRapid MAGLEV) systems. Such a multicriteria approach has proven convenient regarding the general objectives expected to be fulfilled by the above-mentioned HS (High Speed) systems as well as the often conflicting preferences of the increasing number of actors/stakeholders involved in the DM (Decision Making) process (Giuliano 1985; Haimes 1985; Speling 1984).

The main objectives of HS passenger transport systems are as follows:

- Competing against and complementing each other at the same time, particularly in high volume and density demand transport markets (corridors); the consequent effects include redistributing existing and generating new transport demand, improving the efficiency of particular transport modes based on market-driven forces, and revitalizing local, regional, and national transport networks by contributing to their stronger integration in global national and international transport networks. Examples in Europe include TENs (Trans-European Transport Network(s)) (EC 1998a, b);
- Influencing the micro and macro spatial, socioeconomic, and political development of particular regions in which they operate; in general, influence on the structure and "spatial redistribution" of the current and new socioeconomic activities is mainly expected;
- Covering much wider areas in a much shorter time than their conventional counterparts, thus ensuring the more efficient and effective physical mobility of persons and goods and consequently, in addition to telecommunications and information technology, further enhancing globalization, internalization, and integration of the regional and national economies and societies; and

M. Janić, *Advanced Transport Systems*,
DOI: 10.1007/978-1-4471-6287-2_5,

- Playing an important role as promising options for the further sustainable development of the transport sector in terms of supporting further demand growth driven by economic growth on the one hand, while mitigating the negative impacts on the environment and society such as energy consumption and related emissions of GHG (Green house Gases), land use/take, noise, congestion, and traffic incidents/accidents (i.e., safety), on the other.

5.2 High Speed Transport Systems

5.2.1 Definition, Development, and Use

The HS (High Speed) passenger transport systems such as APT (Air Passenger Transport), HSR (High Speed Rail), and TRM (Trans Rapid Maglev) possess the common performance—"high speed"—while operating in the given environment under given conditions. In addition, each of them possesses specific performances in the given context. For ATP these are aircraft capabilities, regulation of operations, airline strategy, and mitigating impacts on the society and environment. For an HSR and TRM these are the infrastructure network, rolling stock, speed, and traffic. They are a priory considered to mitigate impacts on the environment and society by design, construction, and operations.

5.2.1.1 Analysis of Performances

APT (Air Passenger Transport)

1956	The flag carrier of the USSR Aeroflot is the first in the world to sustain regular jet services with the Tupolev Tu-104 aircraft (USSR)
1978	Congress passes the Airline Deregulation Act, thus opening conditions for the development of a competitive (fully deregulated national) air transport market (USA)
1990s	The EU (European Union) air transport market is liberalized, significantly impacting the structure of air transport industry, shifting toward growing LCCs (Low Cost Carrier(s)), partially on the account of traditional national airlines (Europe)

The APT (Air Passenger Transport) system emerged in 1950s and as such, at that time, represented the most advanced transport mode, i.e., the earliest HS system. From its very beginnings it has been constantly modernized and improved in terms of the above-mentioned aircraft capabilities, regulation of operations, airline strategy, and mitigating impacts on the environment and society (particularly over the past decade and a half) (Boeing 1998; Janic 2007).

Aircraft capabilities

Development of "aircraft capabilities" includes speed, take-off weight, payload, and range characteristics. Aircraft capabilities have also influenced the other three

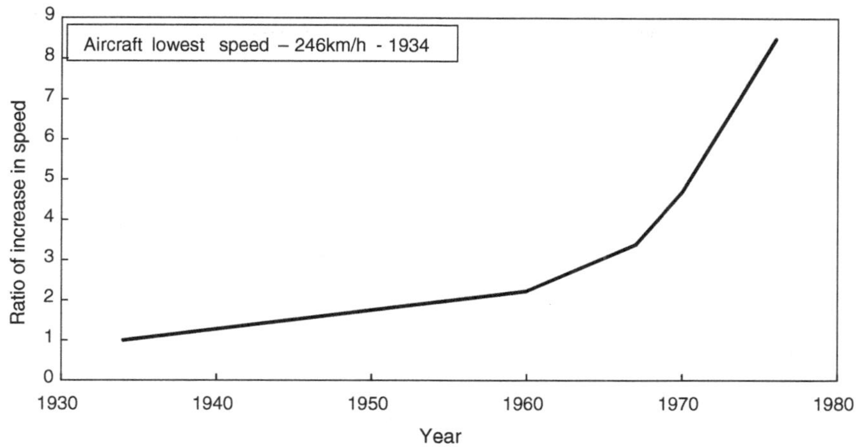

Fig. 5.1 Evolution of aircraft speed over time (Boeing 1998; Horonjeff and McKelvey 1994)

above-mentioned performances, and vice versa. Both speed and payload have contributed to an enormous increase in aircraft technical productivity—by more than 100-fold over the past 40 years (Horonjeff and McKelvey 1994). Specifically, this increase in speed is illustrated by relating the speeds of various commercial aircraft to the speed of slowest (Douglas DC3 (238 km/h)) as shown in Fig. 5.1.

As can be seen, the relatively fast (exponential) technological progress over time has temporarily come to an end with the launch of the Concorde supersonic aircraft in 1974.

In addition, the payload has increased contributing to increasing aircraft technical productivity over longer ranges. Figure 5.2 shows the most recent payload developments in terms of the number of seats per aircraft and range for Boeing and Airbus long-haul aircraft.

As can be seen, with the exception of Airbus A380 and Boeing B747-8 aircraft, the number of seats varies between 200 and 450, and the range between 6,000 and 18,000 km (Some additional figures on the payload-range relationship are also presented in Chaps. 2 and 6).

Regulation of operations

"Regulation of operations" relates to the airline's freedom to enter markets, set airfares, supply transport capacity in terms of the number of seats and flight frequencies, and with regard to ownership structure. For many years, national governments regulated these three elements through national, international bilateral, and multilateral agreements. Under such conditions, the airline industry was regulated through very constrained airline rights dealing with where, when, and how to operate, mainly applying the principles of reciprocity. However, at one moment, regulation became a barrier to the further development of the airline industry and particularly its expected role in supporting globalization of the

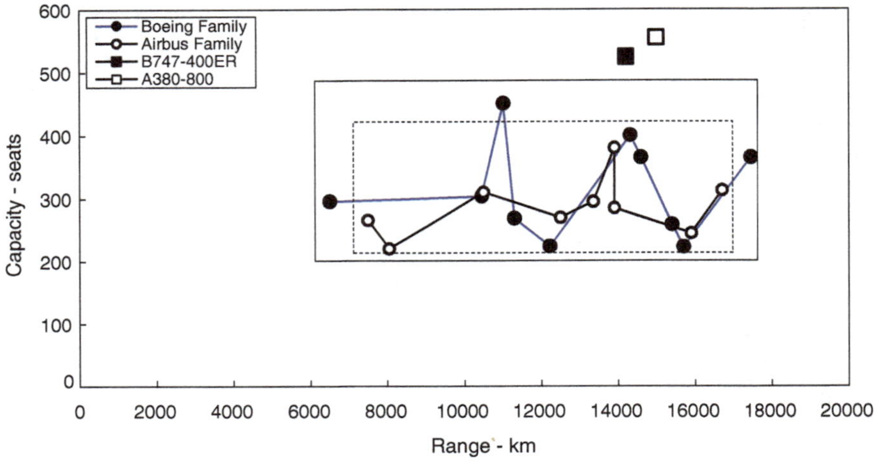

Fig. 5.2 Relationship between the seat capacity and range for the selected series of aircraft (http://www.boeing.com; http://www.airbus.com/aircraftfamilie)

economy and society. Consequently, deregulation of the U.S. airline industry in 1978 and liberalization of the intra-EU air transport market, the later as a gradual process during the period 1987–1992 in combination with deregulation of national markets removed barriers for airlines to freely enter the market(s), set up airfares, and supply transport capacity–seats and flight frequencies. In addition, different "open skies" agreements were concluded in international markets to mitigate international institutional barriers. At the same time, many airlines were privatized, resulting in the removal of any "governmental protection" particularly in subsidizing nonprofitable services.

The above-mentioned developments have had a dramatic impact on the airline industry, airports, and users of air transport services—passengers and goods shippers. For example, in the U.S., many airlines went into bankruptcy while others changed the structure of their operational and business models from the previously and exclusively used "point-to-point" to the "hub-and-spoke" network(s). As a result, airports included in the given airline network started operating as either hub or spoke airports. In general, many airports previously used as airline bases have been converted into hubs. In addition, airports have become the hub of several airlines. Furthermore, the same airline can operate one or several hub airports (BCG 2004). Many previously connected airports have lost regular connections while some others have gained connections after being included in airline hub-and-spoke networks.

In Europe, major national (flag) airlines have mainly remained at their base airports by developing "star-shaped" hub-and-spoke instead of the previously used "star-shaped" point-to-point networks. In addition, many national flag airlines have concluded alliances with their partners from both Europe and other continents. This consolidation process has resulted in three large alliances emerging:

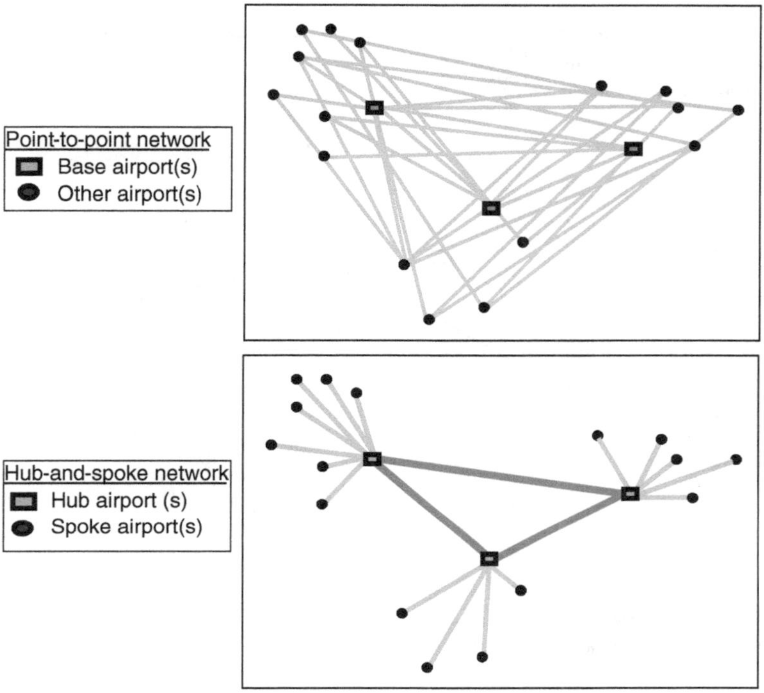

Fig. 5.3 Schemes of airline air route networks

STAR led by Lufthansa (Germany), Oneworld led by British Airways (UK) and SkyTeam led by Air France-KLM (France-The Netherlands) (EC 2003). These alliances have consolidated the hub-and-spoke networks of their leading partners around their hub airports—Frankfurt Main (Lufthansa), London Heathrow (British Airways), and Paris Charles de Gaulle-Amsterdam Schiphol (Air France-KLM), respectively. Consequently, these networks have developed into global mega networks and their hubs into the global mega-hubs.

In addition, in both the U.S. and Europe air transport market, the LCCs (Low Cost Carriers(s)) have emerged as the strong competitors.

Airline strategy

As mentioned above, after deregulation and liberalization of the airline industry, full cost airlines previously exclusively operating 'point-to-point' networks from their base airports have changed their business/operational model into star-shaped "hub-and-spoke(s)" networks. A simplified scheme is shown in Fig. 5.3.

The main characteristics of the newly emerged "hub-and-spoke(s)" networks include increased traffic concentration on routes connecting the hub to spoke airport(s) and consequently increasing proportions of connecting/transfer passengers. Table 5.1 shows an example of such early developments at U.S. major

Table 5.1 Connecting passengers at U.S. major airlines (http://www.transtats.bts.gov/Tables.asp?DB_ID=125)

Year	Proportion (%)
1993	60
1995	59
1998	59
2000	58
2003	62

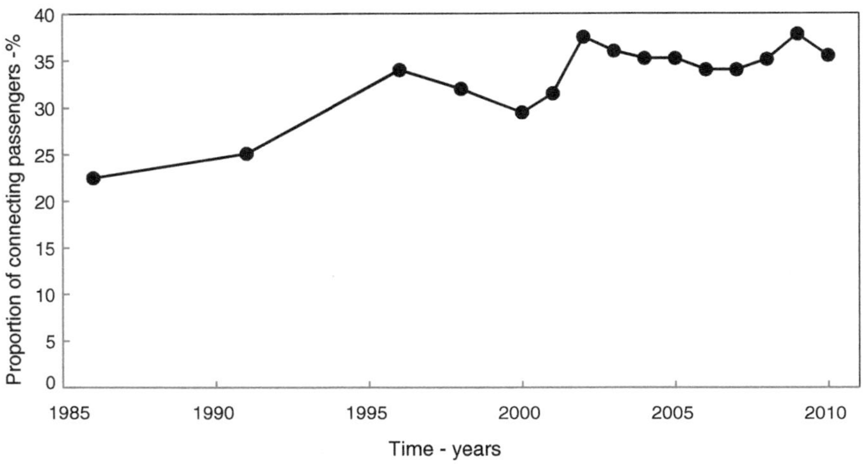

Fig. 5.4 Trends in the development of connecting passenger traffic at London Heathrow airport (UK) (CAA 2011)

airlines: America West, American, Continental, Delta, Northwest, United, and U.S. Airways Airlines.

In addition, Fig. 5.4 shows an example of the development of the share of connecting passenger traffic at London Heathrow airport (UK), which has significantly increased over the past two decades.

This has occurred mainly due to the stronger concentration of Transatlantic flights at Heathrow airport after liberalization of the EU–US air transport market, the abandonment of BA's (British Airways) twin-hub strategy, and the growth of LCCs (Low Cost Carriers) at London Gatwick airport (CAA 2011).

Consequently, the 'point-to-point' network(s) shown in Fig. 5.3 have remained to be mainly operated by LCCs (Low Cost Carrier(s)) as new market entrants. For example, large European LCCs have operated several base airports and associated subnetworks. One of these, Ryanair operates 17 base airports in different European countries, each with 4–8 stationed aircraft. At this and other European LCCs, the nonintended connectivity of flights at the base airport(s) implies negligible if any volumes of transfer passengers. However, some U.S. LCCs have substantial proportions of connecting (transfer) passengers one among the largest ones, Southwest Airlines, has about 25 % of such passengers.

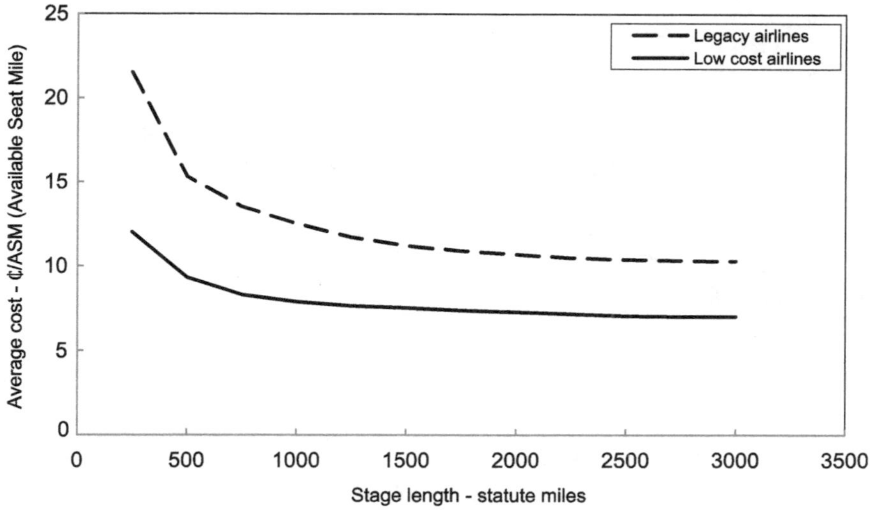

Fig. 5.5 Relationship between the average unit cost and the stage length—the U.S. domestic market (2003) (GAO 2004)

In addition to the route network configuration and structure, their operating costs are the main difference between the two airline categories—business models/ concepts. Figure 5.5 shows an example for the U.S. airline industry.

As can be seen, in the given case, the average unit cost of the conventional (legacy) airlines is higher than that of the LCCs by about 30–40 %. In addition, at both types of airlines, the unit cost decreases more than proportionally as the stage length increases, thus indicating the existence of economies of distance.

Mitigating Impacts on the environment and society

The airline industry and aircraft manufacturers have continuously endeavored to mitigate the impacts of their operations on the environment and society, mainly in terms of fuel consumption and related emissions of GHG, and noise (Janic 2007). At the same time, reducing fuel consumption has contributed to decreasing operating cost. Figure 5.6 shows an example of these efforts by the U.S. airline industry to improve the energy intensity expressed by ASK (Available Seat Kilometers)/kg of fuel over the past 50 years.

The energy intensity of both domestic and international operations has increased in line with increasing of the ASK volumes at a decreasing rate. The EI (Energy Intensity) and its rate of international operations have been overall greater than that of domestic operations by about 20 %. At the same time, volumes of domestic operations are about two times greater than volumes of international operations. Due to their nature, the related intensities of GHG emissions have followed a similar pattern, thus indicating gradual but permanent improvements (Janic 2007) (see also Chap. 2). In addition, improvements through investments in new technologies (aircraft engines and airframes) have mitigated aircraft noise at

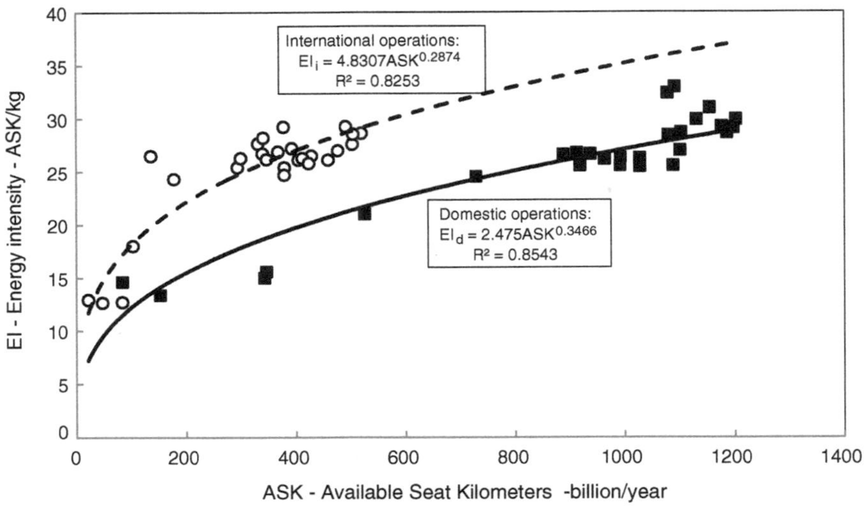

Fig. 5.6 Relationships between EI (Energy Intensity) and available capacity of U.S. certified airlines (1960–2010) (USDT 2012)

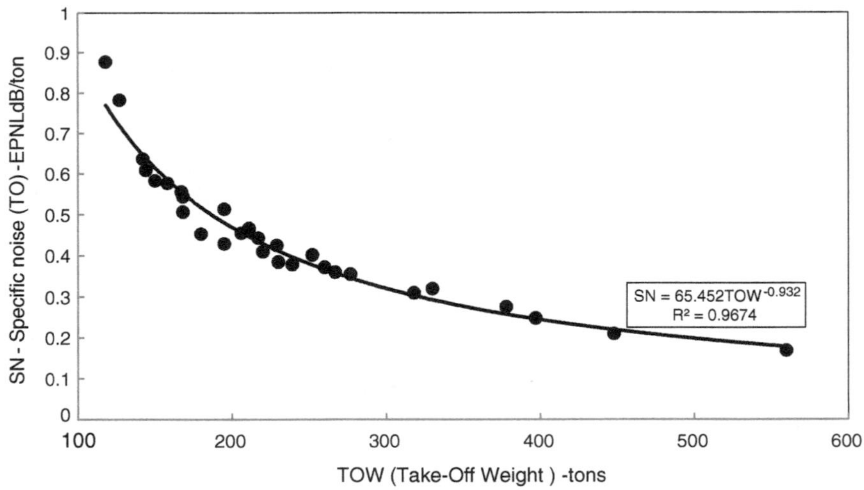

Fig. 5.7 Relationship between the certified T/O (Take-Off) SNL (Specific Noise Level) and TOW (Take-Off-Weight)—heavy aircraft (FAA 1997; EASA 2011)

source. An example of the relationship between SN (Specific Noise) expressed by EPNLdB (Effective Perceived Noise Level in Decibels) per unit of TOW (Take-Off Weight) (ton) and the aircraft TOW (Take-Off-Weight) (tons) is shown in Fig. 5.7. The EPNLdB values represent certified noise levels for Heavy aircraft (TOW > 100 tons) during the T/O (Take-Off) phase of flight.

As can be seen, SN decreases more than proportionally as the aircraft size increases, indicating the substantial achievements in reducing noise while at the same time increasing the aircraft size, i.e., TOW (Take-Off-Weight).

HSR (High Speed Rail)

1933	"Fliegender Hamburger" diesel-powered trains with Jakobs bogies begin regular services between Hamburg and Berlin with a maximum speed of 160 km/h operated by Deutsche Reichsbahn-Gesellschaft, some 50 years before the advent of the TGV train (Train à Grande Vitesse) (Germany)
1964	The first modern Tōkaidō Shinkansen HS (High Speed) trains begin operations between Tokyo and Osaka (Japan)
1981	TGV (Train à Grande Vitesse) HS trains begin operations on the first section of the new Paris-Lyon high-speed rail line (a distance of 418 km in 2.5 h) (France)
1998	HS Direttissima trains begin operations (Italy)
1991	ICE (Intercity-Express) HS trains begin operations on the new Hannover-Würzburg high-speed rail line (Germany)
1992	Class 100 high-speed trains derived from the French TGV begin services on the Madrid–Seville high-speed rail line (Spain)
1994	Eurostar high-speed trains begin operations through the Channel Tunnel between France and the UK (United Kingdom)
2000	Acela Express trains begin operations as the first HS train in North America
2004	Construction of the HSR network begins in China

HSR (High Speed Rail) is characterized by the above-mentioned infrastructure network, the rolling stock, speed, and traffic.

The infrastructure network

One of the main characteristics of an HSR is the network length. In 2011, the length of the global HSR network in operation was 15,231 km, of which Europe shared 43.5 %, Asia 54.0 %, and other countries 2.5 %. In addition, 9,172 km of an HSR network are under construction around the world and nearly 17,594 km are planned. Consequently, by the end of 2025, the length of the world's HSR network is expected to reach 41,997 km, of which Europe will still have a substantial share (UIC 2011).

As far as the length of the network of particular countries is concerned, Japan has the longest HSR network in operation, followed by France, Spain, and Germany. This ranking has been made by taking into account also the HSR lines under construction in 2011. The total length of China's HSR network of 4,576 km includes the Beijing-Tianjin, Wuhan-Guangzhou, Zhengzhou-Xi'an, and Shanghai-Nanjing lines (with the highest operating speed in the world). According to the planning and current construction status, the total operating mileage of an HSR in China will almost triple by 2020 when it will exceed 13,000 km (UIC 2011).

HSR development in Europe is mainly influenced by institutional attainment, namely the decisions of particular EU (European Union) Member States to build the Trans-European HSR network. The planned length of this network is about 29,000–12,500 km of which are new lines. Total costs were estimated at about

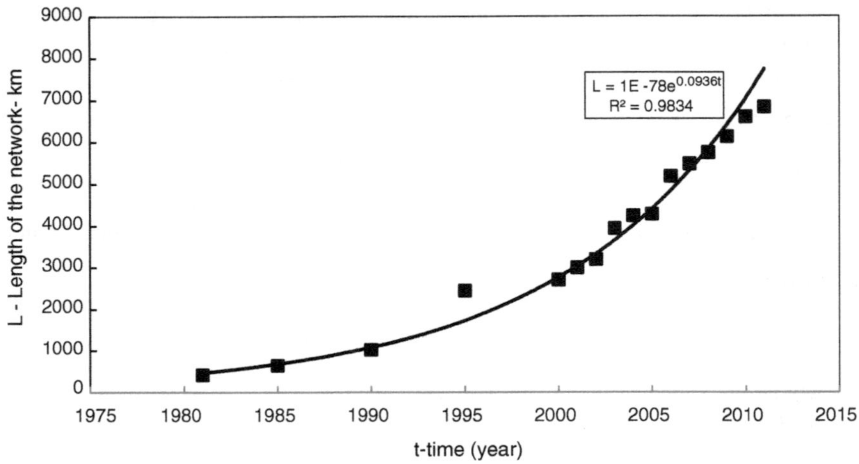

Fig. 5.8 Development of the European HSR network over time (EC 2010, 2012; UIC 2011)

€240Bn (billion), €207Bn of which would be allocated to the rail infrastructure and €33Bn to the rolling stocks. Figure 5.8 shows an example of dynamism of development of the HSR network in Europe.

As can be seen, the network length has increased more than proportionally over time, which appears to be in line with the plans to complete the entire network by 2020–2025 (CEC 1993a; 1995; UIC 2011).

Rolling stock

The HSR rolling stock is characterized by infrastructural, technical/technological, operational, economic, environmental, and social performances given for existing TGV (Train à Grande Vitesse) and forthcoming AGV (Automotice Grande Vitesse) passenger trains in Table 5.2.

The forthcoming AGV train appears superior as compared to its TGV counterpart, mainly with respect to its technical/technological, operational, economic, and environmental performances. It is supposed to operate on the same HSR infrastructure as TGV trains. The principal technical/technological difference is the positioning of bogies between the cars of the train and distribution of the traction system below the floor of cars and not only at the front and the back as in TGV units. This is expected to have an advantage due to maintaining the power of the train independent between units/cars and providing more space for passengers. AGV trains are fully interoperable, which enables them to operate throughout different European countries with quite different power supply systems. Aluminum and steel are mainly used as construction materials, resulting in a lighter and thus relatively more energy efficient train. Thanks to the greater seating capacity and operating/cruising speed, the technical productivity of an AGV train is expected to be higher than that of a TGV train. Since AGV trains are expected to operate on the same

Table 5.2 Performances of selected HS (High Speed) trains (BBVA 2009; http://en.wikipedia.org/wiki/TGV; Alstom 2008; www.alstom.com)

Performance	Type	
	TGV-All	AGV11-14
Infrastructural		
Width of tracks (mm)	1,435	1,435
Maximum radius of curves (km)	≥4	≥7
Maximum gradient (%)	3.5–4.0	3.5–4.0
Technical/technological		
Power supply (kV)	25 kV7 50 Hz AC, 15 kV/16.7 Hz; 5 kV DC	25 kV/50 Hz–15 kV/ 16.7 Hz-3 kV DC-1
Signaling system	TVM/ETCS[a]	TVM/ETCS[a]
Power of locomotive (MW)	6.45–12.24[b]	9.0
Number of bogies	13	11–12
Length of train set (m)	200–394	220–252
Number of cars (cars/train set)	8–10	11–14
Gross weight (tons)	415–816	410–522
Empty weight (tones)	383–665	370–425
Power/empty weight (ratio) (W/kg)	16.84–18.41	24.32
Operational		
Operating/cruising speed (km/h)	270–320	320–350
Minimum scheduled interval (min)	3	3
Transit capacity (seats)	345–750	458–593
Technical productivity (000 seat-km/h)	93.15–240.0	146.56–207.550
Economic		
Investment in infrastructure (mio€/km)	18	18
Direct operating costs (€/seat-km)	0.0935–0.1171	–
Break-even load factor (%)	60–65	60–65
Environmental		
Energy consumption (kWh/train-km)	19.01	16.15
Emissions of GHG (gCO_{2e}/train-km)	401.1	341
Land use/take (ha/km)	3.2	3.2
Social		
Noise (dB(A))/Speed (km/h)	92/300	92/360
Congestion—delays (min)	–	–
Safety (deaths + injuries/billion p-km)	0.00114	–
Policy		
Contribution to sustainability	Yes	Yes

[a] *TVM*–Transmission Voie-Machine (Transmission Track-Machine)—a block of the approximate length of 1.5 km; ETCS (European Train Control System) Level 2; [b] under 25 kV

infrastructure, the corresponding investment costs will be almost equivalent. However, its direct operating costs are supposed to be lower mainly due to the lower energy consumption and related costs, and lower maintenance costs (by about 15 %) at comparable break-even load factors (during breaking, an AGV train returns about

8 MV of energy to the power supply network). Lower energy consumption also implies lower emissions of GHG(Green house Gases). Despite the AGV train being faster than its TGV counterpart, it is less noisy mainly thanks to its lower weight (i.e., higher power/weight ratio) and improved aerodynamics. Furthermore, like its TGV counterpart, it is expected to not cause congestion and delays for AGV and other trains on lines with mixed traffic. Last but not least, AGV trains are expected to be safe, i.e., free of incidents and accidents, again very much like their TGV counterparts. Consequently, it can be said that AGV trains will represent a step forward toward improving the sustainability of the passenger transport sector through its HSR component (Alstom 2008).

Speed

In terms of dynamism, the development of passenger train speeds has been much slower than that of the APT by stretching over more than a century and half. During that time, the speed had increased by more than 10-fold, namely from about 50 to more than 500 km/h as shown in Fig. 5.9.

At present, the maximum operating/cruising speed of TGV (Train à Grande Vitesse) is 320 km/h. The Japanese Shinkansen and German ICE (Inter-City-Express) trains currently operate at a maximum speed of about 300 km/h. However, the forthcoming more advanced European AGV (Automotice Grande Vitesse) and Japanese Fastech 360Z trains are expected to operate at an average speed of 350 and 360 km/h, respectively. AGV trains have also achieved maximum operating speeds of 574.8 km/h, close to half the speed of sound.

Traffic

Development of the HSR network in Europe has been carried out through several phases. It started with giving priority to national projects. Later on, with the introduction of the concept of TENs (Trans-European Transport Networks), international and cohesive projects came to the forefront. As mentioned above, the first HSR line started operations in 1981 when the French railways launched the HSR service on the Paris-Lyon line. One of the main reasons for beginning HSR operations in France was the oil crisis in 1973, which resulted in huge fuel deficits. At that time, electrically powered HSR was expected to reduce the dependence of the transport sector on crude oil and thus overall fuel consumption by replacing air transportation in particular markets. Such development of infrastructure has also driven traffic growth, resulting in HSR amounting for 15 % of the total passenger market share in Europe. Some relationships between development of the HSR infrastructure and related traffic volumes over the 1981–2010 period are shown in Fig. 5.10. As can be seen, traffic volumes have continuously increased, albeit at decreasing rate.

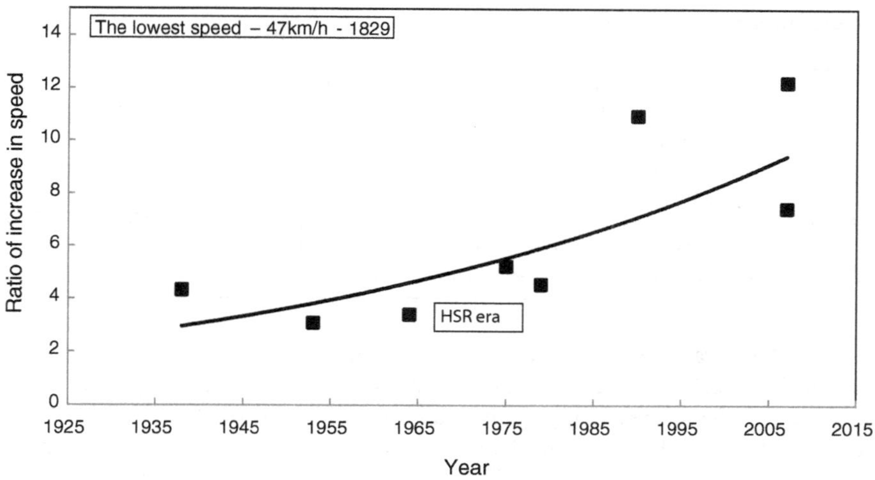

Fig. 5.9 Evolution of the speed of passenger trains over time (UIC 2011)

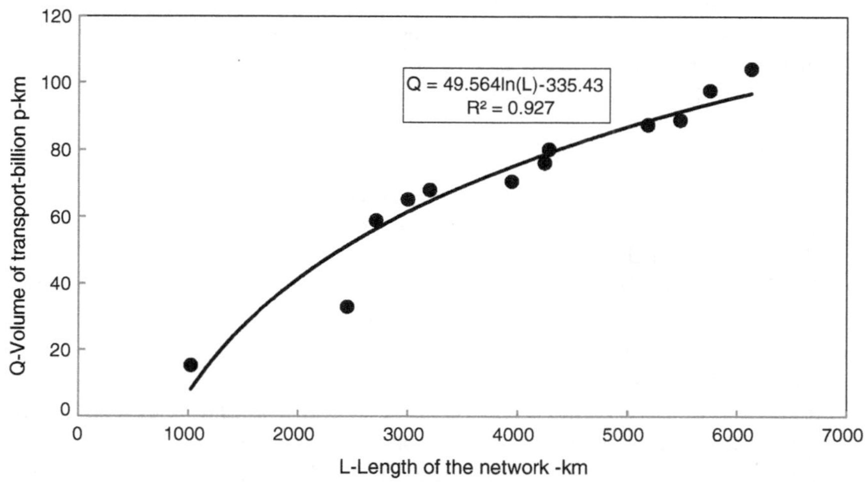

Fig. 5.10 Relationship between the volume of traffic and the length of the European HSR network (EC 2012)

TRM (TransRapid MAGLEV)

1970s	Research on maglev transportation intensifies (Japan, Germany)
1977	The first TRM (TransRapid Maglev) test line of a length of 7 km is built (the test speed achieved is 517 km/h) (Japan)
1993	The TRM test of 1,674 km is carried out (the achieved speed is 450 km/h) (Germany)
1990/1997	The Yamanashi TRM test line of a length of 42.8 km is constructed in 1990 and the first test is carried out in 1997 (EDS (Electro Dynamic Suspension)) (Japan)
2004	The first TRM line between Shanghai and its airport (China) is commercialized

The TRM (TransRapid MAGLEV—MAGnetic LEVitation) system is based on Herman Kemper's idea of magnetic levitation from the 1930s. Magnetic levitations enables the suspension, guidance, and propelling of MAGLEV vehicles by magnets rather than by mechanical wheels, axles, and bearings as is the case in HS wheel/rail vehicles. Two forces (lift and thrust/propulsion), both created by magnets, are needed for operating TRM vehicles. Although the TRM system has been matured to the level of commercialization, the relevant infrastructure has only been partially built, mainly connecting airport(s) with city centers, which is still far from network development comparable to that of HSR (Geerlings 1998; Kertzschmar 1995; Powell and Danby 2007).

Similarly as the HSR, the TRM (TransRapid MAGLEV) system can also be characterized in the given context as mentioned above by its infrastructure, rolling stock, speed, and traffic.

Infrastructure

The main TRM infrastructure is a guideway consisting of concrete (prefabricated) supporting piers and beams. The concrete piers are located at different distances depending on type of the guideway: 31 m—type I, 12 m—type II, and 6.19 m—type III). The beams lying on the solids can have different lengths: 62.92 m—type I, 24.78 m—type II, and 6.19 m—type III. The height of a beam can be: 2 m—type I, 1 m—type II, and 0.4 m—type III. The total height of the guideway can be: 2.2–20.0 m—type I and II, and 1.35–3.5 m—type III. The beam carries the vehicle and provides power to the entire system. Each type of beam has a trapezoidal-like profile with the body and the track on top. The beam can be made of steel, concrete, or hybrid (steel/concrete mixture). The three above-mentioned types of guideway can be single or double (or triple) track constructions. The profile area of the guideway is 1.445 m^2 (Shibo 2008).

Rolling stock

Principles of operations TRM trains differ from the wheel/rail HS trains in the way they operate along the tracks. A scheme of TRM system in Fig. 5.11 shows that the support and guidance systems follow the principles of magnetic levitation. The individually controlled DC (Direct-Current) support and guidance magnets are located at the undercarriage of the vehicle. The three phase armature windings or

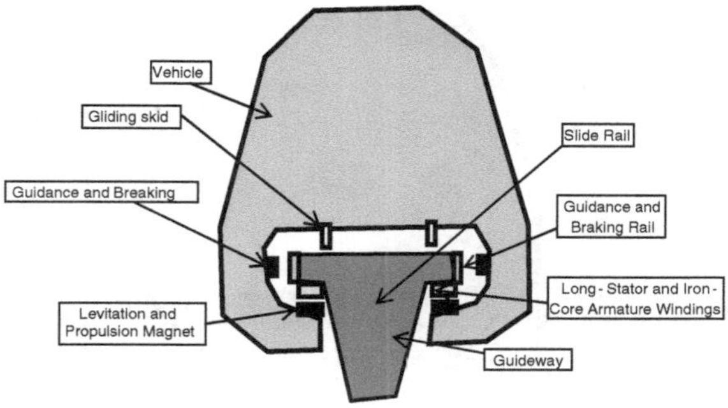

Fig. 5.11 Scheme of the main components of the TRM system (He et al. 1992)

stator windings of the linear electromotor getting power from the modulated inverter system are installed in the slots of iron core, which is mounted on the underside of the gateway. The controlled DC suspension magnets getting energy from the on board power-conditioning equipment also provides excitation fields for the linear motor. The air gap typically 1 cm is maintained as small as possible in order to increase efficiency and power factor of the system (He et al. 1992). Such design enables the TRM vehicles to levitate above the tracks supported by magnets and run on the principle of electromagnetism. As mentioned above, the magnets create two forces—lift and thrust or propulsion for operating the TRM vehicle. The lift force keeps the vehicle above the guideway at a distance of at most 1 cm but usually 10–15 mm during the trip. The propulsion force enables acceleration and deceleration as well as overcoming air resistance during the cruising phase of the trip. Thanks to levitation, there is no friction between the tracks and the vehicle's wheels, thus enabling TRM trains to operate at much higher speeds. Figure 5.11 shows a simplified cross-sectional profile containing the particular components enabling operation. Electric energy is consumed for generating both lift and thrust force, but in much greater proportion for the latter than for the former. In addition, a part of the energy spent during deceleration is returned to the network, thus also indicating possibilities for saving energy similarly as in the case of HSR. Figure 5.12 shows a scheme of the TRM cross-section profile.

Levitation and propulsion Two TRM technologies have been developed to maturity enabling their full commercial implementation: EMS (Electromagnetic Suspension) and EDS (Electro Dynamic Suspension). They enable three basic functions of TRM vehicles: (i) levitation above the track; (ii) propulsion enabling moving forward in terms of acceleration, cruise, and deceleration; and (iii) guidance implying maintaining stability along the guideway.

- *EMS technology* as a wheel-less system enables TRM trains to levitate above the steel beam thanks to magnets attached to the vehicle and oriented toward the beam. This is arranged as a series of C-shaped profiles with the upper part attached to the vehicle and the lower inside part containing the magnets, thus positioning the rail between the upper and inner part. Due to the high fluctuation

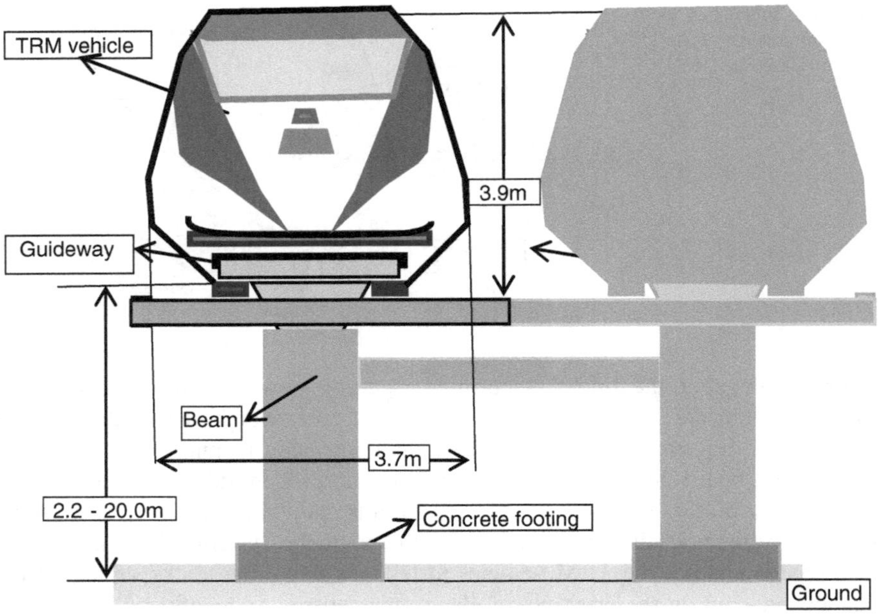

Fig. 5.12 Scheme of the TRM system cross-section profile

of the magnetic field with distance, a highly reliable and redundant electrical control system is installed to continuously maintain the vehicle at a constant distance from the track of 10–15 mm depending on the type. This technology operates at all speeds from minimum speeds of 10–15 km/h to maximum speeds of 400–500 km/h. The thrust of the TRM vehicle for acceleration, cruising, and deceleration is provided by the on board synchronous long-stator linear motor creating a propulsive force moving the vehicle forward.

- *EDS technology* as a wheel-based system uses the magnetic field generated by magnets installed both on the TRM vehicle and on the guideway. The magnetic force in the vehicle is generated by the magnetic field from either superconducting magnets or by a series of permanent magnets. The magnetic force in the track is generated by an induced wire- or conducting strips in the track-based magnetic field. The inherent advantage of this technology is that the levitation is stable and does not require a feedback system (as EMS technology does) to maintain a constant distance between the vehicle and the track during the journey. Since the magnetic field generated at low speeds is not sufficient to maintain the vehicle above the track, i.e., to levitate, trains must have some kind of wheels to support them until they reach higher speeds. Contrary to EMS, this technology does not include a linear motor on board the TRM train for generating the propulsion force enabling moving forward. Instead, propulsion coils alongside the track are used to extract propulsion force, thus playing the role of linear motor. The main principle is as follows: alternating current, whose frequency is synchronized to match the vehicle's speed(s), is pushed through the propulsion coils, thus generating continuously changing magnetic field moving the vehicle forward along the track. Under such conditions, the counterbalance between the strength of the magnetic fields created on board the vehicle and that from the track creates propulsive force moving the vehicle forward. However, this is possible only while operating (cruising) at high speeds. During acceleration and deceleration, at low speeds, a secondary propulsive system based on the conventional electrical linear motor is used in combination with the wheels ('landing gear').

Weight and energy consumption The typical weight of an empty TRM vehicle is about 50 tons, which, combined with the weight of payload (passengers) of about 20 tons, gives a total gross weight of 70 tons. A portion of the vehicle's empty weight represents the weight of on board magnets. If the magnet force to maintain levitation of the vehicle is about 1–2 kW, then the total energy consumed for this part could be 70–140 kW. Table 5.3 gives some specifications of TRM (TranRrapid MAGLEV) 07.

Usually, similarly as at HSR, the energy consumption of TRM is expressed in the relative terms, i.e., either per unit of input—supplied seat or per unit of output—passenger carried. This is often called SEC (Specific Energy Consumption). Figure 5.13 shows an example of the relationship between the SEC and operating speed of TRM07. For the comparative purposes, the corresponding SEC for HSR is also given.

As can be seen, the SEC of both systems increases more than proportionally with increasing of the operating speed, mainly due to increasing the air resistance. In addition, the SEC of HSR is higher than that of TRM for the range of operating speeds (For example, at speed of 300 km/h, this amounts about 33 %).

Control systems Contrary to conventional rail and HSR, TRM systems using either technology do not rely on an outside signaling system. TRM systems have fully automated communication systems controlled by the computer system. This consists of the main computer in the system's command center and that on board the vehicle(s). These computers continuously communicate between each other, thus providing the necessary monitoring and control/management of the driverless vehicles along the line(s). In addition, TRM trains change tracks by using bending switches. These consist of bending beams with drive units installed on every second solid of the bending switch. There are low-speed and high-speed switches. The former are used near and at stations enabling passing between the tracks at speeds of about 100 km/h. The former are used along the main portion of the guideway enabling switching between the tracks at speed(s) of about 200 km/h.

Speed

TRM trains can operate at a range of speeds from about 10–15 km/h to about 400–500 km/h due to the lack of physical contact with the dedicated guideway. This enables them to cover travel distances of between 500 and 800 km, also typical of HSR and short-haul APT system operations, in 1–2 h, respectively. This

Table 5.3 Technical/technological and operational performances of TRM (TransRapid MAGLEV) 07 (He et al. 1992; Janic 2003)

Performance	Value
Carriages per train (linked)	2
Total length (m)	50
Width (m)	3.70
Height (m)	3.9
Weight of a train (unloaded/loaded) (tons)	90/110
Seating capacity (average) (seats)	200
Operating speed (km/h)	400–450
Maximum speed (km/h)	500
Maximum acceleration/deceleration (m/s^2)	0.8–1.5
Lateral tilting angle (^{0})	12–16
Levitation air gap (mm)	8
Maximum thrust (kN)	85
Total tractive resistance (kN)	
Speed (km/h)	
300	22
400	35
500	52
Guideway loads (at 400 km/h)	
Longitudinal (kN)	300
Lateral (kN/m)	14
Vertical (kN/m)	33
Track structure	Single beam elevated Span 12 and 16 m
Technical curve radius (m)	2,825–3,580

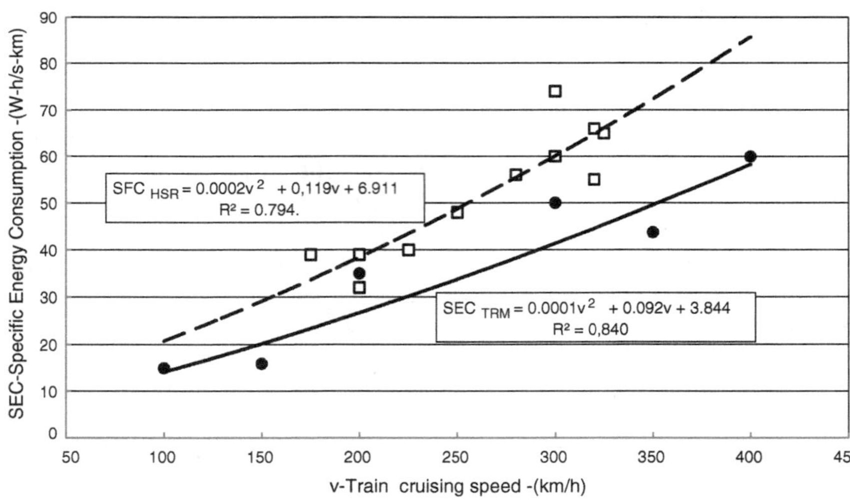

Fig. 5.13 Relationship between SFC (Specific Energy Consumption) and crusing speed of TRM07 and HS train(s) (Lukaszevicz and Anderson 2009; TIG 2012)

performance qualifies the TRM system as both a competitive and complementary alternative to the other two HS systems on comparable distances. Consequently, TRM will be able to connect centers and edges of major cities with airports on the one hand, and provide an intermodal connection with existing APT, HSR, and urban-mass transit and the eventually forthcoming ETT (Evacuated Tube Transport) system, on the other. However, similarly as in the case of HSR, both national and/or international dedicated infrastructure networks for the long-distance TRM may take decades to develop up to a level that will have a significant influence on the existing and prospective transport market(s) (Kertzschmar 1995).

Traffic

Individual TRM lines have been constructed so far for different purposes. In particular, in addition to five testing tracks and four tracks (lines) under construction, only three lines have been commercialized for public use as follows (http://en.wikipedia.org/wiki/Maglev):

- The Shanghai Maglev Train following the German TRM design, which began operations in 2004 in Shanghai (China) by covering the distance of 30 km between the city of Shanghai and its airport in 7 min; this implies an average commercial speed of about 268 km/h;
- The "Urban MAGLEV" also known as the Tobu-Kyuryo Line, namely the Linimo system that started operation in 2005 in Aichi (Japan); the length of the line is 9 km with nine stations, a minimum curve radius of 75 m, and a maximum gradient of 6 %; the trains—almost free from noise and quite resistant to disruptions by bad weather—operate at maximum speeds of 100 km/h, thus offering highly regular and reliable services (nearly 100 %); consequently, nearly 10 million passengers used this system during the first 3 months of operation; and
- The urban MAGLEV system UTM-02 in Daejeon (South Korea) began operations in 2008; the length of the line is 1 km connecting the Expo Park and National Science Museum.

5.2.1.2 Modeling Performances

Background

The DM processes related to HS passenger and other transport systems are mostly based on different ranking methods. Generally, these methods are commonly applied (almost as standard) to ranking alternative systems operating either by the same or different transport modes including already existing (operated) and/or nonexisting (forthcoming) alternatives. Ranking within the scope of the same transport mode is necessary for identifying the "best" or "most preferable" among different available alternatives, while ranking within the scope of different transport modes is necessary for identifying the "most preferable" alternative under conditions when both competition and complementarity by substitution between given HS modes can be expected. Competition is possible in high-density markets (corridors) with sufficient volumes of transport demand supporting efficient and effective operations of the given HS systems. Complementarity by substitution can take place in high-density markets-corridors where one HS system fully or partially replaces/substitutes another aiming at mitigating its current capacity burden (i.e., congestion) on the one hand, and its overall impact on the environment and society, on the other (EC 1998b; Levison et al. 1996; Janic 2003).

Ranking of HS alternatives can be carried out respecting the preferences of the different actors/stakeholders and interest groups involved. Each group can use the specified methodology/method applied to the selected ranking attributes/criteria represented by the indicators and measures of performances of alternative HS passenger transport systems.

Actors/stakeholders and interest groups

In Europe and the rest of the world, different actors/stakeholders and interest groups have acted as DM (Decision Makers) in developing and implementing the above-mentioned HS passenger transport systems. In general, these have been public, semiprivate, and private investors; manufacturers of these systems in the broadest sense; transport operators; governmental authorities at the local (regional and national) and global (international) level; users/passengers; the general public; and local community members. Usually, these actors and groups have different preferences regarding the full commercialization of such systems. On the positive side, they have expected benefits from the new systems. On the negative side, they have always been reluctant regarding their costs. In such a case, a compromise solution had to be found.

Both benefits and costs are of different meaning and importance for different actors. On the benefit side, for example, investors in the HS passenger transport system infrastructure and manufacturers of rolling stock and supporting facilities and equipment expect at least a return of their investments over the planned period of time. Local and central governments expect contribution of these systems to the overall social welfare and the economy. Transport operators expect to profitability. Users/passengers expect to utilize their travel budget more efficiently and effectively. Local community members expect local socioeconomic benefits in terms of new job opportunities and improved accessibility of their regions.

On the cost side, investors, local and central governments can very likely be confronted with extremely high infrastructure costs, long return on investment periods, and permanent scarcity of monetary funds. In addition, governments at all levels and local community members can be faced with increased global and local impacts on the environment and society in terms of energy consumption and related emissions of GHG, land use/take, noise, congestion, and traffic incidents/accidents. The systems' manufacturers can be exposed to an increased risk of economic and social failure of the new concepts. Transport operators need to deploy new models and knowledge in order to run their businesses profitably. Users/passengers can expect increased travel costs in absolute terms. Under such conditions, the choice of the appropriate HS system generally appears to be a very complex process, usually resulting in some kind of compromise solution. DMs use different sets of criteria based on the attributes of performances of particular HS transport systems, which are either already commercialized or in the planning phase.

Attributes/criteria of performances

The attributes/criteria of performances of the particular HS passenger transport systems can be selected, defined, and clustered in light of the preferences of the particular actors/stakeholders involved such as users/passengers, transport operators, local and central government, and local community members.

Users/passengers

Users/passengers are particularly interested in the following attributes/criteria of performances of the HS transport systems:

- **Accessibility**
  This attribute can be quantified by the users-passengers' generalized costs to reach the services of a given HS transport system. These costs depend on the total time the user spends on any of the available urban and suburban transport systems to reach the HS transport station from his or her location (house, office, shop). The HS transport system's station can be either the HSR station usually located nearby the center of an urban agglomeration or at an airport. The fare paid for such movement is another factor influencing the generalized costs. Generally, both the travel time and fare depend on the travel distance. In addition to individual modes (car, taxi), mass transport modes such as urban and suburban bus, stretcher, and various rail-based systems can be used (Vuchic 2007). This attribute/criterion is preferred to be *maximal*, implying that accessibility costs are preferred to be *minimal*.
- **Generalized travel costs**
  These costs consist of the costs of users-passengers' total time spent in a given HS system and the fare paid for the respective trip. The total time spent in the HS transport system consists of any schedule delay and in-vehicle time. The former depends on the departure frequency on a given route during the specified period of time (an hour, day), while the latter depends mainly on the length of route and the vehicle operating speed, as well as on the punctuality of operations expressed by delay(s) as deviations of actual arrivals from the schedule. It should be mentioned that HS passenger transport systems operate as scheduled transport systems at which scheduled delays are significantly reduced by advertising the schedule. Being familiar with the schedule, users-passengers choose to arrive just in time to meet their scheduled departure. The speed of HSR and TRM depends on the technical speed and the number and length of stops along a given route. In APT, the aircraft block speed, which depends on the aircraft type, cruising altitude, weather and traffic conditions on a route, and departing and landing airport, is relevant. In contrast to APT, both HSR and TRM are inherently free of congestion and delays that can exclusively be caused

by imbalances between demand and capacity. In general, these delays can be used for measuring punctuality of services of particular HS transport systems. The fare is primarily dependent on the class of trip (business, leisure), time of trip (high or low season, weekend or working day, time of day, etc.), length of travel distance, and the type of 'yield management system'. This attribute/criterion is preferred to be *minimal*.

- **In-vehicle comfort**
  The internal comfort, choice of services (drinks, meal, newspapers, videos, possibility of working on board, etc.), and feeling of security are important attributes (criteria) for users-passengers of a given HS transport system. They are preferred to be *maximal*.
- **Reliability of services**
  This attribute reflects the rate of canceled departures. The HS system with the lowest rate is preferable. This attribute/criterion is expected to be *maximal*, implying that the number of canceled departures in comparison to the number of realized ones is preferred to be *minimal*.
- **Security and safety**
  Users always expect this attribute to be as high as possible. The choice of HS alternative is always related to the perceived risk for a user to be injured and/or killed due to potential incidents and/or accidents. The preference is for this attribute/criterion (security and safety) to be *maximal*.

Transport operators

Transport operators are expected to consider the following attributes/criteria of performances of HS passenger transport systems:

- **Operating costs**
  Operating costs are primarily dependent on the constructive characteristics of particular HS transport systems, and the volumes and prices of inputs such as labor, capital, and energy. This attribute/criterion is preferred to be *minimal*.
- **Operating revenues**
  Operating revenues depend on the internal and external characteristics of both HS transport systems and their operators. Internal characteristics imply the system's inherent profitability (i.e., some systems are profitable because they are planned to be so, and others are not) and the capabilities and skills of transport operators to successfully use and sell their capacities (use of the 'yield management system'). External characteristics imply the volumes and structure of demand and the prevailing market prices under given conditions. This attribute/criterion is always preferred to be as high as possible, i.e., *maximal*.
- **Utilization of resources**
  The art of utilizing the resources engaged to carry out transport services is reflected by the load factor. This is expressed as the ratio between the number of passengers on board and the number of offered seats by the given transport services on a given route/line during a given period of time. HS passenger transport system operators prefer systems that can easily provide at least the break-even

load factor. Therefore, from their point of view, this attribute/criterion (the break-even load factor) is preferred to be as low as possible, i.e., *minimal*.

- **Technical productivity**
  Operators always prefer to operate systems with higher technical productivity. Technical productivity is expressed by the volume of services carried out per unit of time. Therefore, technical productivity increases in line with the vehicle and line capacity on the one hand, and the vehicle operating speed on the other (Vuchic 2007). This attribute/criterion is preferred to be *maximal*.

Local and central government

Local and central governmental authorities can consider the following attributes/criteria of performances of the HS passenger transport systems (Tabucanon and Mo-Lee 1995):

- **Investments**
  The investments are always carefully considered in evaluating any HS passenger transport system since the related costs are always very high. Therefore, this attribute/criterion is always, independent on the investor, preferred to be *minimal* while the return on investment is preferred to be *maximal*.
- **Energy consumption**
  Energy consumption is important for planning the energetic balance of the country and/or of a given region. In any plan, the proper balance between different types of primary sources of energy/fuel needs to be established in addition to minimizing overall energy/fuel consumption. Therefore, in the case of HS passenger transport systems, planners and policy/decision makers prefer this attribute/criterion to be *minimal*.
- **Congestion**
  Congestion causes delay of transport services and consequently of users-passengers while traveling door-to-door. In general, these delays impose costs on the affected parties. Operators have to engage more vehicles to serve the same volumes of demand under given conditions. Users-passengers loose time, which increases their generalized travel costs. Both increase the overall social costs and generate the need for building additional infrastructure, which may negatively affect the environment through change of land use, increased energy consumption and related emissions of GHG, and an increased risk of traffic incidents/accidents. In order to *minimize* these costs, the local and central government will always prefer the HS transport system that can reduce congestion costs, both directly and indirectly.
- **Externalities**
  Externalities of HS passenger transport systems include energy consumption and related emissions of GHG, noise, traffic incidents/accidents, and land use/take. These are preferred to be *minimal* in both absolute and relative terms.

Local community members

Community member(s) are particularly interested in the following attributes/criteria of performances of particular HS passenger transport systems:

- **Accessibility cost**
  Once accepting a given HS transport system in their neighborhood, community members are always keen to use its services in the most convenient way. On the local scale, they prefer efficient and effective access to services, while on the global scale, thanks to the high speed, the preference is to also access many previously rather distant regions in shorter (more reasonable) time. Thus, this attribute/criterion is preferred to be *maximal*.
- **Social welfare**
  Social welfare expresses some general benefits of the local community arising from the given HS transport system. These can be direct and indirect, part-time, and/or permanent jobs. In addition, these can include the overall increase of attractiveness of a given community (region(s)) for housing, and business and leisure activities. Therefore, in the given context, this attribute/criterion is preferred to be *maximal.*
- **Local externalities**
  Local externalities of HS transport systems comprise the environmental and social impacts such as energy consumption and related emissions of GHG, noise, traffic incidents/accidents, congestion, and land use (take). However, this time they are considered at the local community level. For example, changes to land use due to building HS infrastructure can cause displacement of households, creating physical barriers to daily activities, or increase noise, and similar. Therefore, this attribute/criterion is preferred to be *minimal.*

The multi-criteria ranking methodology

The multicriteria ranking methodology of HS passenger transport systems as alternatives includes the DM (Decision-Making) problem and process, models of ranking attributes/criteria, and the structure of the ranking method(s).

The DM (Decision Making) problem and process

The DM (Decision-Making) problem of choosing one among several HS passenger transport system alternatives based on their ranking in terms of overall feasibility can be structured on three levels as shown in Fig. 5.14.

As can be seen, the first level is occupied by the interest groups. The attributes (criteria) of performances of particular HS passenger transport systems are on the second level, while the ranking alternatives, i.e., HS transport systems, occupy the third and final level.

At least theoretically, membership in particular interest groups is not exclusive, meaning that the same person(s) may belong to different groups simultaneously and he or she is assumed to use different attributes/criteria and their weights while making decisions within each group.

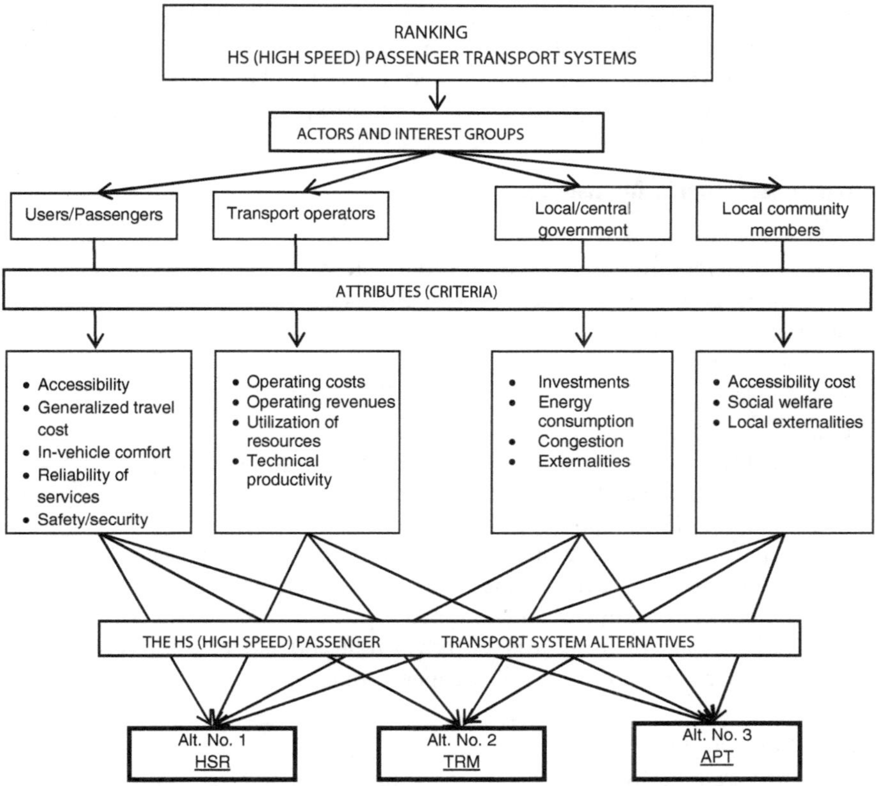

Fig. 5.14 Scheme of the DM (Decision Making) problem and procedure of ranking HS passenger transport system alternatives (Janic 2003)

The DM process of ranking the selected HS transport system alternatives consists of two parts. The first deals with modeling particular attributes/criteria of performances assumed to be relevant for particular interest groups considering the available HS transport systems as alternatives. This enables quantification of the particular attributes/criteria. The second part consists of the multicriteria ranking methodology. In this specific case, the methodology includes the TOPSIS (Technique for Order Preference by Similarity to Ideal Solution) method and the entropy method. The methodology enables particular interest groups to carry out more systematic and consistent evaluation and then selection of the preferable HS system alternative (Hwang and Yoon 1981).

Models of attributes/criteria

Modeling attributes/criteria for ranking the HS transport systems enables their quantification in both absolute and relative terms under given conditions. In absolute terms, they can be quantified for a given route (corridor) or for the entire network of a given HS system, while in relative terms, the attributes/criteria can be quantified per unit of the system's output (seat-km, passenger-km, km of infrastructure, etc.). In both cases, estimates are made for the given traffic scenario implying operations of the particular HS transport systems along comparable (or the same) routes/corridors.

- **Attributes/criteria for users/passengers**

– (A1) Accessibility
This attribute/criterion for a given HS transport system can be expressed by the generalized cost of users/passengers to access its services from their current location as follows:

$$C_a = \alpha * T(d) + d * p(d) \tag{5.1a}$$

where

α is the value of user-passenger time while traveling (€/pass-h);
$T(d)$ is the perceived travel time on the distance d between the user's location and the HS transport system station ($T(d) = d/v(d)$ where $v(d)$ is average speed of a chosen transport mode operating on the distance d) (min; h);
d is the average travel distance from the user's temporal location to the HS transport system station (km); and
$p(d)$ is the fare for using a given urban transport mode to arrive at the HS service (€/p-km; p-km—passenger kilometer).

– (A2) Generalized travel cost
The attribute/criterion of a given HS transport system can be estimated as follows (Janic 1993):

$$C(D) = P(D) + \alpha_0 * [T/4f(T, D) + D/v(D) + T_d] \tag{5.1b}$$

where

$P(D)$ is the fare paid by the passenger for traveling on the route of length D (€/passenger);
α_0 is the value of passenger time while in a given HS transport system (both waiting for departure and in-vehicle time) (€/pass-h);
D is the length of the route (km);
T is the period in which the departures on route D are scheduled (h);
$f(T)$ is the service frequency on the route D in time T (-);
$v(D)$ is the average operating speed on the route D (km/h); and
T_d is the anticipated delay on the route D (min; h).
According to Eq. 5.1b, the generalized travel costs will rise in line with the fare, value of passenger time, and the time of staying in a given HS transport system, the latter depending on the number of departures, length of route, operating speed, and anticipated delay(s).

– (A3) In-vehicle comfort
This attribute/criterion for a given HS transport system refers to its in-vehicle comfort, security/safety, quality of food/drinks, kindness of cabin attendants, possibilities for on board work, etc. This can be estimated either by interviewing the users-passengers or by some quantitative estimates such as for example the number of different services offered. Generally, the number and diversity of these services on shorter routes is smaller than on longer ones.
– (A4) Reliability of services
This attribute/criterion for a given HS transport system is usually expressed by the perceived rate of reliability of services. For the route of length D, this rate can be determined as:

$$R_r = N(D, T_0)/N(D, T_0) \quad (5.1c)$$

where

$N_{cd}\ (D,T_O)$ is the number of canceled departures on the route D in time T_O; and
$N(D,T_O)$ is the total number of realized departures on the route D in time T_O (day, week, month, and year).

– (A5) Safety and security
This attribute/criterion for a given HS transport system is usually expressed by the ratio between the number of events/causalities (incidents/accidents/deaths/injuries) and the volume of output (vehicle-km, or p-km) carried out during a given period of time (usually 1 year).

- **Attributes/criteria for transport operators**

– (B1) Operating costs
This attribute/criterion usually reflects the capital, operating, and maintenance costs of transport means/vehicles of a given HS transport system. The common expressions to estimate these costs can be as follows (Bianco and DiMajo 1991; Janic 1993; ITA 1991):

HSR and TRM

$$C_t(D, S_t) = c_0 * D * S_t \quad (5.2a)$$

where

c_O is the unit cost per unit of output (€/seat-km); and
S_t is the average number of seats per departure (seats).

The other symbols are analogous to those in Eq. 5.1a, b and c.

APT

$$C_a(D, S_a) = a_0 + a_1 * D + a_2 * S_a \quad (5.2b)$$

where

a_i is the coefficient estimated by calibration of the regression equation (i = 0, 1, 2).

The other symbols are analogous to those in Eq. 5.2a. The causal relationship (Eq. 5.2a, b) can also be in the log-linear form. As can be intuitively expected, the cost per departure of the given HS transport systems—HSR, TRM, and APT—increases in line with the vehicle (train, aircraft) seating capacity and route length, and vice versa.

– (B2) Operating revenues
  This attribute/criterion for a given HS transport system can be expressed by the following causal relationship (ITA 1991):

$$P(D) = b_0 + b_1 * D \tag{5.2c}$$

where

b_i is the coefficient determined by calibration of the regression model ($i = 0, 1$).

The other symbols are analogous to those in the previous equations. Evidently, the average fare rises in line with the route length (travel distance).

– (B3) Utilization of resources (break-even load factor)
  This attribute/criterion for a given HS transport system reflects the level of utilization of its operators' resources and is expressed as the break-even load factor. For a single service on a route, it can be determined after equalizing the operating costs and the operating revenues as follows:

$$\lambda_b(D) = C(D)/P(D) \tag{5.2d}$$

where

C(D) is the operating cost per departure on the route *D* (€/dep); and
P(D) is the average fare on the route *D* (€/passenger).

– (B4) Technical productivity
  This attribute/criterion for a given HS transport system can be determined for a single vehicle as well as for the route where a fleet of vehicles operate as follows:

Vehicle

$$P_v = S * v(D) \tag{5.2e}$$

Route

$$P_l = f(T, D) * S * v(D) \tag{5.2f}$$

where all symbols are as in the previous equations.
Both the above-mentioned attributes/criteria are expressed in seat-km/h. The criterion of a single vehicle increases in line with both its seating capacity and operating speed, while the criterion of the route increases in line the technical productivity of a single vehicle and the service frequency of the route.

- **Attributes/criteria for the local/central government**

– (C1) Investments

HSR and TRM
This attribute/criterion for HSR and TRM reflects construction investments and the costs of maintaining the infrastructure of a given HS transport system and can be expressed as follows:

$$C_{ti} = (I_{ta} * D)/Q_t \tag{5.3a}$$

where

I_{ta} is the average annual annuity on the investments per unit length of line (M€/km);
D is the length of route (km); and
Q_t is the total number of passengers expected to travel on route D (pass/per year).

As expressed per passenger, this cost is expected to decrease as the passenger numbers on a given route increase. In addition, if the investment and maintenance costs are additionally divided by the length of route D, the average unit cost per p-km during a given period of time can be estimated.

APT
This attribute/criterion for APT reflects the investment and maintenance costs in airports and ATC (Air Traffic Control System) and can be estimated as follows (Manheim 1979):

$$r_a = (R_a - C_a)/I_a \tag{5.3b}$$

where

R_a is the average annual operating revenues (M€);
C_a is the average annual operating cost (M€); and
I_a is the total investment cost (M€).

– (C2) Energy consumption

HSR and TRM
This attribute/criterion for HSR and TRM is expressed by the energy consumed per p-km on a given route and can be estimated as follows (ITA 1991):

$$SPC_t = EC_t/(\lambda_t * S_t) \tag{5.3c}$$

where

EC_t is the energy consumption of an HSR or TRM train (KWh/km);
λ_t is the average train's load factor; and
S_t is the train's seating capacity (seats/train).

Multiplying Eq. 5.3c by the number of departures *f(D, T)* in time *T*, the route length *D*, and the seating capacity of train S_t, the total energy consumption for a given traffic volume can be estimated.

APT

This attribute/criterion for *APT* is expressed by the quantity of energy/fuel consumed per p-km and can be estimated as follows:

$$SFC_a = \frac{FC_a(D)}{\lambda *_a S_a * D} = \frac{a_0 + a_1 * T_g + a_2 * (\frac{D}{v(D)} + T_d)}{\lambda_a * S *_a D} \tag{5.3d}$$

where

$FC_a(D)$ is the average fuel consumption per flight on the route of length D (tons/flight);

a_i is the coefficient determined for each flight; it is dependent on the aircraft type and length of flight (long, medium, short haul) ($i = 0, 1, 2$); and

T_g is the taxi and/or idle time while the aircraft is on the ground with its engines turned on (min).

The other symbols are analogous to those in Eqs. 5.1b and 5.3a–5.3c. For a given flight, SFC_a decreases more than proportionally as its length and the aircraft size increases. By multiplying Eq. 5.3d by the flight frequency *f(D,T)*, the aircraft seating capacity S_a, and the route length D, the total energy consumption for carrying the given volumes of passengers can be estimated.

- (C3) Congestion

  This attribute/criterion reflects the total costs of congestion and delays including the costs of HS vehicle delays and the costs of user/passenger time losses under given conditions. It can be estimated as follows:

$$C(T_d) = T_d * f(D, T) * [C(D, S) * v(D)/D + \alpha * \lambda * S] \tag{5.3e}$$

where the symbols are analogous to the previous equations.

- (C4) Externalities

Emissions of GHG

This attribute/criterion for a given HS transport system reflects the total emissions of GHG generated on the route D in the time period T. It can be estimated as follows:

$$TE = [f(T, D) * \lambda * S * D] * \sum_{j=1}^{M} SE_j \tag{5.3f}$$

where

SE_j is the specific emission of (j)-th type of GHG (g/p-km) ($j = 1.2, \ldots, M$).

The other symbols are analogous to those in the previous equations.

Land use
This attribute/criterion for a given HS transport system reflects the size and efficiency of land use, which can be quantified by the land use coefficient as follows:

$$\partial(Q, A) = Q/A \tag{5.3g}$$

where
Q is the volume of output (seat-km or p-km); and
A is the area of land acquired for infrastructure of a given HS system (km^2).

Specifically, in the case of HSR and TRM, $A = s_0 D$, where s_0 is the unit area (km^2/km) needed for building the line (D is the length of line). In the case of APT, A can be considered as the size of the airport airside area (runways, taxi-ways, apron(s)).
Noise
This attribute/criterion for a given HS transport system can be expressed as measured—in decibels on the logarithmic scale dB(A).
Incidents/accidents (safety)
This attribute/criterion for a given HS system can be determined similarly as in the case for an individual user/passenger (A5).

- **Attributes/criteria for community members**

- (D1) Accessibility costs
  On the local scale, this attribute/criterion for a given HS transport system can be estimated similarly as that for users/passengers (A1) (Eq. 5.1a). On the global scale, it can be estimated by the number, diversity, and generalized travel costs of connections between the given and other regions.
- (D2) Social welfare
  This attribute/criterion for a given HS transport system reflects the permanent and temporary jobs created by its commercialization. This can be quantified by the number of new jobs created per thousand of active inhabitants in the given region.
- (D3) Local externalities
  This attribute/criterion for a given HS transport system can be estimated similarly as that for local/central governments (C4). In such cases, the externalities relevant for the local community need to be selected.

The multicriteria ranking method

The most common method applied to ranking HS transport systems is the conventional "pure monetary" EAT (Economic Analysis Technique). The EAT evaluates particular alternatives (projects) on the basis of their "revenues" and "costs" during their life cycles. The outputs are expressed exclusively in monetary terms such as NPV (Net Present Value), BCR (Benefit-Cost Ratio), and/or IRR (Internal Rate of Return). In many cases, the application of EAT is not easy due to many reasons. First, taking into account the other nonmonetary attributes such as the environmental and social impacts of particular alternatives adequately and objectively is not always easy. Second, dealing with sometimes quite opposite interests and objectives of the various interested groups involved in the evaluation procedure is sometimes quite difficult, complex, and time-consuming (Giuliano 1985; Tabucanon and Mo-Lee 1995). Consequently, various modifications of the conventional EAT have emerged enabling systematic inclusion of more alternatives and ranking attributes/criteria, the latter reflecting specific interests of particular actors/interest groups involved in the DM process. In academic and professional literature, these are referred to as MCDM (Multi-Criteria Decision Making) methods. In particular, these methods have been proven to cope with the problems of using a large amount of diverse information, clearly identifying trade-offs between conflicting goals, and comparing the available alternatives in a systematic and consistent way (Geerlings 1998; Hwang and Yoon 1981). One of the strongest research-based recommendations for using MCDM methods instead of "pure monetary" EAT was made in the European COST 328 Action (EC 1998a).[1]

The multicriteria method presented for ranking three HS passenger transport systems consists of the entropy method as a commonly used above-mentioned analytical method for assigning weights to attributes/evaluation criteria and the TOPSIS (Technique for Order Preference by Similarity to Ideal Solution) method as one multicriteria analytical method alleviating selection of the preferable among several available alternatives. Alternatively, weights can be assigned to criteria empirically by interviewing representatives of the particular interest groups involved in the DM process (Hwang and Yoon 1981).

- **The entropy method**

The entropy method implies that the data on the decision matrix representing the quantitative estimates of the selected attributes/criteria of the given alternatives are known. Let $A_i (i = 1, 2, 3, \ldots, m)$ alternatives need to be evaluated according to n attributes/criteria X_j $(j = 1, 2, 3, \ldots, n)$. Let X_{ij} be the outcome of the i-th alternative with respect to the j-th criterion. The values of X_{ij} form the decision matrix D containing a certain amount of information. Let p_{ij} be the probability that the alternative A_i is the 'best' per criteria X_j. Then the probability p_{ij} can be estimated as follows:

$$p_{ij} = X_{ij} / \sum_{i=1}^{m} X_{ij}, \forall (ij) \tag{5.4a}$$

Since the choice of the 'best' alternative according to any X_j criterion is related to some measure of uncertainty, the entropy of criteria E_j can be expressed as follows:

[1] The MCA (Multi-Criteria Analysis) is proposed as a useful and convenient method for evaluating projects in the scope of Trans-European Transport Networks (s) (TENs).

$$E_j = -[1/\ln(m)] * \sum_{i=1}^{m} p_{ij} * \ln p_{ij}, \forall j \tag{5.4b}$$

where the expression [1/ln(m)] guarantees fulfillment of the condition: $0 < E_j < 1$. If the DM (Decision Making) has no reason to prefer one criterion over the other, the weight of criteria X_j can be determined as follows (Hwang and Yoon 1981):

$$w_j = (1 - E_j)/\sum_{j=1}^{n} (1 - E_j), \forall j \tag{5.4c}$$

- **The TOPSIS method**

The TOPSIS method evaluates the decision matrix $D[X_{ij}]$ $(i = 1, 2, \ldots, m, j = 1, 2, \ldots, n)$ with *(m)* alternatives each having *(n)* criteria by assuming that each criterion takes either increasing or decreasing utility. This implies that the larger the criterion outcomes, the grater the preference for the 'benefit' criterion and less the preference for the 'cost' criterion. The computation process of the TOPSIS method consists of the following steps:

STEP 1: The dimensions of various criteria are transformed into nondimensional criteria allowing comparison. This allows a normalized decision matrix R to be set up whose elements are determined as follows (Hwang and Yoon 1981):

$$r_{ij} = X_{ij}/\sqrt{\sum_{i=1}^{m} X_{ij}^2} \tag{5.5a}$$

STEP 2: The set of weights $w_j (j = 1, 2, \ldots, n)$ obtained by the expression (5.4c) (or provided by the DM) is accommodated to the decision matrix by multiplying each column of the matrix R with its associated weight w_j. Thus, the normalized decision matrix takes the following form:

$$v_{ij} = [w_j * r_{ij}], \ (i = 1, 2, \ldots, m; j = 1, 2, \ldots, n)$$

STEP 3: After determining the matrix, $V[v_{ij}]$ the positive and negative ideal solution A^* and A^-, respectively, can be estimated as follows:

$$A^* = \{(\max_i v_{ij} | j \in J); (\min_i v_{ij} | j \in J') | i = 1, 2, \ldots, m\} = \{v_1^*, v_2^*, \ldots, v_j^*, \ldots, v_n^*\} \tag{5.5b}$$

and

$$\begin{aligned} A^- &= \{(\min_i v_{ij} | j \in J); (\max_i v_{ij} | j \in J') | i = 1, 2, \ldots, m\} \\ &= \{v_1^-, v_2^-, \ldots, v_j^-, \ldots, v_n^-\} \end{aligned} \tag{5.5c}$$

where J is associated with the 'benefit' and J' with the 'cost' criteria.

STEP 4: Compute the separation of each alternative from the positive and negative ideal solution. A* and A-, respectively as follows:

For the positive ideal solution, it amounts to:

$$S_{i^*} = \sqrt{\sum_{j=1}^{n} (v_{ij} - v_j^*)^2}, \quad \text{for } i = 1, 2, \ldots, m \tag{5.5d}$$

Similarly, for the negative ideal solution, it amounts to:

$$S_{i-} = \sqrt{\sum_{j=1}^{n} (v_{ij} - v_j^-)^2}, \quad \text{for } i = 1, 2, \ldots, m \tag{5.5e}$$

STEP 5: Compute the relative closeness of the alternative A_i with respect to A^* as follows:

$$C_{i^*} = S_{i-}/(S_{i^*} + S_{i-}), \; 0 < C_{i^*} < 1, I = 1, 2, \ldots, m \tag{5.5f}$$

Apparently, $C_i^* = 1$ if $A_i = A^*$, and $C_i^* = 0$ if $A_i = A^-$. An alternative is closer to $A*$ as C_i^* approaches 1. A set of alternatives can now be ranked in descending order of the value of C_i^* .

Application of the multi-criteria ranking methodology

The multicriteria ranking methodology is applied using the attributes of HSR and APT performances estimated through real-life data, and those of TRM estimated though hypothetical (derived) data.

Input

The estimated values of 15 selected attributes as criteria of performances for three HS passenger transport systems are given in Table 5.4 including the preferences of the particular actors and interest groups. Despite not perfectly coinciding with those shown in Fig. 5.14, the selected attributes/criteria are sufficiently relevant for the purpose.

The attributes/criteria in Table 5.4 are expressed in absolute and relative terms and not always by using the above-mentioned quantification models. In addition, the inputs for their estimation have been taken from different sources. Furthermore, one should be aware that these sources may change over time. Under such circumstances, this ranking should be considered as an illustrative example and guidance for future similar work.

As already explained, the first four criteria (1, 2, 3, and 4) are mostly relevant for users/passengers. Operators of HS passenger transport systems are mainly interested in the next three criteria (5, 6, and 7). The remaining criteria (8, 9, 10, 11, 12, 13, 14, and 15) are mostly relevant for the local and central government and community members. Furthermore, among the 15 criteria, five are considered

Table 5.4 Attributes as criteria for ranking the HS passenger systems in the given example (Janic 2003)

Criterion	Objectives for criteria	The HS alternative		
		HSR	TRM	APT
1. Local accessibility (km)	"cost"-(min)	0.5–3.0	0.5–3.0	10-30
2. Generalized travel cost (€/p-km)	"cost"-(min)	0.188	0.257	0.121
3. In-vehicle comfort and convenience[a]	"benefit"-(max)	2	2	3
4. Safety (deaths/100 million p-km)	"cost"-(min)	0.000009	0.00000	0.0400
5. Operating costs (€/p-km)	"cost"-(min)	0.042	0.085	0.055
6. Operating revenues (€/p-km)	"benefit"-(max)	0.069	0.147	0.068
7. Technical productivity (seat-km/hour)	"benefit"- (max)	98,550	132,800	100,200
8. Investments (M€/km)[b]	"cost"-(min)	7.14	15.14	14.94
9. Rate of return on investments (%)	"benefit"-(max)	13.7	13.5	10.0
10. Social welfare[c]	"benefit"-(max)	138/25	30/10	817/1,270
11. Energy consumption (Wh/seat-km)	"cost"-(min)	60[d]	42[d]	230–570
12. Emissions of GHG (g/seat-km)	"cost"-(min)	14.2–25.0[e]	9.9–17.5[e]	140
13. Land use (ha/km)	"cost"-(min)	3.2–3.5	1.2	15
14. Noise (dB(A))	"cost"-(min)	84–105	79–93	46.5–92.5
15. Congestion (average delay) (min)	"cost"-(min)	1	1	11.7

[a] The number of different in-vehicle services; [b] The cost per km of infrastructure; [c] The number of part-time/permanent jobs/km of infrastructure; [d] At the speed of about 300 km/h; [e] Depends on the primary sources for obtaining electric energy

as "benefit" criteria and the other ten as "cost" criteria independently of the actors or interest group involved in the DM. The particular criteria are quantified as follows:

- *"Local accessibility"* is expressed in terms of distance rather than in monetary terms. Since HSR and TRM terminals are usually located in the center of an (urban) agglomeration, it is assumed that the access distance to these terminals may vary from 0.3 to 5 km. The distance of the user-passenger's location to/from an airport is assumed to vary from 10 to 30 km (Lufthansa 1996).
- *"Generalized travel cost"* is determined under the assumption that all three alternatives operate according to the same schedule (i.e., they have equal number of service frequencies along a given route) where all users-passengers have approximately the same value of time. Under such conditions, the generalized travel costs are dependent only on the in-vehicle time and fare. In this example, the cost of in-vehicle time and fare are expressed in monetary terms per unit of travel distance. The average value of travel time is adopted to be 32 €/pass/h. The unit value of travel time per unit of travel distance is calculated as follows: for HSR—32 €/pass/h × 1 km/270 km/h = 0.119 €/p-km; for TRM—32€/pass/km × 1 km/400 km/h = 0.080€/p-km; and for APT—32 €/pass/h × 1 km/600 km/h = 0.053 €/p-km. The fares are 0.069 €/p-km, 0.147 €/p-km and 0.068 €/p-km for HSR, TRM, and APT, respectively (CEC 1995; ITA 1991; Levison

et al. 1996; Witt 1995). Summation of the above values provides the total generalized travel costs for each alternative (Table 5.4).

- *"In-vehicle comfort and convenience"* is expressed by the number of different services provided to users/passengers while on board (SNCF 1996; Lufthansa 1996).
- *"Safety"* is expressed by the number of deaths per volume of output. For HSR, the estimate takes into account the ICE accident of 1998 in which 102 passengers were killed. TRM is assumed to be absolutely safe (no events and no experience). The risk of death in the APT mode is estimated from current ICAO (International Civil Aviation Organization) figures (Corrie 1994).
- *"Operating costs"* are estimated for typical operating conditions of HS passenger transport systems as the product of the average cost per seat-km and the load factor adopted to be 65 % (CEC 1995; ITA 1991; Levison et al. 1996; Witt 1995).
- *"Operating revenues"* are estimated similarly to the "operating costs," but also by assuming that 50 % of first and 50 % of second fare-class passengers use the HSR and APT services. In the case of TRM, only one fare-class is considered (Witt 1995).
- *"Technical productivity"* is estimated for the typical operating speed that may be attained on the competing travel distances of 600–800 km. For HSR this speed is adopted to be 270 km/h and the seating capacity of a train set 365 seats. The typical operating speed of TRM is adopted to be 400 km/h and the seating capacity of a train (the longest version) 332 seats. The typical operating speed of APT is assumed to be 600 km/h and the seating capacity 167 seats (A320-200 aircraft) (CEC 1995; Ellwagner and Wilckens 1994; Geerlings 1998; ITA 1991; Kertzscmar 1995; MVP 1994; Witt 1995).
- *"Investments"* are expressed in M€/km of the infrastructure represented by HSR lines, TRM guideways, and airport runway(s). For HSR, this is determined by dividing the total cost of building and upgrading the HSR network in Europe by the total network length (CEC 1995). For TRM, this is estimated by using data for the Berlin-Hamburg line (Witt 1995). In order to obtain comparable estimates, in the case of APT, the total investments of 25,000 M€ intended to improve the existing and build new airport and air traffic control capacity during the 1994–2000 period are allocated to the total length of runways at 437 airports in EU (European Union) Member States (CEC 1994b).
- *"Rate of return on investments"* represents one of the measures of investment effectiveness where the pure "Economics Analysis Technique" is applied. Essentially, it expresses the gains on the invested capital expressed in percents. This time, this is used as a single attribute (criteria). For the HSR, this is estimated for the entire network (CEC 1995), for TRM for the line Hamburg-Berlin (Witt 1995), and for APT for the airport and air traffic control infrastructure (Doganis 1992).
- *"Social-welfare"* is estimated in terms of the newly generated jobs per km of infrastructure, i.e., for the entire HSR network, single line of TRM, and 437 APT airports in the EU. In the latest case, the jobs at airports, air traffic control, airlines, and supporting airspace industry are taken into account (CEC 1994a; CEC 1995; Doganis 1992; Witt 1995).

- *"Energy consumption"* is estimated for typical operating conditions of particular HS systems. In case of HSR and TRM, the energy consumption increases in line with the operating speed and is constant as the travel distance and train seating capacity increase. The values in Table 5.4 correspond to operating speeds between 200–320 and 400 km/h for HSR and TRM, respectively. The energy/fuel consumption of APT decreases as the nonstop travel distance increases. Also, it varies with changes of the aircraft seating capacity. Typical values for A320-200 aircraft flying on routes of between 300 and 800 km are considered (ITA 1991; Geerlings 1998).
- *"Emissions of GHG"* are expressed by the total amount of GHG emitted per seat-km during the specified period of time. The GHG considered are: CO, NO_x, SO_2, CH, and CO_2 (the share of CO_2 is about 98.8 % in the total). In the case of HSR and TRM, these amounts are estimated as generated from the primary sources for obtaining electric energy in the EU15 Member States, while in the case of APT, the average emissions of GHG are estimated for A320-200 aircraft operating at speeds of about 500–600 km/h.
- *"Land-use"* is estimated per kilometer of HSR (tracks), TRM (guideways), and APT (airport runways) infrastructure. As given, this attribute should be taken with caution since the absolute values of land use (requirements) for the entire HSR and TRM networks and airports can show quite different relationships (ATAG 1996; CEC 1993b; Witt 1995).
- *"Noise"* is estimated for the typical operating conditions of HSR (200–300 km/h) and TRM (200–400 km/h) (Geerlings 1998; Kertzshmar 1995; Levison et al. 1996). In the case of APT, the ICAO noise certification limits for aircraft of different weights while taking-off, landing and at the side line are considered (ATAG 1996).
- *"Congestion"* is expressed by the "average delay" of service operated by each HS transport system. It is adopted to be 1 min for HSR and TRM. In the case of APT it is calculated based on the prevailing demand/capacity ratio of forty EU airports (CEC 1994b).

Results

Two rankings of the three above-mentioned HS transport systems are carried out using the inputs in Table 5.4. The first produces the preferable alternative with respect to all 15 criteria, while the second produces the preferable alternative with respect to the specific set of criteria.

- *The preferred HS transport alternative respecting all criteria* the normalized decision matrix is constructed as the initial step in selecting the preferable HS alternative (Eq. 5.5a) and weights are assigned to particular criteria by using the entropy method (Eq. 5.4) their values are given in Table 5.5. By comparing the values in Tables 5.4 and 5.5, it can be seen that the highest weights are assigned to the attributes/criteria with the greatest difference in absolute values, which emphasizes their relative importance for the dm (decision maker(s)). For example, the "most important" criteria appear to be: (1) "safety", (2) "social

Table 5.5 Normalized decision matrix and weights for particular criteria (Janic 2003)

Criterion	Normalized value/alternative			Weight	Rank
	A_1-HSR	A_2-TRM	A_3-APT		
1. Local accessibility	0.090	0.090	0.990	0.132	4
2. Generalized travel cost	0.552	0.754	0.355	0.011	10
3. In-vehicle comfort and convenience	0.485	0.485	0.728	0.005	12
4. Safety	0.0003	0.000	0.999	0.271	1
5. Operating costs	0.383	0.773	0.550	0.011	11
6. Operating revenues	0.392	0.835	0.386	0.018	8
7. Technical productivity	0.510	0.687	0.518	0.002	14
8. Investments	0.318	0.675	0.666	0.012	9
9. Rate of return of investments	0.632	0.623	0.461	0.002	13
10. Social welfare	0.078	0.019	0.997	0.200	2
11. Energy consumption	0.224	0.585	0.760	0.026	7
12. Emissions of GHG	0.265	0.216	0.940	0.056	6
13. Land use	0.211	0.075	0.975	0.107	5
14. Noise	0.650	0.591	0.478	0.002	15
15. Congestion	0.012	0.012	0.143	0.144	3

Table 5.6 Ranks of the HS passenger transport system alternatives (Janic 2003)

Alternative	Proximity to the ideal solution C_{i*}	Rank
HSR	0.644	1
TRM	0.632	2
APT	0.358	3

welfare", (3) "congestion", (4) "local accessibility", and (5) "land-use".
The ideal and the negative-deal solution and the separate measures are calculated by applying Eq. 5.5b–5.5e to the values in Table 5.5. Then, the relative closeness to the ideal solution is determined using Eq. 5.5f. The results are presented in Table 5.6.
Consequently, the alternatives are ranked according to descending order of C_{i*}. As can be seen, HSR emerges as the preferred in comparison to the other two alternatives (TRM and APT). However, as compared to APT, TRM is very close to HSR.

- *The preferred HS transport alternative respecting the specific attributes/criteria* From the methodological perspective, selection of the preferred HS transport system alternative respecting the specific set of criteria of their performances represents a sensitivity analysis of outputs with respect to changes of inputs, i.e., the number and type of criteria involved in ranking the alternatives. According to the importance of particular criteria for particular interest groups, the following four cases are elaborated:

  – Criteria such as "accessibility," "generalized travel cost," "in-vehicle comfort and convenience," and "safety" relevant for users/passengers and community members while acting as users of HS systems;

Table 5.7 Sensitivity analysis in ranking the HS passenger transport system alternatives (Janic 2003)

Case/criteria	Rank/(C_{i*})		
	A_1-HSR	A_2- TRM	A_3-APT
Case 1/(1, 2, 3, 4)	*1*/0.980	*2*/0.955	*3*/0.450
Case 2/(5, 6, 7),	*2*/0.342	*1*/0.611	*3*/0.234
Case 3/(8, 9, 10, 11)	*1*/0.095	*3*/0.026	*2*/0.073
Case 4/(12, 13, 14, 15)	*2*/0.955	*1*/0.999	*3*/0.009

C_{i*}—*"Closeness" to the ideal solution*

- Criteria such as "technical productivity," "operating costs", and "operating revenues" relevant for transport operators;
- Criteria such as "investments," "rate of return on investments," "social welfare," and "energy consumption" relevant for the governmental, public, semipublic, and private institutional and financial authorities and bodies; and
- Criteria representing externalities of particular HS systems such as "emissions of GHG", "land use," "noise," and "congestion" mainly relevant for the local and central governmental authorities, and local community members.

The results of these subevaluations are given in Table 5.7.

In the first and third case in Table 5.7, HSR emerges as the preferred alternative in comparison to TRM and APT systems. TRM appears the preferable alternative in the second case. APT is the least preferred alternative in the first, second, and fourth case. The score indicates that HSR is a favorable alternative for users/passengers, governmental authorities at both local (regional), national, and international level, as well as for potential investors. Transport operators, local and central governmental authorities, and local community members, who are mostly interested in externalities, favor TRM. However, this alternative appears to be the least preferable for governmental authorities and investors due to its very high investment costs and a rather low contribution to social welfare as compared to the other two alternatives. Nevertheless, this last consequence should be viewed with caution since only the social effects of a single TRM line have been considered rather than those of a developed network.

The above-mentioned scores again confirm that rail-based HS systems are competitive alternatives to APT systems in short- to medium-haul high-density transport corridors/markets. The variations of preferences across the specific sets of criteria indicate why the attitudes of different interest groups must be respected in the DM process, particularly those related to the development of very important HS projects. In addition, the sensitivity analysis reveals the high sensitivity of outputs to changes of inputs. This may highlight the problem of selecting the number, type, and indicators of performances to be included as attributes/criteria in the DM procedure. Furthermore, in addition to application of analytical tools for weighting attributes/criteria, the subjective judgments of representatives/experts of different interest groups need always to be taken into account if at all possible.

5.2.2 Evaluation

The HS (High Speed) passenger transport systems—APT (Air Passenger Transport), HSR (High-Speed Rail), and TRM (TransRapid MAGLEV)—and the methodology for their ranking as alternatives possess both advantages (Strengths and Opportunities) and disadvantages (Weaknesses and Threats).

Advantages and disadvantages

HS systems

- HSR has emerged as the preferable alternative as compared to both TRM and APT with respect to the selected 15 ranking criteria;
- Users/passengers, policy makers, and investors prefer HSR under the given circumstances;
- TRM appears as the preferable alternative of policy makers and community members, both actors/interest groups primarily interested in transport externalities; and
- APT appears as the least preferable alternative for virtually all actors and interest groups.

The multicriteria ranking methodology

- Enables ranking of predetermined transport alternatives (both existing and planned) while respecting the many often conflicting objectives by using attributes/criteria of performances of particular HS systems quantified for the given (assumed) traffic scenario(s);
- Ensures flexibility in including additional indicators of performances as attributes/criteria in the ranking procedure of existing and completely new HS system(s) (still at the conceptual level);
- Can be efficiently and effectively used by particular actors and interest groups involved in the DM process by enabling easier articulating, more transparently presenting, and more exactly quantifying the relevant attributes/criteria of the performances of HS transport alternatives;
- Enables relatively easy verification of the consistency of assigning weights to attributes/criteria including subjective judgments, thus mitigating or completely avoiding some irrational, time-consuming and unexpected activities and outcomes from the DM process; and
- Seems to be the preferable methodology compared to the conventional EAT (Economic Analysis Technique(s)), thus contributing to more efficient and effective DM (Decision Making) process relating to the further development and commercialization of HS passenger transport systems.

Finally, it can be said that HS passenger transport systems can be viewed as advanced with respect to all their performances taken into account in their multicriteria ranking as mutually substitutable alternatives.

References

Alstom (2008). *AGV-a cutting-edge technology integrator*. Berlin, Germany: Alstom Transport Europe, INNOTRANS-BERLIN.

ATAG. (1996). *Aviation and environment*. Geneva, Switzerland: Air Transport Action Group.

BBVA. (2009). *Economic analysis of high speed rail in Europe*. Bilbao, Spain: BBVA Foundation.

BCG. (2004). *Airports-dawn of new era: Preparing for one of the industry's biggest shake- ups*. Munich, Germany: Boston Consultancy Group.

Bianco, L., & DiMajo, D. (1991). Perspectives of high-speed rail transport in short-medium period. *Transportation Research-A, 25A*, 193–202.

Boeing (1998). *Evolution of the world fleet: Time line*. Boeing Aircraft Company. http://www.boeing.com.

CAA (2011) *UK airports market-general context, working paper*. London, UK: Airport Market Power Assessment, Civil Aviation Authority.

CEC. (1993a). *European high-speed rail network: Socio-economic impact study*. Commission of the European Communities, Directorate General for Transport, Final Report, Brussels, Belgium.

CEC. (1993b). *The European high-speed train network: Environmental impact assessment*. Executive Summary, Commission of the European Communities, Directorate General for Transport, Brussels, Belgium.

CEC. (1994a). *Expanding Horizons, A report by Comite de Sages for air transport to the European Commission*. Brussels, Belgium: Directorate General for Transport.

CEC. (1994b). *Progress report on the guidelines for the Trans-European airport network*. Brussels, Belgium: Commission of the European Communities.

CEC. (1995). *High-speed Europe, high level group. The European high speed train network*. Brussels, Belgium: Commission of the European Communities.

Corrie, S. J. (1994). *Potential growth in air travel demands renewed effort to improve safety record* (pp. 7–9). Montreal, Canada: ICAO Journal.

Doganis, R. (1992). *The airport business*. London, UK: Routledge.

EASA. (2011). *Type-certificate data sheet for noise A380, TCDSN EASA, Issue 6*. Koln, Germany: European Aviation Safety Agency.

EC. (1998a). *Integrated strategic infrastructure networks in Europe*. Final Report on the Action COST 328, EUR 18165, European Commission, Luxembourg.

EC. (1998b). *Relationships between high-speed rail and air passenger transport*. Final Report on the Action COST 318, EUR 18165, European Commission, Luxembourg.

EC. (2003). *Analysis of the European transport industry*. European Commission—DG TREN, Final Report, Brussels, Belgium.

EC. (2010). *High speed Europe: A sustainable link between Citizens*. Luxembourg: European Commission.

EC. (2012). *EU transport in figures: Statistical pocketbook 2012*. Luxembourg: European Commission.

Ellwagner, G., Wilckens, M. (1994). European high-speed transport: A service with a future. *International 33*, 2–12.

FAA. (1997). Noise level for U.S. certified and foreign aircraft. Washington, USA: AEE-110Federal Aviation Administration, U.S. Department of Transportation.

Geerlings, H. (1998). The Rise and Fall of new technologies: MAGLEV as technological substitution? *Transportation Planning and Technology, 21*, 263–286.

Giuliano, G. (1985). A multicriteria method for transportation investment planning. *Transportation Research-A, 19A*(1), 29–41.

GAO. (2004). *Airline financial conditions*. Report, The US Government Accountability Office (GAO), Report, Washington DC, USA

Haimes, Y. Y. (1985). Multiple-criteria decision-making: A retrospective analysis. *IEEE Transactions on Systems, Man, and Cybernetics, 15*(3), 313–315.

He, J. L., Rote, D. M., & Coffey, H. T. (1992). *Survey of Foreign maglev systems*. Argonne, Illinois, USA: Center for Transportation Research, Energy Systems Division, Argonne National Laboratory.

Horonjeff, R., & McKelvey, F. X. (1994). *Planning and design of airports* (4th ed.). New York, USA: McGraw- Hill Book Company.

Hwang, L. C., & Yoon, K. (1981). *Multi attribute decision-making: A methods and applications. Lecture series in economics and mathematical systems*. Berlin, Germany: Springer.

ITA. (1991). *Rail/air complementarity in Europe: The impact of high-speed rail services*. Paris, France: Institute du Transport Aèrien.

Janic, M. (1993). A model of competition between high-speed rail and air transport. *Transportation Planning and Technology, 17*, 1–23.

Janic, M. (2003). Multiple criteria evaluation of high speed rail, transrapid maglev, and air passenger transport systems in Europe. *Transportation Planning and Technology, 26*(6), 491–512.

Janic, M. (2007). *The sustainability of air transportation: A quantitative analysis and assessment*. Aldershot, England: Ashgate Publishing Limited.

Kertzschmar, R. (1995). *Transrapid maglev: Prospects for fast regional transportation service. Paper presented at the 1st European workshop on high speed maglev transport system: European prospects*. Padua, Italy: University of Padua.

Levison, D., Gillian, D., Kanafani, A., & Mathieu, J. M. (1996). *The full cost of inter-city transportation—A comparison of high speed rail, air and highway transportation in California. Institute of Transportation, Research Report, UCB-ITS-RR-96-3*. Berkeley, USA: University of California.

Lukaszevicz, P., & Anderson, E. (2009). *Green train energy consumption: Estimation on high-speed rail operations*. Stocholm, Sweeden: KTH Engineering Science, KTH Railway Group.

Lufthansa. (1996). *Timetable*. Frankfurt, Germany.

Manheim, M. L. (1979). *Fundamentals of transportation system analysis (Vol. 1: Basic concepts)*. Cambridge, Massachusetts, USA: The MIT Press.

MVP. (1994). *Riding of flying? TRANSRAPID and its links with the car, railway and aircraft*. Germany: Versuchs-und Plannungsgesellschaft fur Magnetbahnsysteme MbH.

Powell, J., & Dunby, G. (2007). *Maglev: Transport mode for 21st century* (pp. 44–55). Science and Technology: EIR.

Shibo, S. (2008). *Dynamic simulation of the maglev guideway design*. MSc Thesis, Faculty of Civil Engineering and Geosciences, Delft University of Technology, Delft, The Netherlands.

Speling, D. (1984). Assessment of technological choices using a pathway methodology. *Transportation Research-A, 18A*(4), 343–353.

SNCF. (1996). *Timetable-TGV/Thalys*. Paris, France.

Tabucanon, T. M., & Mo Lee, H. (1995). Multiple criteria evaluation of transport system improvements. *Journal of Advanced Transportation, 29*(1), 127–143.

TIG. (2012). *High tech for flying on the ground*. Berlin, Germany: Transrapid International GmbH & Co.

UIC. (2011). *High speed rail and sustainability*. Report, International Union of Railways, Paris, France

USDT. (2012). *National transport statistics*. U.S. Department of Transportation Research and Innovative Technology Administration Bureau of Transportation Statistics, Washington DC, USA.

Vuchic, R. V. (2007). *Urban transit: Systems and technology*. Hoboken, New Jersey, USA: Willey.

Witt, M. (1995). *The transrapid system and the Hamburg-Berlin connection, the 1st European workshop on high-speed maglev transport systems: European prospects*. Padua, Italy: University of Padua.

Chapter 6
Advanced Transport Systems: Future Concepts

6.1 Introduction

This chapter describes five concepts of future advanced transport systems: (i) PRT (Personal Rapid Transit) systems; (ii) UFT (Underground Freight Transport) systems; (iii) ETT (Evacuated Tube Transport) system; (iv) advanced ATC (Air Traffic Control) technologies and procedures for increasing airport runway capacity; and (v) advanced STA (Supersonic Transport Aircraft).

- PRT (Personal Rapid Transit) systems were elaborated in the 1970s and commercialized in the 2010s. They are a fully automated transit systems consisting of dedicated aboveground guideways enabling movement of small passenger car-like vehicles, i.e., capsules. As such, PRT systems are expected to offer competitive—efficient and effective—transit services by substantially replacing conventional passenger cars, motor- and trolleybuses, and LRT (Light Rail Transit) systems in urban and/or suburban areas. At the same time, PRT will contribute to mitigating the overall impacts of urban transit systems on the environment and society in terms of energy consumption and related emissions of GHG (Green House Gases), land use/take, noise, congestion, and traffic incidents/accidents. PRT commercialization is shown to be a rather gradual process mainly due to the high investment costs starting in some niche markets;
- UFT (Underground Freight Transport) systems were elaborated for the first time in the 1970s but particularly in the 1990s and 2000s. They are fully automated systems for moving freight/goods shipments of different sizes such as standard pallets, containers, and/or swap bodies along a network of underground pipes/tunnels. Shipments are loaded onto dedicated vehicles/capsules. Underground pipes/tunnels connect different freight/goods origins and destinations in a given urban or suburban area including large intermodal transport nodes such as ports, airports, and inland freight terminals. UFT systems are expected to substantially replace road trucks in congested urban areas and consequently mitigate their

M. Janić, *Advanced Transport Systems*,
DOI: 10.1007/978-1-4471-6287-2_6, © Springer-Verlag London 2014

impacts on the environment and society. The commercialization of any kind of UFT system is still under consideration due to the high investment costs related to building underground pipes/tunnels.

- Evacuated Tube Transport system was elaborated for the first time in the 1970s, but remained at that stage until recently (Salter 1972). This system is used to transport passengers and freight/goods along long-haul (intercontinental) distances by TRM (TRansRapid MAGLEV) trains running through pipes located below the ocean's floor at supersonic speed.[1] The EET system's eventual full commercialization will likely be the matter of discussion and decisions in the forthcoming decades since if the decision is taken to design such a system, this will certainly not be earlier than around the 2050s or latter. The main issues still to be resolved are choosing the design of infrastructure (pipes) and proving the system's economic feasibility, mainly due to the very high investment cost for building the infrastructure (underwater pipes);
- Advanced ATC (Air Traffic Control) technologies and procedures for increasing the airport runway capacity are currently under development in the scope of the U.S. FAA NextGen (Next Generation Air Transport System) and European-EUROCONTROL SESAR (Single European Sky ATM Research) programs (EC 2007; FAA 2010). The advanced procedures supported by these technologies imply the application of ATC mixed horizontal/vertical, exclusively vertical distance-based, and time-based separation rules instead of the current horizontal distance-based separation rules between landing aircraft. In parallel, ATC can alternatively apply the existing FCFS (First Come First Served) and the advanced PR (Priority) service discipline for landings on a single runway. In addition to maintaining the required level of safety, these advanced technologies and procedures are expected to increase the airport runway system capacity without requiring new land for expanding the infrastructure, i.e., building new runway(s) (LHR 2010). The full commercialization of these technologies and procedures is the matter of a decade or so, i.e., by the year 2020/2025; and
- STA (Supersonic Transport Aircraft) were already in commercial operation as the French-British Concorde and Soviet Union's TY144 supersonic aircraft. Still being at the conceptual level, the aircraft is supposed to operate at speeds higher than Mach 2.0. Their full commercialization is likely to occur in about two to three forthcoming decades (2030/2040). This is because, at present, the main aircraft manufacturers Airbus and Boeing, and other agencies and institutions dealing with the visions of development of the commercial air transport system do not explicitly consider such developments to occur earlier than around 2030 (ACARE 2010; Airbus 2012; Boeing 2012).

[1] The supersonic speed is greater than the speed of sound, which depends on the air ambient temperature. For example, at the sea level with standard temperature of 15 °C, it is 1,225 km/h; at the cruising altitudes of most commercial jets and supersonic aircraft of 11,000–20,000 m with temperature of −57 °C, it is 1,062 km/h.

6.2 Personal Rapid Transit Systems

1972/1975	The WVU PRT Morgantown system begins operations (U.S.)
2001	The first self-balancing PRT (Personal Rapid Transit) system (Segway PT) is launched (U.S.)
2011	The PRT ULTra (Urban Light Transit) system begins operations at London Heathrow airport (UK)

6.2.1 Definition, Development, and Use

One of the most dominant social-economic trends during the twentieth century has been urbanization characterized by growing cities and the consequent increase in proportion of the urbanized population. This has reached about three billion, i.e., about 47 % of the total global population in the year 2000. The current and future trends indicate that about 6 billion or two-thirds of the world's population are likely to live in the urban areas by the year 2050. In parallel, the urban mobility problems in terms of provision of sufficient transit capacity, emissions of GHG (Green House Gases), land use, congestion, and noise have been increasing and becoming more complex. This has often required adaptation of the growing urban forms and structures to the requirements of existing individual and mass urban transit systems, and vice versa (Rodrigue et al. 2006). These urban transit systems referred to here as conventional are considered to be bikes, motorbikes, individual passenger cars, taxis, motor buses, and trolleybuses (including BRT (see Chap. 2)), trams, and LRT (Light Rail Transit) system(s). Future urban developments will certainly require the commercialization of advanced urban transit systems in order to support the above-mentioned conventional ones and satisfy growing demand for urban mobility more efficiently, effectively, environmentally friendly, and safely. Some of these systems/concepts are: (i) ADAS (Advanced Driver Assistance System) for individual passenger cars; (ii) PRT (Personal Rapid Transit) system; (iii) Advanced bus systems; and CTS (Cybernetic Transport Systems) with the "road-based people movers" and "advanced car sharing" subsystems (EC 2007).

Specifically, PRT (Personal Rapid Transit) as urban transport systems offer a potential solution for mitigating urban congestion and improving travel accessibility. It is designed for use in a variety of cases aiming at replacing mainly individual passenger cars and urban motor buses and trolleybuses as:

- An exclusive urban and/or suburban transit system;
- A system feeding other mass urban and suburban transit systems (conventional buses, BRT, LRT, metro);
- An airport ground access system/mode complementing to the existing ones;
- An inter-airport transit system connecting airports belonging to the same airport system (for example, New York and/or London airports);

- An intra-airport transit system at a large airport connecting distant structures such as terminals (the satellite concept) to long-stand parking areas, and vice versa; and
- An exclusive system operating in campus environment such as universities and self-contained business/entertainment parks.

The PRT systems consist of fully automatized small driverless vehicles transporting an individual or small group(s) (typically 1–6) persons along dedicated guideways. These can be elevated, at ground level, and underground. The transit services, available on demand rather than according to a fixed schedule over most or all of the day are provided between particular origins and destinations without stops at intervening stations, i.e., without the need to transfer. The PRT systems can be implemented as a single- and dual-mode systems. The most well-known single-mode PRT systems in the U.S. are the earliest WVU PRT Morgantown (1972/1975), Raytheon PRT2000 (1995), and Taxi 2000 HCPRT (High-Capacity Personal Rapid Transit) (2003). In Europe, the most well-known systems include the UK ULTra and Dutch 2getThere PRT systems. The most well-known dual-mode PRT system is Denmark's RUF (Rapid Urban Flexible) Intelligent Auto/Transit concept. This system is designed to operate PRT vehicles both automatically and manually. The main element of the RUF system is dedicated guideways for the automatic portion of each trip. The system uses small- and medium-sized intelligent electric vehicles, which can move along conventional roads/streets as well as along the automated rail network. Small vehicles with a capacity of 2–4 persons could be in the public or individual private ownership and as such in the latter case parked at the individual's residence and/or workplace. Larger vehicles with a capacity of 10 persons are usually publically owned and parked at the public parking areas similarly to small buses. As such, the concept brings together the convenience of the individual passenger car and the efficiency of urban railways. While operating on roads/streets in manual mode, the vehicles are powered by the electricity from the on-board batteries enabling an average range of about 50 km. While operating in automated mode, the vehicles are powered by electricity obtained from the rail electricity supply system, enabling relatively high operating speeds over longer distances. Such smart integration of operating modes enables the concept/system to offer users/passengers efficient and effective connection between numerous origins and destinations within and through large, dense, and congested urban agglomerations. Consequently, fully automated computer-controlled operations enable traveling free of congestion and potential incidents/accidents (RUF 2008).

To date, three PRT systems have been commercialized: the WVU (West Virginia University) PRT system (designed by Boeing) started operation during the 1972/1975 period; the partially commercialized Cybercab PRT system of the Dutch manufacturer 2getThere operating in Masdar City (Abu Dhabi) started operation in 2010; and the ULTra (Urban Light Transit) system commenced operation at London Heathrow airport (London, UK) in May 2011. The forth, the Vectus PRT system of the Swedish/Korean manufacturer Vectus is planned to start operation in 2013 in Suncheon Bay (South Korea) (EC 2007; http://www.advancedtransit.org). In addition, the ULTra system yet to be commercialized in Amristar (Punjab, India) in 2014 will be the world's first and largest actually urban PRT system. The length of its network is planned to be 8.0 km with seven stations accommodating 240 vehicles expected to serve peak demand of up to 100,000 passengers/day. This completely privately financed system is expected to bring savings in the current journey time of up to about 30 min (http://www.advancedtransit.org). Furthermore, some prototypes of PRT systems such as the LINT PRT system by the manufacturer Modutram (Mexico), the Microrail PRT/Dual Mode system by the manufacturer Megarail Transportation Systems (US), and the Skyweb Express PRT system by the manufacturer Taxi 2000 (US) have been operationalized to the level of test track.

6.2.2 Analyzing and Modeling Performances

6.2.2.1 Background

The generic characteristics of a PRT system are as follows: automatically guided high-pressure rubber tire or steel wheel vehicles with a capacity of 1–6 persons move along exclusive/dedicated concrete steel elevated guideways at relatively high speed for the given (urban) conditions. The vehicles are powered by an electric motor propulsion system controlled by microprocessors, making them reliable, less energy consuming and air polluting, and less noisy, all as compared to individual cars and other urban mass transit systems. The small and relatively light vehicles allow for lighter and consequently less costly guideways. Nonstop services are provided thanks to using off-line stations, which are closely spaced in the network of interconnected guideways, thus eliminating the need for passengers transferring between lines. At stations, a few empty vehicles usually wait (similarly as at taxi stations) to be picked up by users/passengers. First, passengers need to select their itinerary, buy their ticket, and pass the information from the ticket to the vehicle's control system. Then, after the vehicle's door opens they board the vehicle, the door closes, and the computer control system opens a path to the vehicle. Then, the vehicle accelerates and joins other vehicles passing through a given station, and then forward toward the destination. The elevated configuration of guideways reduces the required land only to that needed for setting up posts and stations (Anderson 2000, 2005, 2007).

Similarly as the other advanced transport systems, a given PRT system can be characterized by its infrastructural, technical/technological, operational, economic, environmental, social, and policy performances, all of which are further analyzed and modeled.

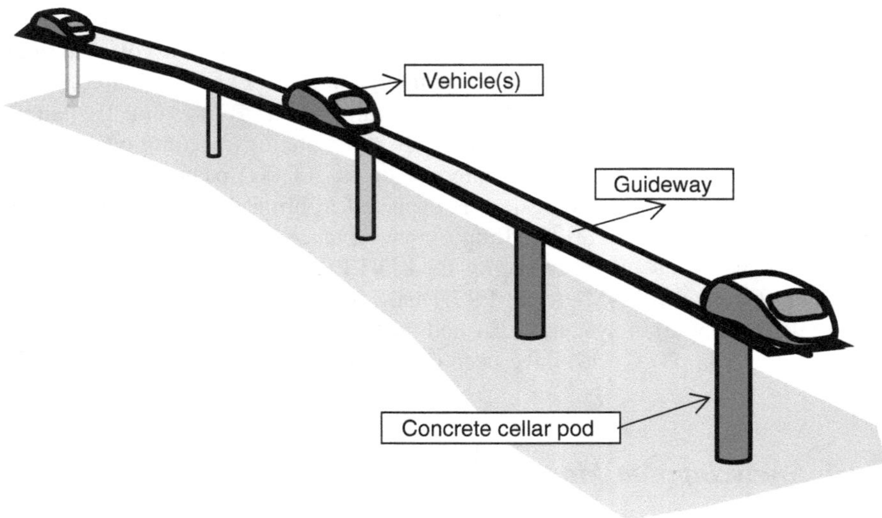

Fig. 6.1 Scheme of the PRT system

6.2.2.2 Infrastructural Performances

The infrastructural performances of a given PRT system refer to its design, geometry, and material of guideways; spatial layout and coverage of the network; layout, configuration, and number of stations; and switching.

Guideways

The PRT system guideway(s) as an elevated construction(s) usually with top-mounted vehicles gives the visual impression shown in Fig. 6.1. A typical PRT guideway is a U-shaped configuration made of concrete laying on a concrete and/or steel/concrete construction. This construction is supported by pods (solid pillars) located at certain distances, thus enabling the relatively uniform distribution of its own weight and the weight of moving vehicles. The cross-section of the U-shape guideway with vehicles determines the square-shaped vehicle dynamic envelope with a height of about 3 m (from the top of a vehicle to the bottom of the guideway aerial structure) and a width of about 5 m for the two-track guideway(s) shown in Fig. 6.2. The height of the pods (solid pillars) from the ground can vary depending on the roadway and/or pedestrian way clearance requirements.

The guideways are also designed with respect to stress (bending and torsion, natural frequency and critical speed) and passenger ride comfort, generally with or without walkways, and with many curves and elevations. Both horizontal and vertical curves need to be designed to provide for the riding comfort of seating passengers according to the standards for the lateral and normal acceleration and jerk of about 0.25 g and 0.25 g/s, respectively. In some cases, covers made of composite materials and a thin aluminum layer inside are attached to the sides of

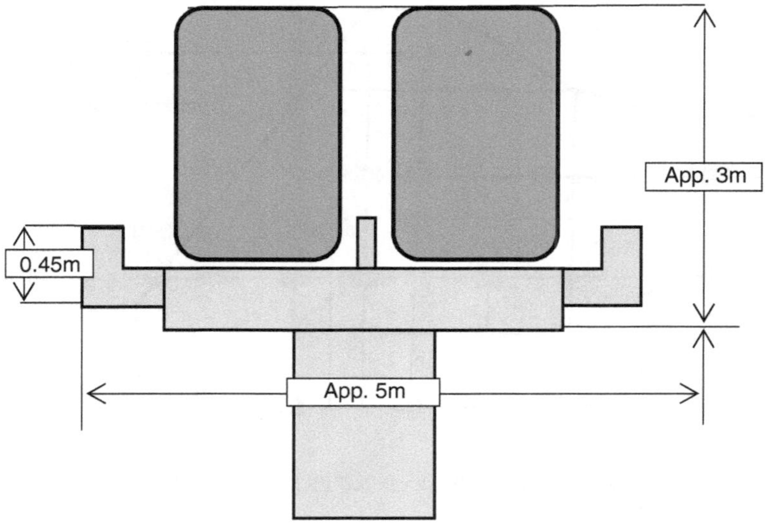

Fig. 6.2 Scheme of the vehicle dynamic envelope for the two-track guideway PRT system

the guideway in order to protect the tracks and tires from extreme weather (frost during cold nights, the accumulation of snow, and solar heating during the summer), and in some cases, from electromagnetism (Anderson 2007).

Network

The PRT network consists of guideways. Its spatial layout is usually adjusted to the spatial density of user/passenger demand. The network can consist of a single line connecting a given origin and destination (for example in the case of airport access or internal transit systems), a small-scale network of lines feeding other urban mass transport systems such as BRT, LRT, and/or metro, and a larger scale network operating in a given urban complex as an independent and/or exclusive public transport mode. In most cases, network development usually begins by building a single guideway and then gradually expanding by adding new ones through adapting to transport needs. Figure 6.3 shows a simplified layout of an exclusive PRT network.

Stations

PRT system stations are designed as off-line guideways located usually between intersections of guideways at distances of about 800 m, or shorter or longer depending on the spatial distribution of user/passenger demand. In the vertical plane, the stations are designed as elevated platforms leveled to the elevated guideways and standing either alone or incorporated into other existing urban

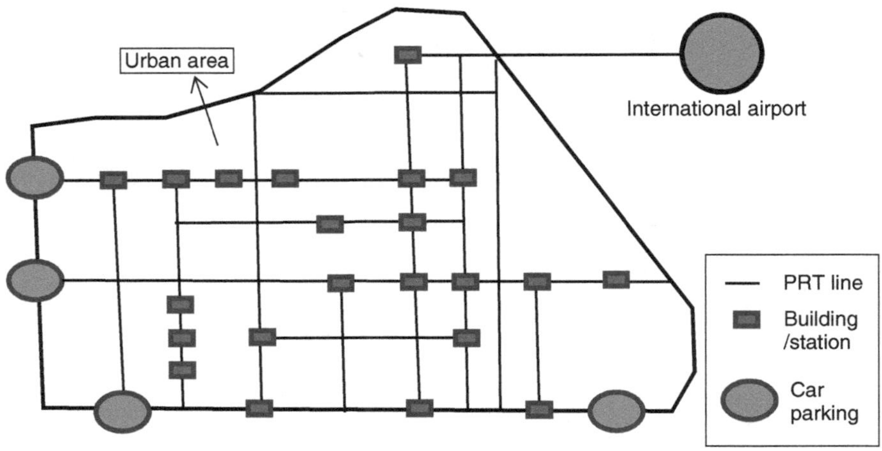

Fig. 6.3 Scheme of the prospective fully developed PRT network (Masdar City—Abu Dhabi) (http://www.advancedtransit.org)

structures. The length of an off-line station can be, depending on the vehicle's operating speed, about 33–39 m, thus providing safe maneuvering space and berths for about 3–5, and, if needed, after extension for up to 12–14 vehicles. Additional berths are also needed for vehicles waiting due to any reason—for example, while vacant during off-peak hours, maintenance, or when all tracks are occupied. Since the passengers are supposed to spend only a short while at the stations during boarding and disembarking the vehicles, the intended space is relatively small, thus reducing the construction costs, and mitigating the visual impacts and crowd-related (social) problems. Such off-line stations enable continuous vehicle movement, thus maximizing the transit speed and minimizing the transit time, while enhancing riding comfort. In addition, stations also provide good spatial accessibility (Anderson 2007).

Switching

Switching is an important component of the PRT infrastructure. At present, switching is based on mechanical back up. In such cases, switches have two stable positions, are self-stable, and can be thrown manually if required. They are unaffected by centrifugal forces, operable under all weather conditions, and powered by low-voltage batteries.

6.2.2.3 Technical/Technological Performances

The technical/technological performances of a given PRT system relate to its vehicles (cars), the system for controlling their movement, and the system for interaction with users/passengers.

Vehicles

The PRT vehicles, designed as small, light, and short as possible—typically 2.7 m long and 1.65 wide, are usually placed above the guideway(s). Each vehicle consists of three parts: a passenger cab, a bogie for interface with the guideway, and a propulsion system. The typical capacity of the passenger cab is four persons plus two children (and their baggage between the seats) including space for wheelchairs, bikes, etc. The bogie has rubber-tired or steel wheels, thus guaranteeing efficient and effective braking at an acceptable noise level. The propulsion system is usually an LEM (Linear Electric Motor) (LSM—Synchronous or LIM—Induction) powered by electricity from on-board batteries or from an external electricity supply system, and controlled by computer microprocessor. Either lead–acid or lithium iron phosphate batteries can be used. The LEM enables weather-independent acceleration and braking, and a reaction and breaking time of a few milliseconds, each as compared to that of humans—0.3–1.7 s, and that of mechanical brakes of 0.5 s, respectively.

The operating speed of vehicles can vary between 30 and 80 km/h depending on the type of PRT system, and the location, geometry, and topography of guideways (the elevation change can be up to 15 % along the distance of 9.0 m).

The vehicle movement control system

The control of vehicles while operating along the guideway is carried out by a fully automated centralized dual-redundant computer system providing high dependability and safety. The control function is set up hierarchically at three levels: (i) vehicle control; (ii) zone control; and (iii) central control. The vehicle control system generally generates the intended routes/paths for each vehicle boarded by an individual and/or group of users/passengers between their intended origins and destinations, and accelerates/decelerates, monitors, and controls the operating speeds and headways between the vehicles. This implies that each vehicle follows the assigned/commanded trajectory. The zone control system enables each vehicle to communicate with the zone's computer controller. Each vehicle transmits information about its position and speed to this controller through the cable in the guideway, and the controller responds back with the maneuvering commands, both in milliseconds. In addition, the merge-zone-control computer receives and maintains information about the current position and speed of all vehicles within a given merging zone, and resolves any potential merging conflicts by sending the relevant maneuvering commands to the vehicles involved. The central control system comprised of the central computer controlling the computers at the other two lower hierarchal levels.

The user/passenger interaction system

The system for interaction of a given PRT system with its users/passengers generally consists of: (i) the information and communication subsystem; and (ii) the fare collection subsystem.

Table 6.1 The subsystem for informing and communicating within a given PRT system

Component	Function
Touch Screen, automated interactive display and audio	Information
CCTV, intercom, emergency assistance request button	Communications initiation
2-way audio intercom and CCTV monitoring	Communications
Application dependent	Destination selection
CCTV, intercom, operator	Emergency response
Facility for ride sharing	Yes

The information and communication subsystem

The components and functions of the subsystem for informing and communicating at both vehicles and berths/stations are given in the self-explanatory Table 6.1.

The fare collection subsystem

The subsystem for collecting fares consists of ticketing machines located at the stations/berths. After selecting the destination and paying by an available means (cash, credit, debit card), the user receives a magnetically encoded card. Then, he/she takes the ticket and goes to the boarding platform where the ticket is passed through the stanchion in front of an available (empty) vehicle. Then, the information about the travel itinerary is transferred to the microprocessor on-board the vehicle. The door opens, the vehicle is boarded, the door closes, and the zone and then the central computer, based on the information received from the given vehicle, generates and then opens the route/path for the vehicle depending on the surrounding traffic. The nonstop transit to the intended destination begins. There, after the vehicle's door opens automatically, the passenger leaves the vehicle.

6.2.2.4 Operational Performances

The operational performances of PRT systems include demand, capacity, service quality, fleet size, and technical productivity.

Demand

The PRT system is designed to replace individual passenger cars, particularly in congested urban areas. Therefore, the primary category of users are passengers abandoning their cars and those using the system for the first time instead of their cars. Some research has indicated that the proportion of car users switching to the PRT system could be up to 30 %. In addition, switching from other urban transit modes could also be expected, but the more exact volumes remain the matter of further investigation on a case by case basis. The PRT system's services are supposed to be equally and fairly available to all categories of individuals or small groups users/urban passengers regarding their gender, social category, trip purpose, and current mode choice.

Capacity

The capacity of a given PRT system generally includes the capacity of vehicles, lines, and stations. The vehicle capacity is expressed by the number of spaces for users-passengers. The capacity of lines and stations can be expressed by the traffic capacity and the transport/transit capacity. The former reflects the number of vehicles and the latter the number of users-passengers handled and transported, respectively, during the specified period of time (usually 1 h) under conditions of constant demand for service.

The traffic capacity of a given PRT system's line can be estimated as follows:

$$\Lambda_{\text{traffic}} = T/h \tag{6.1a}$$

where
T is the period of time (minute, hour); and
h is the minimum headway (seconds, minutes).

From Eq. 6.1a, the transport/transit capacity of the same PRT line can be estimated as follows:

$$\Lambda_{\text{transport/transit}} = \Lambda_{\text{traffic}} * \lambda * n \tag{6.1b}$$

where
λ is the average vehicle's load factor or occupancy rate; and
n is the number of spaces for users/passengers on a vehicle (pass/veh).

The capacity of a PRT system station can be static and dynamic. Its static capacity can be expressed by the required number of berths simultaneously used by PRT vehicles as follows:

$$N_s = \Lambda_s * \tau + N_w \tag{6.1c}$$

where
Λ_s is the intensity of demand for using berths at a given station;
τ is the average occupancy (i.e., dwell) time of a berth at a given station by an active (operating) vehicle; and
N_w is the number of berths for vehicles waiting to be called.

The other symbols are analogous to those in the previous equations.

The dynamic capacity or throughput of a given station is expressed by the number of vehicles processed during a given period of time. For a station with N_s berths, its capacity or throughput (veh/min or veh/h) can be estimated as follows:

$$\Lambda_s = N_s/\tau \tag{6.1d}$$

For example, let the minimum headway in Eq. 6.1a be: $h = 0.5$ s. This implies that the traffic capacity of a line during the time of: $T = 1$ min can be: $\Lambda_{\text{traffic}} =$ 120 veh/min (Similarly, with a headway of $h = 1$ and 2.5 s, this capacity will be:

$\Lambda_{traffic} = 60$ and 24 veh/min, respectively). If each vehicle has $n = 3$ spaces/seats and all are occupied, i.e., $\lambda = 1.0$, based on Eq. (6.1b), the transport/transit capacity of a given line will be: $\Lambda_{transport/transit} = 120 \times 3 \times 1.0 = 360$ pass/min (if the traffic capacity is: $\Lambda_{traffic} = 60$ and 24 veh/min, this capacity will be 180 and 72 pass/min, respectively). After considering a given line as a part of the PRT network by taking into account the more realistic values of the headways and the load factor of vehicle, both capacities will certainly be lower. Nevertheless, the minimum headway must not be shorter than the vehicle breaking distance at a reasonable deceleration/braking rate (1.0–1.5 m/sec^2).

If the intensity of demand for a given station in Eq. 6.1c is: $\Lambda_s = 120, 60,$ and 24 veh/min, and the vehicle dwell time at the berth $\tau = 20$ s, the required number of berths will be: $N_s = 40, 20,$ and 8, respectively. Similarly, it follows from Eq. 6.1d that the maximum throughput of a station with 5, 10, and 20 berths, each occupied by a vehicle for about: $\tau = 20$ s, will be: $\Lambda_s = 15, 30,$ and 60 veh/min, respectively.

In order to enable efficient and effective operations of a given PRT system, the capacities of lines, merging segments, and off-line stations need to be balanced. Consequently, the resulting capacity of a given line with the given number of off-line stations will be the minimum among the capacities of these elements.

Quality of service

Measures of the quality of service of a given PRT system can include accessibility and availability, schedule delay, transit time, punctuality and reliability, and riding comfort.

Accessibility and availability

The PRT system is designed to be highly spatially accessible in terms of walking distance and related time. Nevertheless, locations for joining the system are not as close to the users' residences or workplaces such as those of cars as users have to walk a short distance to one of the PRT stations. Ideally, the distance is about 150 m requiring up to 5 min walking time. In terms of availability, the PRT system is inherently highly available (some concepts are specified to operate 24 h/day).

Schedule delay

PRT is designed as a demand-responsive system where vehicles wait for users-passengers, which is distinctive from other urban mass transit systems where the users/passengers usually wait for the vehicles. However, this is not always true. PRT systems may also incur schedule delays as the time difference between the moment a user-passenger makes a request for a free vehicle and the moment it becomes available. This vehicle can be from a stock waiting at the given station, sent from some other waiting location, or have just arrived and been vacated by its previous user(s). Consequently, the schedule delay can be zero, tens of seconds, and/or a few minutes, the latter two during peak and off-peak periods, respectively. Facilitating the users/passengers and vehicles in the first of the above-mentioned cases confirms the claim that at PRT system(s), the vehicles wait for users/passengers rather than vice versa.

Transit time

The transit time depends on the distance between the selected origin(s) and destination(s), and operating speed. This time consists of the time of boarding the vehicle after buying a ticket, the time for the vehicle's acceleration/deceleration (not higher that 1–1.5 m/s^2), and the time the vehicle cruises at the operating speed mentioned above. Such composed transit time is continuous, uninterrupted by congestion, minimal, and highly predictable under the given circumstances.

Reliability and punctuality

Like other public urban public transit systems, the reliability of a given PRT system can be expressed as the ratio between the realized and planned services during the specified period of time (day, month, year). The reliability to date of currently operating and planned PRT systems has been very close to 100 %. The punctuality of PRT system can be expressed as the proportion of user-passenger trips carried out during the specified period of time, on time, or within the prescribed time deviation(s) (in the latter case it is often referred to as dependability). The punctuality of already operating PRT systems has also been very close to 100 %. In general, both measures are actually expected to be near 100 % all the time and under all weather conditions.

Riding comfort

The riding comfort of a given PRT system is expected to be at a higher level than that of other urban transit systems. This is achieved through the above-mentioned design of the guideways, stations, and vehicles themselves. The guideways enable the smooth "gliding" of air-conditioned vehicles. The size and pattern of supply of transport capacity always guarantee a free seat to users-passengers. Thanks to the system of controlling the flow(s) and spacing between vehicles, their operating speed and acceleration/deceleration are always maintained within the prescribed (comfortable) limits (the later of about 1.0–1.5 m/s^2) excluding any unpredictable strong breaking, for example, due to preventing potential collision(s).

Fleet size

The vehicle fleet of a given PRT system consists of the vehicles in operation and in reserve. The former provide services while the latter are only engaged when the former need to be out of service due to repair and maintenance. Based on Eq. 6.1b, the maximum number of PRT system vehicles in operation can be roughly estimated as follows:

$$M = \Lambda_{traffic} * \upsilon \tag{6.2}$$

where

υ is the vehicle's average turnaround time over the PRT network (min).

The turnaround time υ in Eq. 6.2 depends on the average distance between particular origins and destinations and the related average speed. This implies that serving shorter trips at a higher speed will require a smaller vehicle fleet, and vice versa.

Technical productivity

The technical productivity of a given PRT system can be estimated for a single vehicle and for a fleet of vehicles. For a single vehicle, it is equal to:

$$TP_v = n * v \tag{6.3a}$$

where

n is the number of spaces in a vehicle

v is the vehicle operating speed (km/h).

The other symbols are analogous to those in the previous equations.

For example, for a vehicle with 4 spaces/seats operating at a speed of 40 km/h, the technical productivity will be: $TP_v = 4 \times 40 = 160$ seat-km/h.

Similarly, for a fleet of M vehicles, the technical productivity can be estimated as follows:

$$TP_{fv} = N * n * v \tag{6.3b}$$

where all symbols are analogous to those in the previous equations.

For example, the technical productivity of a fleet of $N = 20$ vehicles, each with $n = 4$ spaces/seats and operating at a speed of $v = 40$ km/h will be: $TP_{fv} = 20 \times 4 \times 40 = 3{,}200$ seat-km/h.

6.2.2.5 Economic Performances

The economic performances of PRT system(s) include the system costs, revenues, and the user/passenger benefits expressed in monetary terms.

Costs

The costs of a given PRT system(s) consists of the capital/investment cost for infrastructure, vehicle and control systems, operating cost for providing transit services, and the total cost as the sum of the annual capital/investment and operating cost.

Capital (investment) costs

The capital (investment) costs relate to the cost for building infrastructure and acquiring the vehicles and control systems for a given PRT system. This cost can be estimated as follows:

$$C_{iv} = c_i * L + c_v * M + C_{cs} \tag{6.4a}$$

where

c_i is the cost of building the PRT system infrastructure (€/km);
c_v is the cost of acquiring a vehicle (€/veh);
C_{cs} is the cost of the control systems (€); and
L is the length of the PRT network (km).

The other symbols are as in the previous equations.

The guideways of PRT systems are mainly elevated constructions for all or part of their length, though the structure can be relatively light. Some estimation of these costs are made for different systems and the averages are as follows: 3.8 million €/km for the infrastructure and systems, and 75,000 €/vehicle (EC 2007).

Operating costs

The operating costs mainly consist of the costs of material, energy, and labor for (daily) operating and maintaining the system. These costs can be estimated directly and indirectly: the former, by estimating the quantities of particular cost issues based on the perceived (or actual) scenario(s) of operating the system; and the latter, by assuming theses costs as a portion of the capital/investment cost.

For example, in the direct case, an estimation is arrived at from a base cost for 5 km of track and a fleet of 25 vehicles (including staff) plus additions for infrastructure per km and per vehicle, and for staff per km and per vehicle. The causal relationship obtained is: $C_o = 1600 + 67.0(L-5) + 11.0(M-25)$ (1000 €/year), where L is the length of the single track guideway (km) and M is the number of vehicles (EC 2007). In the indirect case, typical annual operating costs are estimated at about 3–5 % of the capital/investment costs.

Total costs

The total costs are the sum of the capital/investment costs for infrastructure, vehicles, and control systems, and the operating costs imposed by providing transit services during the period of exploiting the system, i.e., its life cycle, which is 20–30 years. Consequently, the total average unit cost per unit of the system's output expressed in pass-km can be estimated as follows:

$$c = [C_{iv} * (1 + q)]/[m * 365 * D * s] \tag{6.4b}$$

where

q is the average proportion of the investment costs taken as operational costs;
m is the depreciation period of the given PRT system (years);
D is the user/passenger demand (trips/day); and
s is the average trip length (km).

The average cost c in Eq. 6.4b expressed in €/pass-km linearly increases with increasing of the capital/investment and operational costs and decreases more than proportionally with increasing of the daily demand (passenger trips) and length of the system's depreciation period.

Revenues

The revenues of a given PRT system can be direct and indirect. The former are gained from passenger fares, the potential transportation of freight/goods, advertising, and subsidies from local authorities and/or particular interest groups. The latter are generally wide and distinctive such as savings in the land use/take combined with increased land values and improved accessibility, increased rents for commercial space located near the system, reduction in traffic congestion, savings in the user-passenger travel time, increased productivity of staff, increased convenience for third party partnerships (e.g., hotel user access), operational savings, and some financial benefits. In any case, both direct and indirect revenues should cover the total costs of the given PRT system.

6.2.2.6 Environmental Performances

The environmental performances of a given PRT system include the energy consumption and related emissions of GHG (Green House Gases), and land use/take.

Energy consumption

PRT systems consume electric energy for their operations. The electricity consumption per unit of output (pass-km) depends on many interrelated factors such as: propulsion efficiency, operating speed, load factor, vehicle weight, drag and rolling resistance, the number of stops along the route/path, and requirements for air-conditioning. Some estimates for U.S. transit systems indicate that the electricity consumption of a PRT system can be about 0.404 kWh/pass-km, which is lower than that of trolleybuses (1.09), passenger gasoline/diesel cars (1.06), and motor buses (0.62). Consequently, in the above-mentioned case, PRT system commercialization resulting in taking over trolleybus, individual car, and motor bus users/passengers could contribute to energy consumption savings of about 0.69, 0.66, and 0.22 kWh/pass-km, respectively. The total savings can be obtained by multiplying these figures by the volumes of pass-km carried out during the specified period of time. This advantage of PRT systems is believed to be mainly derived from substantially reducing or completely eliminating intermediate stops and increasing the average daily load factor (Anderson 2007).

Emissions of GHG

The emissions of GHG depend on the primary sources for obtaining the electric energy used for powering PRT systems, which are mainly country/region specific. However, some proposals for installing solar panels on the sides and the top of guideways for producing the electric energy needed for PRT vehicles have been made. In such a case, emissions of GHG would be minimal or even nonexistent. Consequently, in this respect, a given PRT system could be considered as completely environmentally neutral.

The savings in emissions GHG can be estimated similarly to those of energy consumption.

Land use

The land used by a given PRT can be also direct and indirect. Directly, land is needed for the posts (solid cellars) and stations supporting the elevated guideways. However, the land needed is very small compared to the land taken for building the infrastructure of other urban transit systems. Specifically, the land underneath the guideways (about 4.5–5.0 m wide) can be used for walking and biking paths, as well as for vehicle and pedestrian crossings. Indirectly, PRT systems can save land through reducing the need for operating and parking individual passenger cars, trolleybuses, and motor buses in the scale and scope to which they are replaced.

6.2.2.7 Social Performances

The social performances of a given PRT system include noise, congestion, traffic incidents/accidents (i.e., safety and security), and social welfare.

Noise

The PRT system is supposed to be a relatively "silent" urban transit mode as compared to its counterparts—individual passenger cars, trolleybuses and motor buses, and LRT (Light Rail Transit) systems. This is because the vehicles run on rubber tires along the concrete guideway's floor and the electric motors do not have moving components. Some measurements have shown that passing PRT vehicles generate noise of less than 50 dBA at a distance of 2.5 m and 32 dBA at a distance of about 10 m from the observer (EC 2007).

Congestion

PRT is by design a congestion-free system. This implies that the vehicles run rather independently without interfering with each other, thus providing highly punctual services, i.e., without significant deviations of the actual times at destinations from those planned. In particular, PRT can contribute to mitigating road congestion by replacing individual passenger cars, trolleybuses, and motor buses. For example, if the average occupancy rate of an individual passenger car is 2 pass/car, and that of a PRT vehicle 4 pass/vehicle, then each PRT trip can substitute about two car trips. Such substitution removes these and other (substituted) cars from the road traffic flow and thus contributes to mitigating congestion and related overall impacts. Some estimates indicate that about 30 % of present car users could switch to the PRT system, if an equivalent efficiency and effectiveness of services under given circumstances was provided (Anderson 2005, 2007).

Safety

PRT systems are designed to be absolutely safe implying that incidents/accidents do not occur due to known reasons. The following facts support such expectations: (i) the vehicles operate on dedicated, elevated guideways separated and isolated from other interfering urban transit systems; (ii) the spacing between vehicles is maintained automatically by the three-level computer-based control system particularly controlling their acceleration/deceleration, speed, and breaking; (iii) the vehicles are automatically stopped and the user/passengers allowed to get-off in cases of system component failure and/or emergency; (iv) vehicle boarding takes place individually (or as homogenous groups), thus preventing entry of undesired persons; and (v) the stations are relatively small and monitored, thus preventing collection of too many people, and consequently reducing risk of exposure to antisocial behavior that could compromise both individual and collective security.

Social welfare

The social welfare resulting from a given PRT system consists of urban and social effects

Urban effects

The urban effects can be considered at the macro- and microscale. On the microscale, these generally imply creating more livable high density urban structures and the system's apparent ability to support deurbanization of large urban agglomerations and/or the self-contained urban complexes efficiently, effectively, sustainably, and safely (see Chap. 2). On the macroscale, this implies an efficient use of land under the system's guideways and stations for some other purposes.

Social effects

Social effects could generally include the following: savings in land for individual car infrastructure–roads and parking spaces; reducing the frequency of incidents/accidents including the local third party impact; reducing the overall energy consumption and emissions of GHG; reducing noise emitted by individual cars; making accessibility to the green zones in urban areas on the one hand, and also to the public activities and services—schools, offices, stores hospitals—on the other easier; transporting also freight/goods and mail; and being permanently available and accessible for everybody at acceptable costs. The social impacts could include: reducing the number of employees mainly due to operating driverless vehicles; and suffering from unprofitability, i.e., an inability to cover the costs, thus requiring subsidies.

6.2.2.8 Policy Performances

The policy performances of PRT systems mainly refer to the existing and prospective barriers to their faster commercialization on a wider urban scale(s). Due to their mutual interrelations, the barriers to a given PRT system are summarized in no specific order as follows:

- Some researchers and academics express skepticism toward the system's controversy in combining two mutually incompatible components, namely elements of individual passenger cars and elements of other urban mass transit system(s): very small vehicles with very complex infrastructure are considered infeasible and incapable of handling greater volumes of user/passenger demand in dense urban areas due to insufficient capacity and network spatial coverage; such small vehicles seem to be very convenient in low density suburban areas due to them offering sufficient capacity and network coverage but the user/passenger demand is insufficient to be served efficiently and effectively;
- The planners and prospective system operators are faced with much uncertainty in predicting/forecasting the volumes of user/passenger demand. This makes evaluation of the system's overall social/economic feasibility complex, unreliable, and consequently difficult to make investment decisions. Furthermore, the necessary investments in a PRT system are very high by design;
- Urban authorities are usually concerned with finding funding/investors for the system, with its ability to coexist with existing transit systems, and its vulnerability to weather conditions. This also relates to the system's ability to solve all or only some of the existing urban mobility problems, and its technical reliability including the consequences of technical failures. Furthermore, there is always the inherent concern that introduction of fully automated systems like PRT can rise local unemployment and consequently compromise local GDP (Gross Domestic Product). Last but not least, skepticism could arise from the perceived negative impacts on the visual appearance of a given urban complex after system construction;
- Users/passengers may be skeptical due to their lack of sufficient knowledge about the system they are supposed to use. They could particularly feel uncomfortable when being driven automatically without any ability to control the running vehicles themselves. This also relates to being permanently monitored, traced, and tracked while in the system. The question whether all social groups in terms of age, gender, education, and technical knowledge will be able to use the system appropriately remains a matter of specific concern; and
- Standardized and harmonized legislation for driverless systems, specifying the responsibilities during operation/use of PRT services, is lacking in many countries.

6.2.2.9 Performances of Commercialized PRT Systems

The quantitative performances are illustrated for three already commercialized (WVU, ULTra, and Cybercab), and one forthcoming PRT system (Vectus).[2]

[2] The information is collected from different sources including websites, which are not particularly cited.

- The *WVU PRT system* was commercialized, i.e., started operations, under the same name between 1975 (phase I) and 1979 (phase II) by connecting five WVU (West Virginia University) campuses (Morgantown, West Virginia, USA). The system provides transit services for the University staff and students, as well as local residents.
- The *ULTra PRT system* has been operating from May 2011 under the name "Heathrow pod" at London Heathrow airport connecting Terminal 5 and the Business Car Park. There are plans to extend the network throughout the airport.
- The *Cybercab PRT system* started operations in Masdar City (Abu Dhabi) in 2010 under the name "Masdar PRT" connecting the North Car Park at Masdar City and Phase 1a of the Masdar Institute of Science and Technology. The original plan was to develop the PRT system along with an inter-city LRT (Light Rail Line), both also connecting to the Abu Dhabi airport, which would exclude access of all passenger cars to the city center. These would be parked at the borders of the area covered by the PRT lines as shown in Fig. 6.3.
- The *Vectus PRT system* is planned to start operations as "Suncheon Wetland PRT" *in 2013 connecting* Suncheon City Garden Expo site and the world famous Suncheon Coastal Wetlands Park. As such, the system is expected to serve mainly visitors to the sites.

The quantitative estimates of indicators and measures of particular performances of the above-mentioned PRT systems are given in self-explanatory Tables 6.2, 6.3, 6.4, 6.5, and 6.6.

Infrastructural performances

Specifically, Table 6.2 indicates that the length of the network typically being a single line is relatively short—the longest is that of the earliest WVU PRT system. The guideways are concrete and steel/concrete mainly elevated constructions. They have a similar width (up to 2 m) and maximum gradient (10°), and quite distinctive turning radii. The area of the square dynamic envelope is between about 4 and 8 m^2.

The system for maintaining the guideways operational during bad weather is case specific. Each case has off-line stations enabling the vehicles level entry. At these stations, the berth concept prevails either angled independent (30°) or straight in-line. The minimum berth width is about 4 m. The three recent and forthcoming cases are only with the beginning and end station and the similar dynamic capacity, i.e., throughput.

Technical/technological performances

In particular, Table 6.3 indicates that the vehicles of particular PRT systems are of a similar size, payload, and space capacity with the exception of the earliest WVU system.

In addition, all vehicles have rubber wheels except the forthcoming Vectus PRT system, which has rail wheels. The power is supplied differently to the vehicles of particular systems. For example, in the earliest WVU system, electricity is

Table 6.2 Infrastructural performances of the selected PRT systems

System/Network Performances/ Indicators and measures	WVU PRT Morgantown (USA) (1975)	Cybercab PRT Masdar City (Abu Dhabi) (2010)	ULTra PRT LHR[a] (UK) (2011)	Vectus PRT Suncheon Bay, (South Korea) (2013)
Guideway(s)				
Length of network (km)	13.9	1.7	3.9	5.3
Track width (at grade/elevated) (m)	2.5/2.5	1.75/1.85	1.75/2.10	1.40/1.40
Maximum track gradient (%)	10	10	10	10
Positioning profile	Mainly elevated (65 %)	Mainly elevated	Mainly elevated	Mainly elevated
Square dynamic envelope (width × height) (m)	2.40 × 3.30	1.75-1.85 × 2.30	1.75–2.10 × 2.00	2.10 × 2.50
Pavement (type)	Concrete	Concrete	Concrete on a steel structure	Steel rails on concrete/steel structure
Turning radii (m)	9.1	6.1	5.2	5.0
Height (m)	As necessary	As necessary	5.7	As necessary
Spacing of pods/(concrete pillars) (m)	N/A	N/A	18	30/50
Maintaining guideway operational	Heating	No need	Snow and ice vehicle	N/A
Stations				
Number	5	2	2	2
Spacing (km)	3.3	1.7	3.9	5.3
Berths per station	6/22[b]	4–6	2–4	4
Berth length (m)	5.0	4.3 at 30°	3.2	4.5
Beth capacity (veh/h)	68/210[b]	120	120	160–200

[a] London Heathrow Airport
[b] End-of-line station/Off-line station
N/A Not Available

provided through a 575 V AC bus bar system, which runs along the side of the track and is electrically heated (similar to the principle of metro trains). In the following two systems, the vehicles are equipped with batteries, which can be charged at stations—either at berths or in dedicated charging rooms at the operation facilities. In the forthcoming Vectus PRT system, power is to be provided to vehicles by a current collection system installed along the guideway(s). The vehicles are centrally driverless (WVU) or self-managed (others) and controlled by the central or distributed control system (Vectus PRT). Each vehicle of the above-mentioned systems has an independent CAS (Collision Avoidance System), as well as two-way communications between operator and passenger(s) on-board including CCTV, air conditioning, and LCD screens displaying journey status and other useful messages for passengers.

Operational performances

Table 6.4 indicates that most operational performances of the given PRT systems are quite distinctive. The exceptions are partially the minimum headway, the maximum speed, the maximum speed in curves, acceleration/deceleration rates, and reliability and punctuality of services.

Economic performances

The economic performances of the selected PRT systems given in Table 6.5 are relatively similar. This particularly relates to the unit investment and operational costs, and savings in passenger time. The question remains if these systems are economically feasible without subsidies.

The environmental and social performances

The environmental and social performances of the considered PRT systems in Table 6.6 are characterized by lower overall relative impacts on the environment and society as compared to those of individual passenger cars and motor buses. The energy consumption is roughly proportional to the vehicle size (capacity), while the emissions of GHG depends on the energy consumption and the primary sources for obtaining electricity.

Specifically, electricity is obtained from different primary sources staring from predominantly coal in West Virginia (U.S.), a mixture of primary sources in the UK and South Korea, and exclusively solar panels in Abu Dhabi. Thanks to the predominantly elevated network of guideways, the above-mentioned cases confirm the inherent neutrality of the PRT systems in terms of land use/take. In addition, their noise levels are similar under comparative conditions. Last but not least, they all are absolutely free from occurrence of traffic incidents/accidents, and related property damages, injuries, and losses of lives.

The performances of the above-mentioned PRT systems indicate that they are, at present, mainly designed and commercialized to serve specific—niche—urban transit demand by partially replacing individual passenger car, motor-, trolleybus, and sometimes LRT system use. As such, the case-specific and quite distinctive

Table 6.3 Technical/technological performances of the selected PRT systems

System/Network Performances/ Indicators and measures	WVU PRT Morgantown (USA) (1975)	Cybercab PRT Masdar city (Abu Dhabi) (2010)	ULTra PRT LHR[a] (UK) (2011)	Vectus PRT Suncheon Bay (South Korea) (2013)
Vehicles				
Length/Height/Width (m)	4.72/2.67/2.01	3.90/1.46/2.01	3.70/1.47/1.80	3.60/2.08/2.42
Weight (empty/full) (tons)	3.78/5.40	1.40/2.05	0.85/1.30	1.5/2.5
Payload (tons)	1.62	0.65	0.45	1.0
Capacity (spaces) (adults + children)	20	4 + 2	4 + 2	4 + 4
Suspension/ support	Rubber wheeled	Rubber wheeled	Rubber wheeled	Rail wheeled
Power supply (V/kW)	575 V AC/52	Batteries (Lithium ion)/-	4 × Batteries (Acid) 48/2	500 VDC/5
Engine type	Electric-linear	Electric motor driving vehicle wheels	Electric motor driving vehicle wheels	Electric-linear
Control systems				
Guidance—vehicle	Passive presence detectors	Reference magnetic markers	Electronic control steering	Captive to steel track
Control—vehicle	Central system	Central system	Central system	Distributed system
Operations—vehicle	Driverless/centrally managed	Driverless/ self-managed	Driverless/ self-managed	Driverless/ self-managed

[a] London Heathrow Airport

Table 6.4 Operational performances of the selected PRT systems

System/Network — Performances/ Indicators and measures	WVU PRT Morgantown (USA) (1975/1979)	Cybercab PRT Masdar City (Abu Dhabi) (2010)	ULTra PRT LHR[a] (UK) (2011)	Vectus PRT Suncheon Bay (South Korea) (2013)
Demand				
Passengers (000/year)	2250	255–365	500	3000
Capacity				
Fleet size	71	10	21	40
Traffic capacity[b] (veh/h/dir)	240	720	400	720–900
Transport capacity[b] (pass/h/dir)	4800	2880	3600	5760–7200
Minimum headway (sec)	15	5	6	4–5
Maximum speed (km/h)	50	40	40	70
Maximum speed in curve (radius 20/50/100 m)	N/A	16/26/36	18/28/40	-/36/43
Typical/maximum acceleration/deceleration (m/sec)	1.0/5.0	0.8/4.7	1.25/2.5–5.0	1.2/5.0
Technical productivity (000 pass-km/h)	240.0	76.8	64.0	91.9
Quality of service				
Accessibility (Walking distance-/time-min)	100–150/1.5–5	100–150/1.5–5	100–150/1.5–5	100–150/1.5–5
Availability[c] Wd/Std/Snd (h/day)	16/7.5/0	24/24/24	22/22/22	N/A

(continued)

Table 6.4 (continued)

System/Network Performances/ Indicators and measures	WVU PRT Morgantown (USA) (1975/1979)	Cybercab PRT Masdar City (Abu Dhabi) (2010)	ULTra PRT LHR[a] (UK) (2011)	Vectus PRT Suncheon Bay (South Korea) (2013)
Schedule delay (min)	1–5	0–2	0.3	N/A
Transit time (min)	11.5	2.5	6.0	7.0
Reliability (%)	98.5	99.0	99.7	N/A
Punctuality (delays)	Negligible	Negligible	Negligible	Negligible

[a] London Heathrow Airport
[b] Based on the minimum headway
[c] *Wd* Weekday, *Std* Saturday, *Snd* Sunday, *N/A* Not Available

Table 6.5 Economic performances of the selected PRT systems

System/Network — Performances/Indicators and measures	WVU PRT Morgantown (USA) (1975/1979)	Cybercab PRT Masdar City (Abu Dhabi) (2010)	ULTra PRT LHR[a] (UK) (2011)	Vectus PRT Suncheon Bay (South Korea) (2013)
Costs				
Investment cost (million €/km)[b]	6.55	3.8[c]	6.15	6.50–10.90
Operating cost (million €/year)	2.75	N/A	1.33	N/A
Subsidies (%)	35–40	N/A	N/A	N/A
Operating cost (€/pass)	0.9–1.1	N/A	1.38	N/A
Revenues				
User charge (€/pass)	0.35/free	Free	0.70	N/A
Savings of passenger time (min)/(€/min)	11.2/1.65	2.0/N/A	10.8/4.43	N/A

[a] London Heathrow Airport
[b] One-way guideway
[c] EC 2007
N/A Not Available

Table 6.6 Environmental and social performances of the selected PRT systems

System/Network —— Performances/Indicators and measures	WVU PRT Morgantown (USA) (1975/1979)	Cybercab PRT Masdar City (Abu Dhabi) (2010)	ULTra PRT LHR[a] (UK) (2011)	Vectus PRT Suncheon Bay (South Korea) (2013)
Energy consumption[b] (kWh/km)	0.40	0.19	0.13	0.24
Emissions of GHG[c] (gCO_{2e}/pass-km)	12.8	Nil	11.4	12.9
Land use	N/A	N/A	N/A	N/A
Noise (dBA)	N/A	N/A	<50 at 2.5 m	<50 at 2.5 m
Safety (Incidents/accidents)	Nil	Nil	Nil	Nil

[a] London Heathrow Airport
[b] Full vehicle
[c] Emission rates: UK- 527gCO_{2e}/kWh, South Korea 430gCO_{2e}/kWh (mixture of primary sources); West Virginia (US)—642gC_{O2}/kWh (mainly coal); Abu Dhabi—0gCO_{2e}/kWh (Solar)
N/A Not Available

performances indicate the lack of standardization of PRT systems. This suggests that the future eventual commercialization of PRT systems on a wider urban scale will need some standardization in order to, at least, improve their economic performances, i.e., reduce the overall costs.

6.2.3 Evaluation

PRT system(s) possesses both advantages (Strengths and Opportunities) and disadvantages (Weaknesses and Threats) as compared to individual passenger cars and other urban mass transit systems, which can be relevant for users/passengers, system operators, and local/regional urban communities and authorities.

Advantages

Users/passengers

- High availability closely following the pattern of demand in terms of time and volumes;
- Convenient riding comfort regarding the availability of space(s)/seat(s) and exposure to different forces;
- Acceptable travel cost (fares) as compared to those of other urban transport alternatives including individual passenger cars;

- Savings in the journey time including high punctuality of services, i.e., a high certainty of arrival at the selected destination(s) at the scheduled time; and
- Improved physical (spatial) accessibility as compared to that of other urban mass public transit systems.

System operator(s)

- Low operational costs due to operating driverless and fully automated vehicles and needing generally few employees; and
- High quality of transport services delivered very closely matching the expectations of users/passengers.

Local/regional community and authorities

- Contributing to mitigation of the overall impacts of urban transit systems on the environment and society in terms of energy consumption and emissions of GHG, land use/take, noise, congestion, and traffic incidents/accidents, and
- Contributing to increasing/improving the social/economic/business attractiveness of particular locations in a given urban area(s).

Disadvantages

Users/passengers

- Slightly diminished physical accessibility for those users/passengers switching from individual passenger cars;
- Concerns of particular categories of users/passengers due to their unfamiliarity with approaching and using the vehicle(s)/services; and
- The still present, although at a much lower scale, risk of exposure to some elements of antisocial behavior.

System operator(s)

- Setting up the system is costly mainly on account of high infrastructure costs;
- Covering the costs exclusively by operations is an inherent weakness; and
- The inherent uncertainty of providing sufficient transit capacity comparable to that of the other cooperating and/or competing urban mass transit systems.

Local/regional community and authorities

- Compromising the environment's visual impression due to the elevated infrastructure;
- Potentially requiring subsidies for covering operational and other costs; and
- Inability to resolve all urban transit problems as some authorities may expect.

The experience so far shows that only three of more than 50 conceptual, prototype, and fully developed PRT systems have been commercialized. This is mainly because of the above-mentioned policy–social barriers to their implementation where the total system operating costs seemingly dominate. Nevertheless, in the present context, PRT possesses all attributes of an advanced urban transit system, which certainly should be carefully considered for gradual implementation respecting its overall socioeconomic feasibility. This certainly will be a gradual process on a case by case basis continuing with niche applications. On the one hand, such development will continue to demonstrate and prove the expected advantages. On the other, it will contribute to the further maturating of PRT systems and consequently increased interest in their commercialization on a wider scale. Finally, the trends of both strengthening and weakening of the existing level of urbanization will certainly contribute to the wider use of PRT systems either as a complement or even an exclusive urban transit system.

6.3 Underground Freight Transport Systems

1970s	A British national tube transportation system for general commodity freight (similar to SUBTRANS) is launched by the British Hydro-mechanics Research Association (BHRA) (UK)
1984	The concept of long-haul freight/goods transport in capsules propelled by linear induction motors through a tube of a diameter of about 2.0 m called SUBTRANS is patented (U.S.)

6.3.1 Definition, Development, and Use

Large urban and suburban agglomerations and large multimodal transport nodes such as ports, airports, and other inland intermodal terminals located close to these urban agglomerations have increasingly suffered from severe traffic congestion with the following impacts: (i) lost time of passengers while traveling to and from their workplace and delays in delivery of freight/goods to their final destinations; (ii) incidents/accidents caused by trucks and individual passenger cars resulting in injuries and death, and damage and loss of property (vehicles); (iii) increasing energy consumption and related emissions of GHG (Green House gases) by all vehicles (trucks and cars) while in congestion; and (iv) local noise caused by congested slowly moving traffic. One of the prospective solutions to mitigate such impacts and their further escalation due to the prospective growth in freight transport volumes can be to partially replace trucks used for freight/goods deliveries by the UFT (Underground Freight Transport) systems. The UFT system concept of is not new. One of the first such systems was The Mail Rail System in London (UK) operating from 1927 by Royal Mail for moving mail across the city. At present, some UFT systems have been successfully used in Japan to transport bulk material (Nippon/Daifuku and Sumitomo Electric Industries). In addition, two UFT systems have been operating in Georgia (Tbilisi) for the movement of crashed rock and in Russia (Petrograd) for moving garbage. The others are two automated capsule systems—one for containers and the other for pallets—still at the conceptual level with some pilot trials conducted under laboratory conditions (Rijsenbrij et al. 2006).

The concept is rather simple: instead of moving freight/goods consolidated into shipments of different sizes such as standard pallets and/or containers aboveground by trucks, they are moved through underground pipes/tunnels on automatically controlled dedicated vehicles. The pipes/tunnels, which can be partially aboveground, connect different freight/goods origins and destinations in the given urban and suburban area(s). The vehicles/capsules, loaded with freight/goods shipments, move completely automatically though the pipes/tunnels. At the beginning, the pipes were expected to have a diameter/width of about 1 or 2 m mainly for transporting smaller shipments up to the size of pallets. Later on, pipes/tunnels with a much wider diameter/width have been considered for moving containers, swap bodies, and semi-trailers. This has resulted in emerging two fully automated distinguished concepts: (i) the Automated Capsule System for Pallets such as CargoCap (Germany), Subtrans (Texas, U.S.), and MTM (U.S.); and (ii) the Automated Capsule System for Containers such as CargoCapContainer (Germany) and SAFE Freight Shuttle (TTI, Texas, U.S.). Both concepts, but especially the latter one(s), imply interoperability while connecting to existing (conventional) freight transport systems.. All these systems require building completely new underground infrastructure—tubes/tunnels, which could be very costly if developed as networks. This is one of the main reasons why none of these concepts have been either privately or publically commercialized as yet, despite some positive feasibility studies (Liu 2004; Rijsenbrij et al. 2006).

6.3.2 Analyzing and Modeling Performances

6.3.2.1 Background

The UFT systems are characterized by their infrastructural, technical/technological, operational, economic, environmental, social, and policy performances. In the given context, the infrastructural performances relate to guideways, i.e., pipes or tunnels, the network, and stations. The technical/technological performances refer to vehicles/capsules, the system for their driving, movement control, and vertical transportation. The operational performances relate to demand, capacity, and quality of services, vehicle/capsule fleet size, and technical productivity. The economic performances imply the costs and revenues. The environmental performances, similarly as at other systems, refer to the energy consumption and related emissions of GHG (Green House Gases) and land use. The social performances include noise, congestion and safety (traffic incidents/accidents).

6.3.2.2 Infrastructural Performances

The infrastructural performances of a given UFT system relate to the characteristics of gudeway(s), i.e., pipes/tunnels, network, and stations/terminals.

Guideways (Pipes/Tunenls)

In general, UFT pipes/tunnels can be circular or rectangular/square. The former is recommended in cases when the pipes/tunnels are built deeper underground in order to withstand higher internal pressure.[3] Depending on the location, they need to be positioned sufficiently below the surface to avoid any conflict with existing underground structures such as building foundations, water and gas pipes, metro systems, etc. This could be from 6–8 to 30–50 m or even deeper. In addition, for pipes with a diameter of less than 1.2 m, the construction costs could be lower due to the possibility of using standard commercially available steel-made elements. Figure 6.4 shows a scheme of such circle-shaped design of the above-mentioned Automated Capsule System for Pallets.

As can be seen, to the certain extent, the shape of vehicles/capsules is adapted to the shape of the pipe/tunnel.

Rectangular/square pipes/tunnels appear to be more convenient respecting the shape of freight shipments, usually in the format of boxes/containers and pallets. In addition, the rectangular/squared tunnels can be built by the reinforced concrete similarly as the large concrete canals, which could substantively reduce the construction costs. Figure 6.5 shows an example of the cross-sectional shape of pipes/tunnels of the Automated Capsule System for Containers (Standard container: Length—12.2 m (40 ft), Height—2.60 m (8.5 ft), Width—2.44 m (8 ft); this is an equivalent of 2TEU (Twenty-foot Equivalent Unit) (Liu 2004).

As can be seen, the rectangular shape of pipe/tunnel enables them to more efficiently and effectively fit with the shape of the container(s) and other similar sized loading units (swap bodies and semi-trailers). Nevertheless, the fitting rate is usually determined by determining the proper blockage ratio defined as the quotient between the cross-sectional area of the vehicle/capsule and the inside cross-sectional area of the pipe/tunnel. A higher blockage ratio causes the aerodynamic drag on the vehicle/capsule to increase requiring increased electricity consumption. At the same time, it reduces the risk of collision of vehicles/capsules in case of power failures.

[3] In general, the internal pressure can be estimated as: $\sigma = z\gamma$, where γ is the average specific weight and z is the depth of the soil. This implies that the internal pressure increases linearly with increasing of the depth of soil. For example, if the average specific weight of soil is: $\gamma = 20$ kN/m^3 and the depth: $z = 30$ m, the resulting vertical pressure on the upper wall of a pipe/tunnel located there will be: $\sigma = 20 \times 30 = 600$ N/m^2 (Pa). In addition, the weight of surface constructions (pavements, tracks, buildings, etc.,) need to be added. The other is the lateral pressure component by the surrounding soil at a given depth.

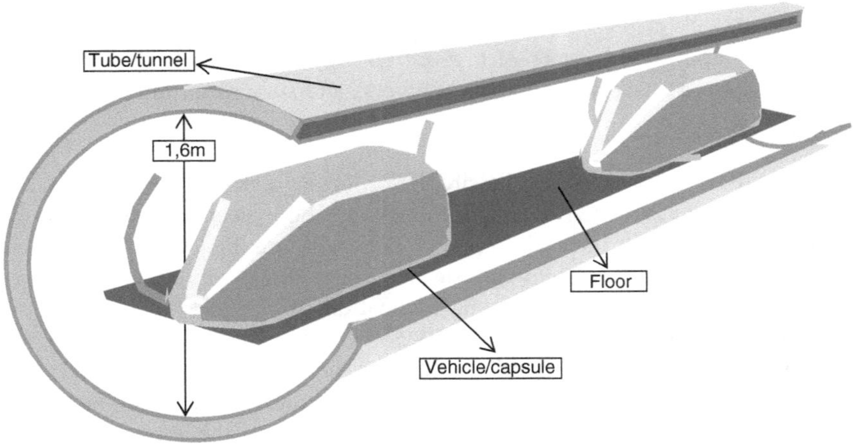

Fig. 6.4 Scheme of the UFT Automated Capsule System for Pallets—CargoCap (CC 1999)

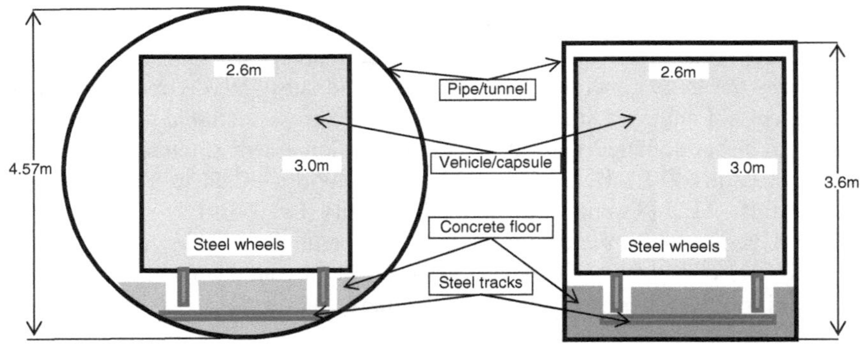

Fig. 6.5 Scheme of the possible cross-sectional shapes of pipes/tunnels at the UFT system for containers—one direction (Liu 2004)

In case of setting up bidirectional tracks within a single pipe/tunnel, its diameter/width will have to be at least doubled. Alternatively, twin pipes/tunnels each with a single track can be a solution. In both cases, different vehicles/capsules (full and empty) can move simultaneously in both directions of a given line. As mentioned above, the main material for building these larger pipes/tunnels can be steel and concrete, which depends on their shape and size as well as underground depth. The floor of pipes/tunnels would be made of concrete or steel with the steel tracks enabling movement of steel-wheeled vehicles/capsules. The alternative can be just a concrete floor enabling the movement of vehicles/capsules with rubber tires (as is the case in the current applications in Japan).

Network

In general, a given UFT system can be designed to operate as a single line or as a network of lines. In both cases, the lines connect network nodes/stations, which act as the entries and exits of the vehicles/capsules to/from the system, respectively. In order to properly cover an urban or suburban area, the UFT network needs to have a sufficient number of nodes/stations, each covering the specified "service area" (few hundred meters around the station at the surface, i.e., street level). In the horizontal plane, these stations/locations should be at or as close as possible to the intermodal terminals where the transfer of shipments between UCT and other conventional freight transport systems/modes, and vice versa, takes place. For example, in case of Automated Capsule System for Containers, these transshipment locations can be the port and inland container terminal(s). In case of Automated Capsule System for Pallets, the locations can be large logistics and shopping centers. This implies that the UFT networks for transporting containers could cover a much wider area than those transporting pallets and other smaller shipments. However, the density of nodes/stations of the latter can be much higher, mainly depending on the size and structure of activities of a given urban area. In any case, both types of systems are expected to operate at short- to medium-haul distances, i.e., up to 350–400 km. The main reason is that the greatest volumes of freight/goods are transported along such distances, thus guaranteeing sufficient demand for these systems.

Stations

The stations of the UFT network as its nodes are underground structures designed as platforms with several tracks for handling the incoming and outgoing vehicles/capsules. The area of floor of a given station depends on the number and length of tracks. The latter is influenced by the length of a single vehicle/capsule or the length of a "train" consisting of few units, i.e., vehicles/capsules. The floor level of these stations is usually located above the pipe/tunnel level, which enables using gravity for acceleration of the incoming and deceleration of the outgoing vehicles/capsules, respectively. Each station can have a set of short tunnels at the same level, each ending with an elevator (if the station is deep underground), which enables lifting vehicles/capsules to/from the street level. From there, the vehicles/capsules are delivered to their final destinations, and vice versa. Figure 6.6 shows a horizontal layout of a station/terminal of a UFT system for pallets. The pipes/tunnels connect the given station/terminal to the rest of the network. They all are underground. Four elevators deliver vehicles/capsules with or without pallets vertically to/from the neighboring streets (Liu 2004).

6.3.2.3 Technical/Technological Performances

The technical/technological performances of a given UFT system relate to the characteristics of its vehicles/capsules, and the systems for their driving, movement control, and vertical transport.

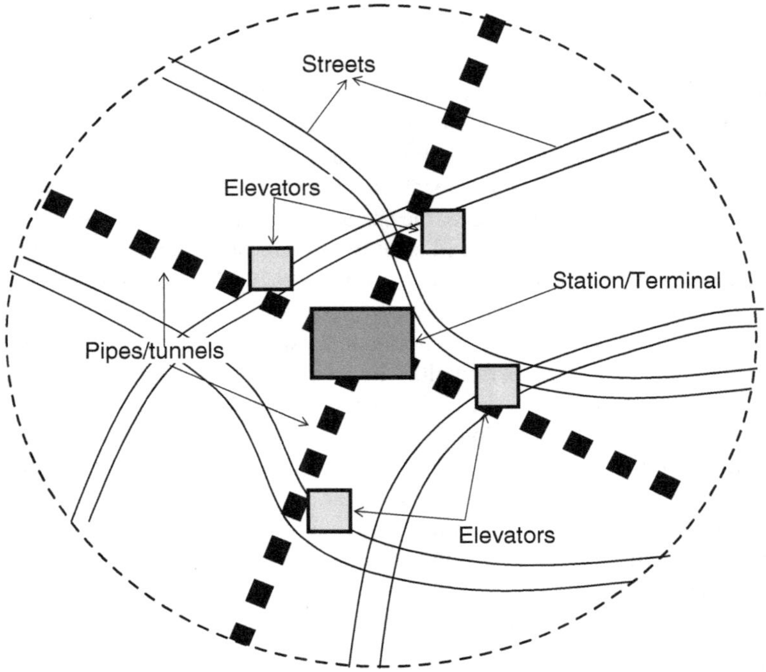

Fig. 6.6 Scheme of the station/terminal of a UFT system for pallets (Liu 2004)

Vehicles/capsules

Depending on the UFT system, the vehicles/capsules can carry (standard) pallets or containers, swap bodies and/or semi-trailers. Their cross-sectional shape would be closely adapted to the shape of pipes/tunnels and their length to the length of the boxes with freight shipments. Their steel wheels mounted on the bottom would support the gross weight, i.e., the weight of the vehicle/capsule and that of the freight/goods onboard. The steel wheels would enable movement along the steel tracks, thus considerably reducing the rolling resistance as compared to rubber tires (up to 0.002) and consequently contributing to saving energy. In addition, so-called side wheels can be mounted on the vehicles/capsules in order to provide their stability during moving through pipes/tunnels and reduce friction between them and the walls of pipes/tunnels.

The vehicles/capsules can operate as single units or be coupled as a "train" of two or three units. The typical capacity of a single vehicle/capsule of a UFT system for pallets usually corresponds to 2 Euro pallets (800 × 1,200 × 1,500 mm) or to about 2 tons of freight/goods. The dimensions and carrying capacity of a single vehicle/capsule of a UFT system for containers and swap bodies usually corresponds to one container or a swap body equivalent to 2TEU. Table 6.7 gives some characteristics of the vehicles/capsules of a UFT system for pallets and containers.

Table 6.7 Characteristics of UFT system vehicles/capsules for pallets and containers (CC 1999; Liu 2004)

Characteristic	UFT system	
	CargoCap (Germany)	New York city (USA)
	Pallets	*Containers*
Length/Width/Height (m)	4/1.4/1.6	12.81/2.6/3.0
Empty weight (tons)	0.8	25
Payload (maximal) (tons)	1.5	64
Gross weight (tons)	2.3	89
Track gage (mm)	800	1,453

The vehicles/capsules are mainly made from steel with thin outer aluminum cladding of 1–2 mm enabling driving by LIM (Linear Induction Motor). The loading/unloading of vehicles/capsules can be carried out at both underground and aboveground stations by using the conveyor technique. The roll conveyors on the vehicles/capsules automatically connect to those at the stations enabling loading units to pass between them.

The system for driving vehicles/capsules

While moving through pipes/tunnels, the vehicles/capsules are supposed to be most efficiently and effectively driven by one or more LIMs. The LIM can be considered as an electric machine that converts electrical energy directly into mechanical energy in translational motion. It consists of two components: a primary and a secondary. When powered by three-phase electric current, the primary winding generates a moving flux. Then the current induced in the secondary reacts with this flux and produces mechanical force. The LIMs are placed at convenient distances along particular lines and particularly at the stations of a given UFT system. At so-called LIM stations, usually two LIMs powered by three-phase current are installed at each side of a pipe/tunnel, acting as the motor's stator, i.e., the primary windings. Their shape should be adapted as close as possible to the shape of the vehicles/capsules. The primaries of such LIMs interact with the vehicles/capsules acting as the secondary while passing through, which generates electromagnetic and then mechanical force pushing them as a pump down the pipe(s)/tunnel(s). In order to act as the secondary, the walls of the vehicles/capsules are made of two layers—the inner of steel as a ferromagnetic material and the outer of aluminum as a good conductor. In order to ensure rather good efficiency, the air gap between the wall of the vehicle/capsule and the LIM must be about 1 cm at most. In general, the power, i.e., thrust force, of a given LIM (MW) can be estimated as follows (Liu 2004):

$$P_{in} = E * V * A * \Delta p \tag{6.5}$$

where

E is the LIM's efficiency (%);

V is the air speed in the pipe/tunnel under steady-state conditions (m/s);

A is the size of the inner area of a pipe/tunnel (m^2); and
Δp is the pressure drop along the entire length of the system under steady-state conditions (Pa(N/m^2)).

The thrust P_{in} in Eq. 6.5 is supposed to be controlled most effectively by changing the input current frequency, which enables the acceleration/deceleration, cruising at constant speed, and breaking/stopping of the vehicles/capsules. Alternatively, the vehicles/capsules could have breaking systems similar to those of conventional rail wagons, i.e., based on the use of compressed air as the energy source to activate brakes. Each vehicle/capsule would have its own tank with compressed air, whose discharging, and thus activating breaks would be achieved through an automatic remote control system and onboard PLC (Programmable Logic Controller). The brakes would be activated whenever needed—while approaching the stations until complete stop and/or when preserving a safe distance to the vehicle/capsule ahead.

For large LIMs driving UFT systems for containers and swap bodies, the required thrust could reach a couple of hundred MW and efficiency E should be about 70–80 %.

In addition, thanks to using rail-based technologies, the steel-wheeled vehicles/capsules could be switched between the steel tracks at stations/terminals or intersections/branches of the pipes/tunnels similarly to conventional rail wagons. The switches could be activated by the central computer when needed.

The system for controlling movement of vehicles/capsules

The vehicles/capsules need to be identified and sorted while moving through a given UFT network of pipes/tunnels. For such a purpose, the corresponding systems would be installed at the locations such as the entry and leave station(s) and inside the pipes/tunnels at each intersection/branching location. In the former case, the type of freight/goods is identified including shippers and receivers. In the latter case, the vehicle/capsule destination is identified in order to activate the appropriate switches. In general, UFT systems for pallets and containers are recommended to use a RFID (Radio Frequency Identification) system. This generally consist of a transported (identification tag) attached to each vehicles/capsule, an antenna and stationary reader both attached to the wall of the pipe/tunnel, and central computer located on the ground. As in the case of aircraft, the reprogrammable transponder sends a high frequency radio signal, which is picked up by antenna, read by the reader, and forwarded to the central computer for processing (Liu 2004).

The system for vertical transport of vehicles/capsules

When the pipes/tunnels of a given UFT system are located deep underground (a few tens of meters), the vertical transfer of both loaded and unloaded vehicles/capsules between the street level and the system needs to be provided. A simple straightforward solution is to use common elevators. They should be designed according to the purpose (UFT system case) with a carrying capacity of one or a few vehicles/capsules and a reasonable vertical speed of about 1–3 m/s. If the tunnel/pipes, particularly those of a UFT system for pallets, are not so deep underground, the entry/exit stations/terminals are elevated above the natural street level, thus enabling gravity force to be used for accelerating and decelerating incoming and outgoing vehicles/capsules, respectively.

6.3.2.4 Operational Performances

The operational performances of a given UFT system are demand, capacity, quality of service, the vehicle/capsule fleet size, and technical productivity.

Demand

Demand for a given UFT system can be expressed in terms of the number of pallets, containers, and/or swap bodies expected to be served during a given period of time (usually one year). In most case studies, UFT systems have been expected to considerably affect the conventional truck transport system in given urban or suburban areas by counting on taking over the volumes of demand that would ensure their economic feasibility over their life cycle. This implies that UFT systems would act as complete substitutes and not as complementary systems to conventional truck transport. In general, demand has shown to substantially influence the economic performances of UFT systems.

Capacity

The capacity of a given UFT system can be expressed by the maximum volumes of freight/goods consolidated into pallets, and/or containers or swap bodies processed during a given period of time under given conditions, i.e., constant demand for service. For example, the capacity of a pipe/tunnel expressed in the weight of freight/goods served can be estimated as follows:

$$C_p(T) = f(T) * n * w \tag{6.6a}$$

where

T is the time period under consideration (hour, day, week, year);

$f(T)$ is the frequency of vehicles/capsules passing through a given location in the pipe/tunnel in a single direction during the time period T;

n is the number of vehicles/capsules per train; and

w is the weight (or the number of units) of freight/goods on-board the vehicle/capsule (tons).

The frequency $f(T)$ in Eq. 6.6a can be estimated as follows:

$$f(T) = T/\tau \tag{6.6b}$$

where

τ is the minimum inter arrival time between two successive vehicle/capsule trains (s).

The time interval τ can be estimated as:

$$\tau = \delta / v \tag{6.6c}$$

where

δ is the minimum distance between any two successive vehicle/capsule trains (m); and

v is the average operating speed of a vehicle/capsule train in the given pipe/tunnel (m/s).

The typical average operating speed of UFT systems for pallets and containers is planned to be about v = 30–40 km/h and the maximum speed about 80–90 km/h.

In addition, the capacity of a given UFT system station in terms of the weight of freight/goods served during a given period of time can be estimated analogously as in Eq. (6.6a). In such case, the time interval τ represents the average injection interval between successive vehicle/capsule trains at a given station (s). The capacity of a pipe/tunnel and a station in terms of the volume of freight/goods served can be estimated similarly as in Eqs. 6.6a–6.6c.

Quality of service

Quality of service is expressed by indicators and measures such as accessibility and availability of services, delivery time, and reliability and punctuality of services.

Accessibility and availability of services

The accessibility and availability of services depends on the spatial coverage of a given UFT system and its operating regime. In particular, accessibility would be high if there were a sufficient number of entry and exit stations of the system in order to provide access of customers from the given urban or suburban area it serves, as well as at a reasonably close distance from their door. This implies that, in the optimal case, this distance should be about a couple of hundreds of meters. Availability implies that the system needs to be accessible in order to accept and deliver freight/goods shipments at the time convenient for the (quite different) users/customers. Therefore, most concepts consider 24-h operation, i.e., availability of access, every day of the year.

Delivery time

The delivery time of freight shipments by a given UFT system depends on the length of pipes/tunnels, i.e., lines, the operating speed of vehicle/capsule trains, and the time needed for their vertical transport between the system and the ground (street) level, and vice versa. This time can be estimated as follows:

$$t_d = 2(h/V) + L/v \tag{6.6d}$$

where

h is the vertical distance between the system and the ground (street) level with the entry/leaving stations (m);

V is the average speed of the elevators at both ends of a given pipe/tunnel moving vehicles/capsules vertically (m/s); and

L is the length of a given pipe/tunnel (m).

The other symbols area s in the previous equations.

Reliability and punctuality of services

UFT systems are expected to provide reliable and punctual services by design. This would be possible thanks to their full automation based on the centralized computer system monitoring and controlling vehicle/capsule train movements.

Fleet size

The fleet size, i.e., the number of vehicles/capsules, can be estimated as follows:

$$N(T) = \max\{Q(T)/w;\ f(T) * [t_l + 2(h/V + L/v) + t_{ul}]\} \tag{6.6e}$$

where

$Q(T)$ is the demand oforf freight/goods to be transported during time (T) (tons); and

t_l, t_{ul} is the average loading and unloading time of a vehicle/capsule at shippers and receivers, respectively (min, hours).

The other symbols are as in the previous equations. Equation 6.6e assumes that each vehicle/capsule, in addition to the time of delivering goods, also spends loading and unloading time at the shippers and receivers of freight/goods shipments, respectively; in addition, each individual vehicle/capsule or train moves full between freight/goods shippers–receivers and empty back, which implies utilization of the available capacity $N(T)w$ of about 50 %. This could be improved by reasonably widening the diversity of freight/goods shipments transported.

Technical productivity

The technical productivity of a pipe/tunnel of a given UFT system can be estimated from Eq. 6.6a as follows:

$$TP_p(T) = C_p(T) * v \tag{6.6f}$$

where all symbols are as in the previous equations. This indicates that, as intuitively expected, container UFT systems have much higher technical productivity than those for pallets, again explaining the shift in design from the latter to the former. For example, let a specific UFT system deal with containers and swap bodies. Each vehicle/capsule carries one container or swap body equivalent to 2TEU at an average speed of $v = 35$ km/h. If they were injected into the system at stations/terminals every: $\tau = 30$ s, during the time: $T = 1$ h, the technical productivity would be: $TP_p(T) = (3600/30) \times 1 \times 2\text{TEU} \times 35 = 8{,}400$ TEU-km/h.

6.3.2.5 Economic Performances

The economic performances of a given UFT system include its costs and revenues. In assessing the feasibility of particular UFT system concepts, both have usually been expressed per year over the period of the system life cycle of about 30 years.

Costs

The costs of an UFT system include the investment costs and operating costs.

The former includes the costs of building the system infrastructure (pipes/tunnels and stations), and the costs of acquiring: (i) supporting facilities and equipment (rails, elevators, LIMs, equipment in stations, control and communication system, etc.); and (ii) vehicles/capsules. The latter includes the costs of staff operating the system, the costs of electricity consumption, and the other (miscellaneous) costs. By dividing the total annual costs by the total quantity of freight transported, the average unit costs can be obtained as the basis for setting the prices of services. For example, some studies have indicated that the average unit costs of the UFT system for pallets serving New Your City (U.S.) could be: 0.177 $US/ton. The costs of the UFT system for containers in the same city would be: 17.2 $US/TEU (Liu 2004).

Revenues

Revenues are obtained by charging for services provided by a given UFT system. Usually rates are set to at least cover the unit total cost, thus guaranteeing feasible operation. Otherwise, subsidies would be needed to cover the negative difference between the costs and revenues. In the above-mentioned examples of the UFT systems serving New Your City, these rates would need to be competitive to those of conventional truck services: 20 $US/ton for the UFT system for pallets and 30 $US/TEU for the UFT system for containers (Liu 2004). This implies that both systems would be profitable even without considering their contribution to reducing externalities.

6.3.2.6 Environmental Performances

The environmental performances of UFT systems include energy consumption and related emissions of GHG (Green House Gases), and land use/take.

Energy consumption and emissions of GHG

UFT systems consume electrical energy mainly for powering the LIMs pushing vehicles/capsules through the pipes/tunnels. The power of a single LIM based on the power consumed by the flow of air and vehicles/capsules can be estimated as in Eq. 6.5. In general, the total electricity consumption of a given UFT system would be proportional to the product of the number of LIMs deployed, the power of each of them, and their operating time. Since all three parameters could be high at large-scale UFT systems, it is realistic to expect that they will, operating continuously, consume a substantial amount of electricity. In relative terms, i.e., the energy consumed per transported shipment, these systems are supposed to be more efficient that their conventional truck counterparts serving the same urban and suburban areas.

The total emissions of GHG of UFT systems is generally proportional to the product of the quantity of electric energy consumed and the emission rates of the primary sources for obtaining this energy. Again, these emissions with respect to UFT systems are expected to be lower than those of conventional trucks in both absolute and relative terms.

The achieved savings in energy consumption and related emissions of GHG by replacement of truck transport under given conditions could thus be used as one of the important criteria for assessing the overall social-economic feasibility of UFT systems. This could particularly be justifiable if emissions of GHG were internalized, i.e., considered as externalities.

Land use

UFT systems do not use/take additional land since they are predominantly underground constructions. Some very limited land may be used/taken at street level for building access to entry/leave stations. By replacing trucks, the requirements for new land for eventually widening the roads and streets including providing parking spaces for trucks would be significantly reduced and not completely eliminated. This would be again taken into account in assessing the overall social-economic feasibility of UFT systems.

6.3.2.7 Social Performances

The social performances of a given UFT system include noise, congestion, traffic incidents/accidents (safety and security), and social welfare comprising both urban and social effects. In contrast to conventional trucks, UFT systems are by design free of impacts such as noise, congestion, and traffic incidents and accidents, the latter causing injuries and loss of life. Even more, they are expected to be beneficial in relieving a given urban or suburban population from these impacts by trucks expected to be replaced. Thus, they contribute to overall social welfare, which should be, similarly as above, taken into account in assessing their overall social-economic feasibility. However, since such completely automatized systems do not require substantive employment, employment in the truck transportation sector due to rather substantial substitution could be affected, thus resulting in an increase of the total unemployment in the given context.

6.3.2.8 Policy Performances

The main policy performance of UFT systems is their attractiveness for the prospective public or private financiers in terms of full commercialization under conditions of completely liberalized freight transport markets (Visser 2010). In many cases, this will need additional clarification of the role of private and public entities interested in commercializing these systems. In addition, such clarification would enable overall changes in the perception of both society and DM (Decision-Making) institutions/bodies at different institutional levels (local-urban, regional, and national). Specifically, the government as the public entity and DM can be involved in commercialization of a given UFT system at four levels: (i) fully through public initiative; (ii) through public–private partnership; (iii) through supporting private initiative; and (iv) through providing a legal framework for a fully private initiative.

In addition, in order to start participating more actively in the eventual commercialization of UFT systems, the government, in general, needs to make three mental shifts (Winkelmans and Notteboom 2000). The first is the cognitive shift implying a change in perception toward UFT systems by considering them as a solution for some freight transport-related problems. The second is the strategic shift implying defining the strategic policy objectives related to UFT systems. The last is the operational shift implying acting according to the specified policy objectives. However, to date, the above-mental shift has not appeared to have fully taken place yet. For example, in the Netherlands, UFT has been the matter of private and public-funded research; governmental-policy consideration including setting up a test site have been carried out since mid-1997, but the whole initiative was put on hold in 2002. In Japan, many private research and commercialization activities were carried out on UFT systems in the 1980s and 1990s, but none with the strong government support. The currently commercialized UFT systems developed by Sumitomo Metal Industries for transporting different materials such as solid waste, minerals, and construction material are owned by private companies. In the UK, in the early 2000s, an initiative to commercialize a UFT system in London based on the Mail Rail tunnel network failed to receive government support at any level: local, regional, or national. In the USA, different initiatives for commercializing UFT systems have been either fully or partially sponsored by the states' or central government since 1990. However, these systems including the above mentioned for pallets and containers in New Your City have not yet received government support for final commercialization. In Germany, despite being co-sponsored by the government, it is still very uncertain whether the CargoCap systems for pallets and containers will obtain government support for full commercialization.

An additional policy performance of UFT systems is represented by access for users/customers acting as systems operators. The question is whether access will be granted to one or a few mutually competing operators, like it was in the past and is at present, respectively, within the railway freight transport sector. This seems to be strongly dependent on the ownership, i.e., on whether the systems will be private or public goods. In the former case, the privately built and operated UFT system would allow access to a single user/operator. In the latter case, access would be given to a few users/operators.

In summary, it seems that the above-mentioned policy performances of UFT systems will need additional time for the overall mental shift to occur, thus opening the path toward their commercialization under given (market) conditions.

6.3.2.9 Quantitative Performances of Selected UFT Systems

The performances of UFT systems are illustrated by two concepts: Automated Capsule System for Pallets—CargoCap (Germany) and Subtarns (USA), and Automated Capsule System for Containers—CargoCapContainer (Germany) and Freight Shuttle (Texas, USA). They are given in the self-explanatory Tables 6.8 and 6.9, respectively (Kersting and Draganinska 2005; Rijsenbrij et al. 2006).

Tables 6.8 and 6.9 indicate that UFT systems for pallets and UFT systems for containers and swap bodies, despite being designed for application under different micro conditions, could be standardized similarly as railways, particularly regarding some of their infrastructural, technical/technological, and operational performances.

6.3.3 Evaluation

The UFT systems for either pallets or containers possess both advantages (Strengths and Opportunities) and disadvantages (Weaknesses and Threats).

Advantages

- The technology is known and available, thus making such systems for both pallets and containers technologically feasible, i.e., sustainable;
- Some improvements in the social-economic feasibility could be achieved by operating 24 h/day and fully automatizing the processes within the logistics chains these systems serve;
- Mitigating the impacts of freight transport on the environment and society in a given urban or suburban area(s) by direct and indirect savings in the energy consumption and related emissions of GHG (Green House Gases), local noise, congestion, traffic incident/accidents, and land use/take—all as compared to the equivalent volumes of the presumably substituted transport by road trucks; and
- Flexibility in improving the overall social-economic feasibility by including direct and indirect impacts on the society and environment– in case of internalizing externalities in the freight transport sector in a given urban or suburban area(s).

Disadvantages

- Requiring completely new, dedicated predominantly underground infrastructure (lines and networks of pipes/tunnels);

Table 6.8 Performances of UFT systems: Automated Capsule System for Pallets—CargoCap/Subtrans (Liu 2004; Rijsenbrij et al. 2006)

Performances/type/indicator	Value of indicator
Infrastructural	
Pipe/tunnel (shape)	
Circular (diameter) (single line) (m)	1.2/2.1
Pipe/tunnel	
Length (km)	80/644
Construction material	Steel or concrete Steel
Tracks	
Station(s) (number/m)	-/1/610
Network (type)	Point-to-point (corridor) (Limited number of freight/goods origins and destinations)
Technical/technological	
Vehicle/capsule	
Body/Wheels	Steel
Carrying capacity	
Euro pallets/vehicle/capsule	
(W × L × H = 800 × 1200 × 1500 mm)	2/6
Maximum load (tons)	2/2
The vehicle/capsule driving system	LIM/LIM
The vehicle/capsule braking system	LIM or compressed air based
Operational	
Demand (tons/day)(000—estimated)	-/-
Capacity (tons/day)(000—estimated)	113/205
Speed (km/h)	
Average	36/40
Maximum	40/90
Operating time (h/day)	24/24
Economic	
Average operating cost (€/ton-km)	0.72/0.54
Average total cost (€/ton-km)	1.51/0.67
Environmental	
Energy consumption (kWh/ton)	-/0.17
Emissions of GHG (Green House Gases)	Depending on the primary sources for producing electricity
Social	
Noise	Free/Free
Congestion	Free/Free
Safety (traffic incidents/accidents)	Free/Free
Policy	
Availability of technology (Yes/No)	Yes/Yes
Level of development	Prototype without tests/LIM tested in the laboratory
Perception/acceptance level	Low-Fuzzy/Low-Fuzzy

Table 6.9 Performances of UFT systems: Automated Capsule System for Containers—Cargo-CapContainer/Freight Shuttle (Liu 2004; Rijsenbrij et al. 2006)

Performance/type/indicator	Value of indicator
Infrastructural	
Pipe/tunnel (shape)	
Rectangular (two lines) (width, height) (m)	10, 7
Circular (diameter) (single line) (m)	8
Pipe/tunnel	
Length (km)	-/116
Construction material	Concrete
Tracks	Steel
Station(s) (number)	-/2
Network (type)	Point-to-point (Corridor)
Technical/technological	
Vehicle/capsule	4-axes automotive rail wagon
Body/Wheels	Steel
Carrying capacity (TEU/vehicle/capsule)	2
The vehicle/capsule driving system	LIM (three-phase current)
The vehicle/capsule braking system	LIM or Compressed air based
Operational	
Demand (TEU/day) (000—estimated)	10/30
Capacity (TEU/day) (000—estimated)	-/30.2
Speed (km/h)	
Average	40
Maximum	80–90
Operating time (h/day)	24
Economic	
Average operating cost (€/TEU-km)	0.03–0.08
Average total cost (€/TEU-km)	0.04–0.35
Environmental	
Energy consumption (kWh/TEU)	-/112.8
Emissions of GHG (Green House Gases)	Depending on the primary sources for producing electricity
Social	
Noise	Free
Congestion	Free
Safety (traffic incidents/accidents)	Free
Policy	
Availability of technology (Yes/No)	Yes/Yes
Level of development	Concept/No prototype or demonstration
Perception/acceptance	Low-Fuzzy

- Being still at the conceptual level of commercialization—despite the availability of the technology, only a few prototypes have been built and limited demonstrations carried out;
- Uncertainty in the overall socioeconomic feasibility due to high infrastructure costs such as the tunnel-building costs and the highly uncertain average operational costs;
- Requiring additional costly consolidation/deconsolidation of freight/goods shipments (the systems for pallets) to enable smooth interoperability with other conventional freight transport systems/modes—particularly road and rail;
- Requiring rather substantial and stable medium- to long-term demand for economically feasible operations under competitive conditions; at present such demand exists only in short-distance markets implying the potential feasibility of these systems in these markets;
- Inflexibility, i.e., inherent vulnerability, in adapting to changes of the structure of given urban and suburban area(s) due to changing locations of the businesses of particular service users/consumers (i.e., users leaving their locations); and
- Complex and time-consuming decision-making, planning, and building processes prolonging and increasing the uncertainty in time of the return on investment; in addition to difficulties to provide sufficient demand in particular markets, these facts make prospective investors hesitant to act.

Finally, it can be said that the UFT systems are still at the conceptual level of development and commercialization. At present, prospective investors seem to be hesitant to commercialize them mainly due to the systems' high infrastructure costs and consequently high uncertainty in the overall social-economic feasibility. Although the technological barriers have been overcome, the policy barriers related to the role of particular actors involved in investment, ownership, access, and position in the liberalized freight transport markets in the urban and suburban areas they serve need further clarification. Nevertheless, it seems very likely that urbanization, the related conventional freight transport, and consequent environmental and social impacts, particularly those of road trucks, will continue to grow. Under such circumstances, UFT systems will have to be reconsidered in all abovementioned aspects as an alternative and complementary solution for mitigating the impacts in the medium- to long-term future.

6.4 Evacuated Tube Transport System

1910s	The modern concept of a Vactrain with evacuated tubes and maglev technology explored by U.S. engineer Robert Goddard (USA)
1914	A Vactrain concept offered in the book Motion without friction (airless electric way) of Russian professor Boris Weinberg (Russia)
1970s	A series of elaborate engineering articles about Vactrains published in the year 1972 and 1978 by R. M. Salter of RAND (USA)

(continued)

(continued)

1980s	The experimental German TRM train to operate in the large underground tunnels with reduced air pressure equivalent to that at the altitude of 68,000 ft (21,000 m) proposed by the Swissmetro (Switzerland)
1980s	The transoceanic tube floating above the ocean floor anchored with cables proposed by F. P. Davidson (the chairman of the Channel Tunnel project) and Y. Kyotani (a Japanese engineer) (UK, Japan)

6.4.1 *Definition, Development, and Use*

The future economy and society until and beyond the year 2050 will likely mainly be characterized by demography, resources, and global development (ACARE 2010). Demography will be characterized by: (i) continuing growth: the world's population is expected to reach 9–10 billion; (ii) an aging population: by 2,100, the average age will be about 40 years in all countries; the largest population group will be that of about 50 years old; and both will likely increase the economic pressure on the health system; (iii) growing developing economies: the strengthening "middle" class in countries like Brazil, India, Russia, China will certainly create new increasing demand for mobility; and (iv) urbanization: by 2025 about two-thirds of the world's population will live in cities, many of which are to be developed into mega-cities.

The resources will be characterized as follows: (i) energy: the cost of energy will continue to increase particularly after the peak of crude oil production, which is expected around 2030; alternative sources will need to be developed in any case simply because the conventional ones will start to run out; (ii) technology: ICT (Information and Communication Technologies) will continue to influence and lead to an increasingly dematerialized economy; nanotechnology will be able to create new materials with a wide range of applications including in transport; and (iii) environment: the main development will seemingly be a change of the earth's climate with its perceivable consequences; certain parts of the world will face limited availability of fresh water.

Global development will be characterized as follows: (i) geopolitics: the rise of new economic and political powers such as China, Brazil, India, and Russia will lead to a multipolar world with the likely gradually decreasing influence of the USA and (ii) economy and trade: the character of economic development in the long-term will continue to be 'growth-leading to-growth' despite impacts of economic/political crises from time to time; the above-mentioned new emerging economic powers will continue to strengthen their influence.

The future transport system will be characterized by continuous growth of demand mainly sustained by demographic and economic development (growth). The main attributes of the transport system serving such demand are expected to be as follows: (i) connecting people: connecting large urban agglomerations and markets thus further fostering globalization of economic and trade; (ii) providing transport services of refined quality at reasonable costs: transport services will have to be adapted to the very differentiated users' needs (business and leisure) in terms of sufficient capacity, travel time, reliability, punctuality, safety, and security, i.e., in terms of the acceptable generalized door-to-door travel costs; (iii) further diminishing the impacts on the environment and society: the energy consumption and related emissions of GHG (Green House Gases), local noise, congestion, traffic incidents and accidents, and waste will have to be further reduced by advanced technologies and operations; and (iv) contribution to the national and global economics: this will continue to be achieved through employment and expanding, i.e., through synergies of new technologies from other fields/areas.

The ETT (Evacuation Tube Transport) or 'Vactrain' defined as a very high-speed long-haul transportation system could potentially be an advanced future transport system. The ETT system was elaborated for the first time in the 1910s by the U.S. engineer Robert Goddard. In general, the concept assumed that the trains would operate at speeds of about 1,600 km/h in vacuumed[4] underground tunnels/tubes. In addition, Russian professor Boris Weinberg designed a 'Vactrain' concept in 1914. The rather well elaborated concept again emerged in the 1970s (Salter 1972). In the late 1970s and early 1980s, Swissmetro considered operating experimental German TRM (TransRapid MAGLEV) trains in large underground tunnels with the pressure reduced to the altitude of 68,000 ft (21,000 m). In the 1980s, Frank P. Davidson, founder and chairman of the Channel Tunnel project, and Japanese engineer Yoshihiro Kyotani proposed a system consisting of a tube floating at least 300 m below the ocean surface anchored with cables. The depth of floating would be sufficient to prevent impacts of water turbulence.

In these concepts of the 1970s, the range of operating speeds of ETT system trains was supposed to be several thousand km/h (6,400–8,000 km/h, which is about 5–6 times the speed of sound at sea level and standard conditions—1,225 km/h). As such, in terms of operating and commercial (travel) speed, the ETT system would be superior to conventional and past supersonic (Concorde, TY144) APT, and quite comparable to the forthcoming STA (Supersonic Transport Aircraft).

After being fully conceptualized, the ETT system would be able to take over substantial numbers of passengers and volumes of freight/goods from the APT (Air Passenger Transport) system and thus, as an environmentally friendlier system/mode, contribute to mitigating the negative impacts on the environment and society in terms of energy consumption from nonrenewable resources (crude oil) and related emissions of GHG (Green House Gases), and noise around airports.

[4] 'Vacuum' is defined as an air-free or almost the air-free space.

6.4.2 Analyzing and Modeling Performances

6.4.2.1 Background

The main components of a given ETT system are its vacuumed tunnels/tubes, TRM trains, and supporting facilities and equipment for energy supply, maintaining vacuum in the tunnels, train/traffic control/management systems on-board the vehicles and on the ground, and fire protection systems. They all influence the system's performances, and vice versa.

Similarly to other transport systems, the ETT system is characterized by its infrastructural, technical/technological, operational, economic, environmental, and social/policy performances. In general, even under the present conditions of no prototype system existing, they could be analyzed and modeled using the "what-if" scenario approach and bearing in mind the preferences of the particular actors/stakeholders involved such as providers/managers of transport infrastructure, manufacturers of vehicles/trains, users and providers of transport services, governing bodies at different institutional levels (local, regional, national, and international), and local communities.

6.4.2.2 Infrastructural Performances

The infrastructural performances include design of the individual tubes and their network.

Tubes

The infrastructure of the EET system would be designed according to two concepts. The first implies building underground tunnels under the sea floor, while the alternative implies installing undersea floating tubes anchored to the seabed by steel cables. The latter concept would consist of two designs: (i) two transport and one separate service/maintenance tube; the separate tube would be also shared with pipelines for oil, water, gas, electric power transmission lines, communication lines, etc. and (ii) a single tube divided vertically into three sections—the main section with the train lines, the maintenance section above, and the emergency section below the main section. Figure 6.7a, b shows these designs where TRM trains would be used (Salter 1972; Sirohiwala 2007).

The floating tubes could be made of two types of materials: either pure steel guaranteeing air-proof at a rather moderate cost or composite materials with steel and concrete where the inner thin layer of steel would be the inside and the concrete the outside wall of the tube. Due to the need for eliminating heating from the tube mainly coming from train operations, materials with good thermal conductivity would have to be used. For example, common steel would be better than stainless steel. Locating tubes deep underwater (300 m) in combination with water pipes inside the tubes could be a solution for the heating problem (Zhang et al.

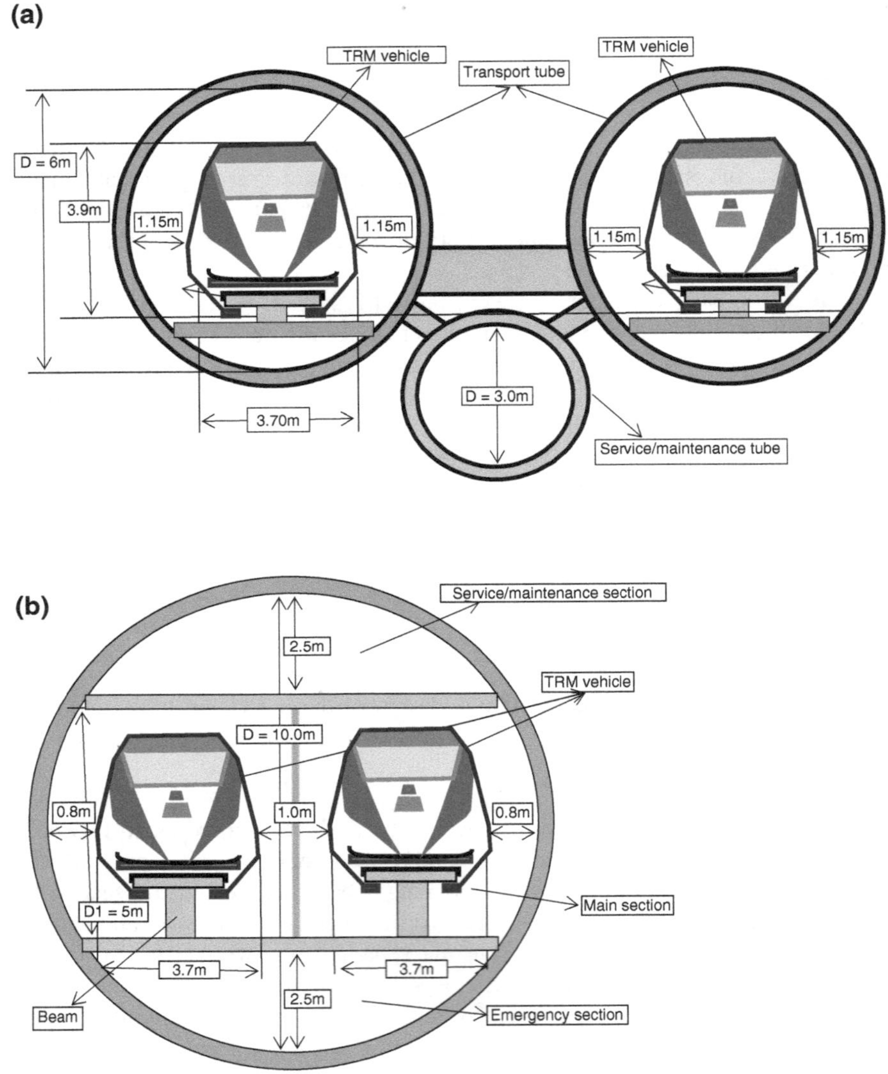

Fig. 6.7 Schemes of the tube design for an underwater ETT system, **a** Double tube, **b** Single tube

2011). The thickness of the tubes' walls would be sufficient to sustain the water pressure at a given depth from the outside and almost zero pressure from the inside (at the depth of 300 m the outside pressure is about 30 atm, i.e., the pressure increases by 1 atm for each 10 m of depth (atm—atmosphere)). The tubes would be composed of prefabricated sections joined together in order to compose an airtight tube. Alternatively, an interlocking mechanism would be incorporated into the sections in order to keep them assembled. The vacuum-lock isolation gates at the specified distance would be constructed in order to evacuate air from particular sections of the tubes more efficiently and thus prevent the spreading of potentially large-scale air leakages throughout the entire tubes. These gates would consist of vertically up and down moving doors, which could also function as part of the fire protection system. These doors would be closed during initial evacuation of air from the tubes and in the cases of large-scale leakages, and opened otherwise (Salter 1972).

The question arises whether such built and equipped tubes incorporating additional TRM guideway(s) would be able to float at the given depth. This will depend on the resultant buoyant force W_b as the difference between the weight of the tube W_p (positive) and the weight of the displaced water W_w (negative) as follows:

$$W_b = W_p - W_w = M - \rho_0 * V \tag{6.7a}$$

where

M is the mass (weight) of the tube(s) (kg);
ρ_O is density of sea water (tons/m^3); and
V is the volume of displaced water equal to the volume of tube(s) (m^3).

If $W_b = 0$, the tube(s) will float at the surface; if $W_b < 0$, the tube(s) will be pushed upwards implying that they would need to be anchored to the ocean floor by a cable system in order to stay at the given depth; if $W_b > 0$, the tube(s) will sink to the sea floor.

The mass (weight) of the tube can be determined as the product of the volume of the shell of a tube ΔV and the specific gravity of the material s_m. The volume of the shell of a tube can be calculated as follows:

$$\Delta V = \pi * (R_2^2 - R_1^2) * L \tag{6.7b}$$

where

R_1, R_2 is the inside and outside radius of the tube (m) ($R_1 < R_2$); and
L is the length of the tube (m).

The mass (weight) of the tube is as follows:

$$M = \Delta V * s_w * f = \pi * (R_2^2 - R_1^2) * L * s_m * f \tag{6.7c}$$

where

f is the factor of increasing the total mass (weight) of the tube due to its internal content.

The mass (weight) of the displaced water is equal to the product of the internal volume of the tube V and (sea) water density ρ_O as follows:

$$\rho_0 V = \rho_0 * \pi * R_1^2 * L \tag{6.7d}$$

where all symbols are as in the previous equations.

As shown in Fig. 6.7a, in the case of the two transport and single service/maintenance concept, the inside and outside diameter of each transport tube would be about $D_2 = 2R_2 = 6.2$ and $D_1 = 2R_1 = 6.0$ m, and that of the service tube $D_{s2} = 2R_{s2} = 3.2$ and $D_{s1} = 2R_{s1} = 3.0$ m, respectively. The diameter of transport tubes is specified regarding the profile of TRM trains of a width and height of 3.70 and 4.16 m, respectively (Table 5.4, Chap. 5). In addition, these tubes would accommodate "at grade guideways" for TRM trains (currently, their height is 1.25–3.5 m with a support by of reinforced concrete piers set up at a distance of 2.75 m; the height of the steel hybrid guider is 0.4 m (Chap. 5). For example, let the specific density of ocean water be: $\rho_O = 1.027$ tons/m^3; the dimensions of the tubes are as above; the factor $f = 2$ for installing guideway(s) and other systems inside; the average specific gravity of the tubes' material is: $s_m = 5.67$ tons/m^3 (this is the 60/40 % mix of steel (specific gravity: $s_s = 7.85$ tons/m^3) and concrete (specific gravity: $s_c = 2{,}400$ tons/m^3)). Then, for a given length of tube of $L = 1$ m, based on Eqs. 6.7a–6.7c, the buoyant force for the transport tube would be: $W_b = 21.72–29.02 = -7.3$ kg < 0, which implies that the tube would float and must thus be anchored to the ocean floor. By carefully adjusting the values of the parameters, particularly factor f, more accurate results could be obtained. In addition, the strength of the buoyant force would be used to specify the need for anchoring cables.

In the above-mentioned calculations, the thickness of walls of all tubes has been adopted to be 200 mm. But would this be necessary? Theoretically, the minimum thickness of a given tube (t_m) can be estimated as follows (Antaki 2003):

$$t_{\min} = \frac{3}{4} P_a * \frac{D_2}{B} \tag{6.7e}$$

where

P_a is the maximum allowable external pressure on the wall of the tube (MP$_a$); and

B is the strength of the material the tube is made of (MP$_a$).

The tube(s) would be evacuated and thus without any significant internal pressure. In the above example, the strength of the material (60/40 steel/concrete mix) is: $B_m = 1{,}240$ MP$_a$, which is at the same time the collapse pressure P_c ($B_s = 2{,}000$ MP$_a$; $B_c = 100$ MP$_a$; MP$_a$—Mega Pascal; 1 atm = 0.101325 MP$_a$). Then the maximum allowable pressure for this mixture of materials would be: $P_a = (1/3)\ P_c = 1{,}240/3 = 413.3$ MP$_a$. In addition, the actual pressure at a depth of 300 m where the tubes would be placed is: $P_a = 30$ atm (i.e., ≈ 3.04 MP$_a$), which is much lower than the above maximum allowable pressure. Then, based on Eq. 6.7e, the minimum required thickness of the tube wall would be:

$t_m \geq 11.03$ mm. For example, at a depth of 500 m, this would be: $t_m \geq 18.39$ mm. Both values are far below the adopted wall depth of 200 mm.

Taking into account the length of the link/line of about 5,664 km (the distance between London (UK) and New York (USA)), the quantity of material used to build two transport and one service/maintenance tube with 200 mm thick walls and the specific gravity of the mixture of material (5.67 tons/m^3) would amount to about 152 million tons, In addition, building the given link/line would take at least 20 years.

As compared to underground tunnels, this concept seems to be more attractive, mainly due to the nature of the construction work and supporting activities, and the overall investment costs. In addition, it seems to be more resistant to earthquakes and other kind of large-scale disruptive events on the one hand, but also more exposed to the eventual terrorist and other attacks on the other. The other details on the construction process could be found in the reference literature (Salter 1972; Sirohiwala 2007; Zhang et al. 2011).

Network

The individual tubes and intermodal terminals would compose the EET system network, which would consist of two subnetworks as follows:

- The subnetwork at the higher level consisting of links/lines connecting large (hub) intermodal passenger terminals at the coasts of particular continents. The links/lines as the tunnels/tubes would lie mainly under the sea level with a short portion as an underground construction (between the coast and the inland intermodal passenger terminal(s)); and
- The subnetwork at the lower level consisting of links/lines as tunnels/tubes connecting continental large urban agglomerations and coast terminals; these would be completely underground constructions.

The two subnetworks would meet each other at intermodal terminals near the coast(s). The links/tubes of the higher network would be predominantly under water. One of the characteristic cases could be the link/line under Atlantic Ocean connecting Europe and North America (e.g., London and New York). The tubes would begin and end at the intermodal passenger terminals enabling exchanging passengers between the ETT system and its feeder systems under the same "roof". Figure 6.8 shows the simplified sheme of an intercontinental higher level ETT system/network with a single link/line.

6.4.2.3 Technical/Technological Performances

Technical/technological performances relate to the vacuum pumps, vehicles, and traffic control/management system.

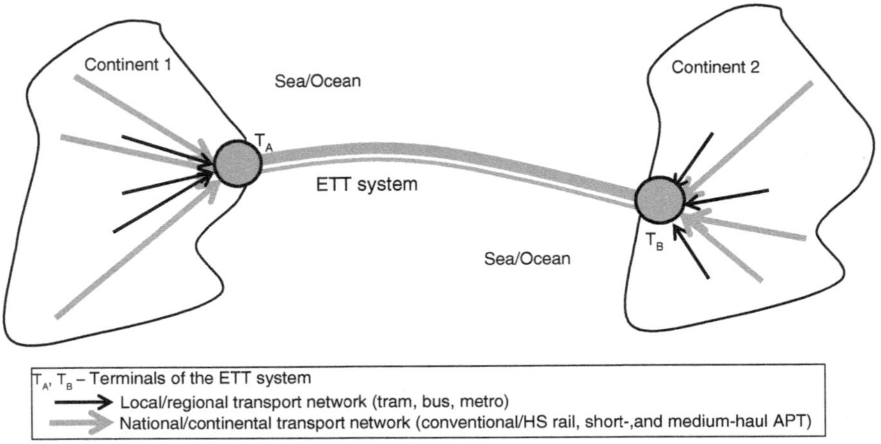

Fig. 6.8 Scheme of an intercontinental ETT system network with a single link/line (tube)

Vacuum pumps

After assembling the tubes, MAGLEV guideways, energy system for both vehicles, vacuum pumps, and other supporting facilities and equipment will need to be installed. The vacuum pumps would be applied to initially evacuate and later maintain the required level of vacuum inside the tunnels/pipes. In particular, creating the initial vacuum consists of two steps: (i) large-scale evacuation and (ii) removal of smaller molecules near tunnel walls using heating techniques. In addition, any potential leakages of air would need to be dealt with. These require powerful vacuum pumps consuming a lot of energy, at least during the initial step. An alternative could be to create a vacuum at very low pressure, for example, at about 0.01–0.0001 atm (equivalent to the altitude of about 50 km). How would the pumps work? At the initial stage, they would be operating until achieving the required tube evacuation level. Once the required vacuum conditions were set up, the pumps would automatically stop and the vacuum-lock isolation gates would be opened. In cases of air leakage in some section(s), the corresponding gates would be closed and the pumps activated again. The pumps would be located along the tube(s) in the required number depending on the volumes of air to be evacuated, available time, and evacuation capacity. For example, along the Atlantic line, about 200 units each with a vacuuming capacity of 100 m^3/min and energy consumption of 260 KW would be located at a distance of about 28 km. The volume of air to be evacuated from the two tubes with a diameter of 6.0 m and one tube of 3.0 m, and a length of 5,564 km would be about 354 million m^3. The initial evacuation would take about 12.3 days (Sirohiwala 2007).

Vehicles and propulsion

As mentioned above, the EET system would use TRM (Trans Rapid Maglev) type vehicles. Such as, for example, German TRM07 trains whose technical/technological performances are given in Table 5.3 (Chap. 5). The exceptions from those in this table are the operating speed and acceleration/deceleration rate(s), which would be, thanks to moving through vacuum, much higher—the operating speed would be about 6,400–8,000 km/h and the horizontal acceleration/deceleration to/from these cruising speeds about 3.0–5.0 m/s^2. The trains would be pressurized similarly to modern commercial aircraft (about 1 atm) (Zhang et al. 2011).

The ETT TRM trains would very likely be powered by some type of rocket engine, mainly due to the requirements for the very high supersonic speeds in the vacuumed tubes/tunnels. This is because these engines do not rely on the atmospheric oxygen. These rocket engines would presumably be powered by nuclear energy as a propellant. Most of the energy would be consumed during acceleration and deceleration phase of a journey. During cruising phase, due to the vacuum conditions in the tube/tunnel, the TRM trains would move thanks to the inertial force gained after acceleration. Consequently, the rolling and aerodynamic resistance during cursing phase of a journey as the two additional main causes for the energy consumption would be completely eliminated. In addition, operating in the airless tinnel(s)/tube(s) at the very high speeds would eliminate, on the one hand, the shock waves at the moment of breaking the sound barrier (important for passing trains in a single tunnel/tube concept), and on the other air friction and the consequent heating of trains. Even under conditions of the very low air density in the tubes, the latter two would be negligible. Nevertheless, heat shields would be incorporated in the trains in order to protect them from overheating caused by unpredictable air leakages. Under such circumstances, if, for example, a mass of the two-car TRM train was 110 tons (as in Table 5.3), then based based on the Newton's laws, a rocket engine of about 733 MW would be needed to accelerate/decelerate the train to/from the cruising speed of 8,000 km/h.

Traffic control/management system

The traffic control/management system would be fully automated. The main reason is the very high speed of TRM trains, which can be monitored but not controlled by their drivers. In addition, under such conditions, the driver simply would not have time to react in unpredicted events. Consequently, the trains would be controlled (guided) automatically analogously to modern unmanned flying vehicles (UAV)[5] , and managed (separated) along the link/line according to TRM principles.

6.4.2.4 Operational Performances

The operational performances relate to demand, capacity, and quality of service, vehicle fleet size, technical productivity, and a "what-if" operating scenario.

[5] At present, these are designed pilotless aircraft, but due to transmitted electronic means and/or their autonomous on-board flight management control system, they do not need active intervention by a pilot-flight controller.

Demand

The passenger demand for the above-mentioned higher level ETT subnetwork would be collected and distributed from the lower level ETT system, short-haul and medium-haul Air Passenger Transport (APT), High-Speed Rail (HSR), TransRapid Maglev (TRM), conventional rail and bus systems, local urban mass transit systems (subway, tram, bus), and individual passenger cars. The corresponding modal-based freight transport systems would be used for collecting/distributing the freight/goods shipments. For the lower level EET subnetwork, the demand would be collected and distributed in a similar way.

The most important attribute of choosing the ETT system would be the total door-to-door travel time. This consists of the accessibility time and the time on-board the ETT system. The time on-board consists of the scheduled delay, i.e., the average waiting time for a departure, depending on the departure frequency, and vehicle/transit time. If the value of passenger time is taken into account, the generalized cost function can be formulated as follows:

$$U(d,T) = \alpha\left[\theta + 1/2\frac{T}{f(T)}\right] + \beta\left[\frac{v}{g^{+}} + \frac{d}{v(d)} + \frac{v}{g^{-}}\right] + p(d,T) \tag{6.8a}$$

where

- α is the value of passenger time of assessing the ETT system and while waiting for a departure (€/min);
- θ is the time of accessing the EET system (min);
- β is the value of passenger time while on-board the ETT system (€/-min);
- T is the period of time (day, h);
- $f(T)$ is the service frequency during the period (T) (dep/T);
- $v(d)$ is the cruising speed of the ETT train along the route/line (d) (km/h);
- g^{+}, g^{-} is the acceleration/deceleration rate of the EET trains (m/s^2); and
- $p(d, T)$ is the fare for a trip on route/line (d) during time (T) (€/pass).

A similar form of generalized cost function could be used for the ETT system's main competitor—APT (Air Passenger Transport). Then, the probability of choosing the ETT system as mode (1) instead of APT as mode (2) at time (T) can be estimated as follows:

$$p_1(T) = \frac{e^{-U_1(T)}}{\sum_{i=1}^{2} e^{-U_i(T)}} \tag{6.8b}$$

The number of passengers choosing mode (1-ETT) at time (T) can be estimated as follows:

$$Q_1(T) = p_1(T) * Q(T) \tag{6.8c}$$

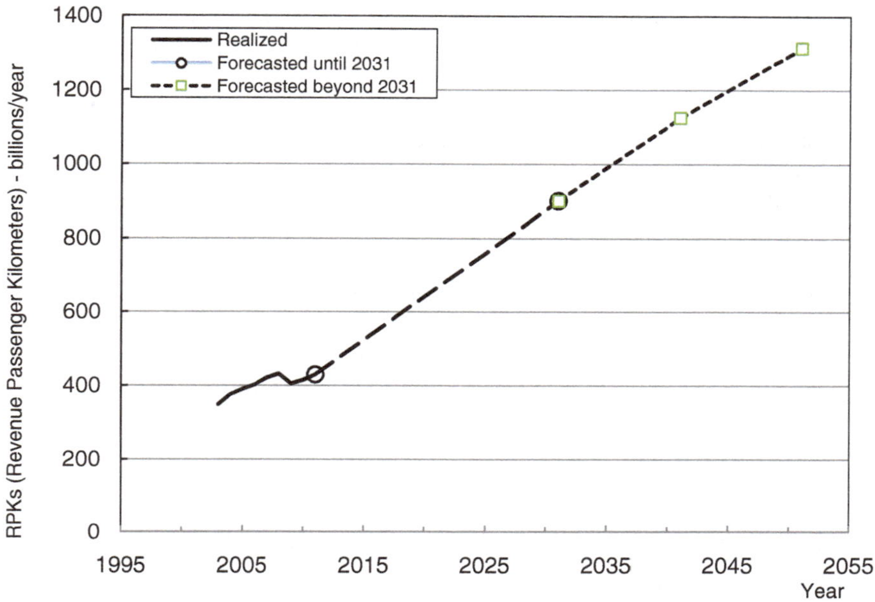

Fig. 6.9 Possible development of APT (Air Passenger Transport) demand between Europe and North America (Boeing 2012)

where

$Q(T)$ is the total number of passengers intended to travel along the given route/line at time (T) by competing modes (1-ETT) and (2-APT).

Given the above-mentioned scenario of the future long-term development of the transport system, operating between large urban (continental and intercontinental) agglomerations appears the most convenient for the ETT system as it would be able to generate sufficient passenger and/or freight/goods demand. This implicitly implies competition with the long-haul APT (Air Passnger Transport) system. One such prospective intercontinental passenger market could be the one between Europe and North America (Airbus 2012; Boeing 2012). Figure 6.9 shows development of APT demand in this market for the 2011–2050 period. Its share in the total global APT[6] demand of about 8.3 % in 2011 would decrease to about 6.5 or 5.4 % in 2031, thus indicating its increasing maturity over time (Boeing 2012; Airbus 2012).

The development of APT demand in Fig. 6.9 is based on the average annual growth rate of 3.8 % over the 2011–2031 period. Beyond the year 2031, this rate is

[6] Generally, for the period (2011–2030/31), both Boeing and Airbus predict annual APT growth in terms of RPK (Revenue Passenger Kilometers) of about 5 % (Airbus 2012; Boeing 2012).

assumed to be 2.5 % for the 2031–2041 period, and 1.75 % for the 2041–2051 period, thus indicating further maturation of the market and weakening of its demand-driving forces on both sides of the Atlantic.

Beyond the year 2050, the EET system is supposed to be in operation and take over part of the APT demand. The attracted demand would consist mostly of business passengers and shippers/receivers of high-value time-sensitive freight/ goods, both considering the transport time as one of the most important attributes for choice of the transport mode/system (currently they use the APT system). Later on, the ETT system could become increasingly convenient for more massive use also by "middle class" users-passengers and other categories of freight/goods shipments. This implies that the main driving force for development of the ETT system, in addition to sufficient demand, would be the intention to travel at the substantially higher speeds and consequently reducing travel time, in an economically viable, environmentally acceptable, and safe way, all as compared to the fastest counterpart—the APT system.[7]

Capacity

The service frequency of the ETT system on a given route/line $f_l(T)$ during time (T) satisfying the expected demand can be determined as follows:

$$f_1(T) = \frac{Q_1(T)}{\lambda_1(T) * n_1(T)} \tag{6.8d}$$

where

$\lambda_1(T)$ is the average load factor of an ETT train departing during time (T); and

$n_1(T)$ is the seating capacity of an ETT train departing during time (T) (seats).

Quality of service

Passengers on-board the ETT TRM trains would be exposed mainly to the horizontal acceleration/deceleration, lateral, and vertical acceleration force g (g = 9.81 m/s^2). The horizontal force would be: $F = ma$ (a = 3–5 m/s^2). The impact could be mitigated through the design of the ETT tubes (preferably as straight as possible in both horizontal and vertical plane) and appropriate arrangements of seats within the trains. The former would be rather complex to achieve particularly in the vertical plane since the long intercontinental tubes have to align with the Earth's curvature. In the horizontal plane, the straight line shortest (Great Circle) distances would very likely be followed.

Thus, the lateral component of the g-force would be minimized, and only the other two-horizontal and vertical—would remain. Under such conditions, if the trains were accelerating/decelerating at the rate of a = 3–5 m/s^2, and the vertical

[7] Economic viability implies the system's ability to cover its costs. Environmental acceptability implies at least not additionally contributing to impacts on environment and society. Safety means operating without incidents and accidents due to known reasons.

component was always 9.81 m/s^2 (i.e., 1 g), the resulting force would be: 1.04–1.12 g, which could be acceptable for most passengers due to not particularly affecting their riding comfort.

Fleet size

Given the service frequency $f_1(T)$ in Eq. 6.8d, the fleet size of the given ETT system can be estimated as follows:

$$N_1(T) = f_1(T) * t_{tr} \tag{6.8e}$$

where

t_{tr} is the ETT train's turnaround time (min).

The turnaround (t_{tr}) in Eq. 6.8e can be estimated as follows:

$$t_{tr} = 2\left(t_s + \frac{v}{g^+} + \frac{d}{v(d)} + \frac{v}{g^-}\right) \tag{6.8f}$$

where

t_s is the train's stop time at the beginning and end terminal (min).

The other symbols are as in the previous equations.

Technical productivity

Technical productivity for the ETT system can be determined as the product of the offered capacity per given period of time and the operating speed.

Train/vehicle

$$TP_{1/v}(T) = n_1(T) * v_1 \tag{6.8g}$$

Route/line

$$TP_{1/r}(T) = 2f_1(T) * n_1(T) * v_1 \tag{6.8h}$$

where all symbols are as in the previous equations.

"What-if" operating scenario

The "what-if" operating scenario is developed for the year 2050/2051 when the EET system connecting Europe and North America is supposed to be in place competing with the APT system. It is assumed that the APT system will not yet operate STA (Supersonic Transport Aircraft). According to the scenario in Fig. 6.9, the APT system is supposed to carry out about 1,350 billion RPKs in 2051. The ETT system operating at a speed about nine times higher and having higher prices of about 100 % would be able to attract about 45 % of this volume. Assuming the travel distance of 5,564 km (that between London and New York), the annual number of trips in both directions attracted by the ETT system would be about 109.184 million. The daily number amounts to about 149,567 per direction (these numbers do not include the induced demand, i.e., that using the system as the first choice from the beginning). If served by TRM trains of a seating capacity of $2 \times 200 = 400$ seats and load factor 0.80, the daily service frequency would be 456 dep/day or 16 dep/h per direction, i.e., every 3.75 min. Assuming that the train's running time along the route/line of length of 5,564 km would be 54 min (acceleration/deceleration at a rate of 3 m/s and cruising at a speed of 8,000 km/h), the number of trains simultaneously operating in each direction would be $17 \times (54/60) \approx 15$, separated by a distance of about 378 km. Respecting the train's running time of 54 min and the stop time at both ends of the route of 15 min, the required fleet for service frequency of 16 dep/h would be $2 \times (54 + 15)/60 \times 16 \approx 39$ trains. At each end terminal, the required number of tracks to handle departing and arriving trains would be $2 \times 16 \times (15/60) \approx 8$ tracks. The length of each track would be about 150–200 m, in order to enable comfortable passenger embarking and disembarking of train(s).

The technical productivity of a single train during cruising phase of a journey would be 400 (seats) $\times$ 8,000 (km/h) $\approx$ 3.2 million s-km/h. The technical productivity of the system during 1 h would be 2×16 (dep/h) $\times$ 400 (seats) $\times$ 8,000 (km/h) $\approx$ 102.4 million s-km/h.

6.4.2.5 Economic Performances

The economic performances of the EET system include costs, revenues, and a "what-if" economic scenario.

Costs

Infrastructure

The costs include expenses relating to the infrastructure and rolling stock. The infrastructure costs would consist of investment, maintenance, and operating costs. The investment costs generally include the costs of building tubes (2 + 1), TRM guideways, and terminals at both ends of the given tube/tunnel. They also include the costs of facilities and equipment such as vacuum pumps, the power supply system, traffic control system, communications, and fire protection system (http://tunnelbuilder.com). The maintenance costs include the costs of regular and capital maintenance of infrastructure and supporting facilities and equipment. The operational costs mainly include the costs of labor and energy for maintaining vacuum in the tubes.

Rolling stock

The cost of rolling stock would consist of the investment and operational cost. The former relates to acquiring the TRM train fleet. The latter includes the cost of maintenance, material, labor, and energy to operate the TRM fleet under given conditions.

Revenues

Revenues would be mainly obtained from charging passengers and freight/goods shippers. In addition, the savings in externalities such as energy consumption and related emissions of GHG, noise around airports, congestion, and traffic incidents and accidents by substituting APT could be taken into account, particularly when the EET system is considered from the broader international social/policy perspective. Furthermore, there could be savings in the costs of passenger time compared to those of the APT system.

"What-if" economic scenario

According to the "what-if" economic scenario, the ETT system connecting Europe and North America is assumed to provide a return on investment over a period of 40 years. At present, the investment costs for building tubes appear to be very uncertain but some estimates indicate that they could be about 13–18 million €/km (i.e., 72–103 billion € for the entire link/line of length of 5,564 km including terminals at both ends) (http://www.popsci.com/scitech/article/2004-04/trans-atlantic-maglev). The cost of building the TRM guideway in the tube in a single direction would be similar as at the today's TRM—about 15 million €/km (i.e., for two tracks this gives the total investment cost of: (5,564 km) × (2 × 15) million €/km = 167 billion €) (Table 5.4 in Chap. 5). Thus, if the system was built over the 20-year period between 2030 and 2050, the total infrastructure investment cost would be about 239–270 billion €, or without taking into account the interest rate(s), 5.975–6.750 billion €/year (the cost of facilities and equipment—vacuum pumps, power supply system, traffic control system, and fire protection system are included). As an illustration, the share of the investment costs in the cumulative GDP of Europe (EU) (544.05 trillion €) and North America (USA, Canada) (688.76 trillion €) during that period would be about 0.04–0.05 %, and 0.035–0.039 %, respectively.[8]

The costs of operating infrastructure would amount to about 10 % of the investment costs, which would give total infrastructure costs of about 6.573–7.425 billion €/year. Assuming that the passenger demand in each year of the investment-returning 40-year period would be at least at the same level as in 2051, i.e., about 592 billion RPKs, the investment costs would be about: 0.0111–0.0125 €/RPK or 0.0139–0.0157 €/s-km (the load factor is 80 %).

The operational cost of a TRM (Transrapid Maglev) system would be about 0.085 €/RPK (similarly as in Table 5.4, Chap. 5). As a result, the total costs of an EET system would be 0.0911–0.0925 €/RPK and 0.11388–0.11568 €/s-km. The minimum return fare between Europe and North America covering the costs would be about 1014–1029 €/pass, which would be reasonably competitive with today's APT business fares in the given market.

[8] The base for estimating GDP in Europe and North America during the period 2030–2051 has been GDP in 2011 (CIA 2012). The average annual growth rate of GDP in both areas is assumed to be 3 % over the 2011–2051 period.

Savings in the costs of passenger time could also be achieved; thanks to the difference in travel speed of both systems. For example, if the average value of passenger time would be rather low, say about 32 €/pass h (Chap. 5), and the difference in travel time about 6 h (0.7 h by the ETT system and about 6.7 h by APT), the total savings in the cost of passenger time in 2051 would be about 20.5 billion € [592 (billion RPK)/5,564 (km) × 6 (h) × 32 (€/h)].

6.4.2.6 Environmental and Social Performances

The environmental and social performances relate to impacts of a given EET system in terms of: (i) energy consumption and related emissions of GHG (Green House Gases) and land use/take (environmental); (ii) noise, congestion, safety (traffic incidents and accidents) (social); and (iii) a "what-if" environmental scenario.

Energy consumption and emissions of GHG

Energy consumption of an ETT system includes the energy for operating TRM trains, setting up and then maintaining vacuum in the tubes, and powering other supporting systems, facilities, and equipment. The amounts of energy would be substantive. For example, only for a single trip (service) the energy consumed for accelerating a 110 tons TRM train from 0 to 8,000 km/h would be about 75.445 MWh. If in this latest and all above-mentioned cases the energy was obtained from renewable (sun, wind, and hydro) and nonrenewable nuclear sources, this would mean almost zero emissions of GHG. Even if a LCA (Life Cycle Analysis) was considered, these emissions would be almost negligible (IBRD 2012).

Land use

ETT system would occupy additional land only for their coast terminals should they not already be incorporated into the surrounding urban structures.

Noise

ETT system would not generate any noise disturbing the population around the begin and end terminals. This is because of operating TRM trains at the low speeds in their vicinity.

Congestion

Due to the nature of operations, the ETT system would be free from congestion along the links/lines. In addition, regarding the intensity of operations, automated traffic management systems would have to provide precise guidance in order to achieve almost perfect (in terms of seconds) matching of the actual and scheduled departure and arrival times. However, while relieving airports from congestion by taking over APT demand, ETT system could contribute to increasing congestion in the areas around the begin and end terminals simply due to the increased intensity of mobility described above.

Safety

ETT system is expected to be at least as safe as its APT counterpart. This implies that incidents and accidents should not occur due to the known reasons. However, special attention would have to be devoted to the external security of infrastructure (tubes), for example, to ways of monitoring and preventing eventual terrorist attacks.

"What-if" environmental scenario

According to the "what-if" environmental scenario, the propellant/energy used by the ETT system would be completely obtained from nonpolluting nonrenewable (nuclear) and renewable (solar, wind, and water) sources. In addition, this implies that the emissions of GHG from burning these propellants would be negligible compared to that from burning today's kerosene. Under such conditions, taking over the passenger demand from the APT system would reduce the volumes of its operations and consequently the overall impacts on the environment and society. Again, it is assumed that the APT system will not operate STA at that time. Let's assume that an aircraft type similar to the B787-8 would be mostly operated in the market in 2051. Its average fuel consumption is about 0.0262 kg/s-km or 0.0328 kg/RPK (the load factor is assumed to be 80 %) (Chap. 2). The emission rate of JP-1 fuel is 5.25 kgCO_{2e}/kg of fuel, thus giving the emission rate of the aircraft of about 0.138 kgCO_{2e}/s-km or 0.172 kgCO_{2e}/RPK. For the annual volumes of APT traffic taken over by the ETT system of 592 billion RPK, the total saved fuel consumption and related emissions of GHG would be about 19.4 and 102.0 million tons, respectively.

The cost of CO_{2e} emissions from APT have been estimated according to different scenarios. In the first so-called BAU (Business As Usual) or "high" scenario it is 0.045 €/RPK, which gives the total cost savings of 26.64 billion € (IWW-INFRAS 2004). If NASA's "Subsonic Fixed-Wing Research Fuel-Reduction Goals" program is realized by the years 2030–2035, the above-mentioned unit emissions could be reduced by about 50–70 % and consequently the unit costs would fall to 0.009 €/RPK ("low" scenario). Consequently, the savings in CO_{2e} emissions in the given case would be about 5.32 billion € in 2051 (GAO 2009; IBRD 2012; IWW-INFRAS 2004).

In addition, the total APT externalities would be: 0.07 €/RPK (emissions of GHG, noise, incidents and accidents, and land use), thus enabling the ETT system to achieve annual savings of about 41.44 billion € in 2051 (if being commercialized) (IBRD 2012).

6.4.2.7 Policy Performances

The ETT system could have significant policy performances both of the national and international (global) character. On the national scale, the main performance would be its contribution to creating an integrated transport system. On the international scale, the main performance would be contribution to creating an integrated global international very high-speed nonair transport system, which would even more strongly contribute to the further globalization of the already highly global economy and society. The former is explained above. The latter, on the one hand, would imply engaging transport/logistics operators to manage door-to-door services, i.e., operate the entire intermodal system in which the ETT subsystem would have the crucial role. On the other, the ETT system project(s) would need substantive knowledge, expertise, and funding, which could be provided efficiently and effectively only through international cooperation. In such respect, the implementation of the system would be a platform for integrating mankind and turning it toward similar future projects for a better future.

6.4.3 Evaluation

The described EET system possesses both advantages (Strengths and Opportunities) and disadvantages (Weaknesses and Threats) as follows:

Advantages

- More or at least equally efficient and effective as the conventional existing and future APT (Air Passenger Transport) system it is assumed to compete with;
- More environmentally and socially friendly as compared to the conventional existing and future APT system;
- Contribution to creating an integrated global transport system for both passengers and freight on both national and particularly on the international-global scale;
- Economically, socially, and politically important on the national and particularly on the international scale due to contributing to further: (i) concentration of existing and developing new resources (knowledge, skills, materials, building/construction processes, mechanization, etc.); (ii) globalization of economies through the project funding sources (the national and particularly international), and (iii) economic, social, and political integration of the societies involved.

Disadvantages

- The system's inherent uncertain real social/economic feasibility (who would really need to travel at the offered supersonic speeds and under what conditions?);
- Very high investment costs, but as a very small proportion of the GDP of the areas involved (this implies that in the given case the principal funding would come from EU and North America (ECB (European Central Bank), U.S. Federal budget, IMF (International Monetary Fund), World Bank, and large private enterprises interested in the system/project);
- Long lead time for commercialization (at present the lead time seems to be no shorter than 20 years from when the works begin);

- The unknown technical/technological barriers in developing components, processes, and techniques, and natural barriers (both could substantively increase the planned funding and lead time for commercialization);
- Uncertain and inherent instability of the political, economic, and social conditions during implementation and operationalization, and latter during full commercialization of the system;
- Inherent vulnerability to potential terrorist attacks; and
- Facing other emerging and presumably more feasible systems and technologies in the meantime, i.e., during the lead time for commercialization.

Finally, before the eventual further elaboration of the feasibility of the ETT system for its eventual commercialization, the above-mentioned advantages and disadvantages should be carefully considered. Some numerical figures provided in this section could change in absolute but not substantially in relative terms including their relationships. They indicate that the ETT system might be, under certain conditions, a new promising very high (supersonic) speed long-haul transport system.

6.5 Advanced Air Traffic Control Technologies and Operations for Increasing Airport Runway Capacity

1921	ATC (Air Traffic Control) is introduced for the first time at London Croydon Airport (UK)
1950s	VOR's and later VORTAC's (Very High Frequency Omni-directional Range and Finding), ILS (Instrument Landing System), and approach lighting systems are implemented (USA)
1950s	Radars are widely deployed to control commercial air traffic; aircraft continue to fly along fixed air corridors (USA)
1960s	Aircraft begin to carry radar beacons/transponders identifying them, and thus making the radar more efficient (USA, Europe) ATC (Air Traffic Control) remains based on radars and flight corridors, but becomes increasingly computerized (USA, Europe)
2000s	Plans to replace radars with satellite-guided systems emerge (U.S. NextGen and EU SESAR Programs) (USA, Europe)

6.5.1 Definition, Development, and Use

The ultimate capacity of a given airport runway system is defined as the maximum number of aircraft that can be accommodated during a given period of time (usually 1 h) under conditions of constant demand for service. At many airports, this capacity and particularly its landing component is considered as the main constraint to their future growth (Janic 2006, 2007a). In general, aircraft landings always have ultimate priority over take-offs when they request service at the same runway at the same time. This however depends on the pattern of demand for these particular operations on the one hand and the number of runways used on the other. At airports with a single runway, the landings and take-offs can interfere with each other when the demand for service reaches the runway ultimate landing or take-off capacity. In such cases, take-offs are usually inserted between successive landings or vice versa, both at the appropriate time gaps. London Gatwick (London, UK) and NY LaGuardia airport (New York, USA) are characteristic examples of such runway operations. At airports with several runways, these can be separately used for landings and take-offs simultaneously or after some time. In this case, the two types of operations do not interfere with each other despite demand for each type reaching the corresponding ultimate runway capacity during a given period of time (usually one to a few hours). Operating two independent parallel runways in segregated mode (one exclusively for landings and another exclusively for take-offs at London Heathrow Airport (UK) at the level of their ultimate capacities during almost the whole day is an illustrative example. This and some other airports such as Paris Charles de Gaulle, (Paris, France) and Amsterdam Schiphol (Amsterdam, the Netherlands) airport (hosting the hub-and-spoke network(s) of the SkyTeam alliance), and Frankfurt Main (Frankfurt, Germany) airport (hosting the mega hub-and spoke network of STAR alliance) in Europe are examples of the airports where "waves" of landings are followed by "waves" of take-offs several times during the day. The incoming "waves" are exclusively served on a given runway(s) operating at its ultimate landing capacity. The "waves" of departing flights are then exclusively served using the same runway(s), this time at its take-off capacity. In each of these cases, the runway landing capacity appears critical because of two reasons. First, it influences the number of landings and consequently the number of take-offs latter on both during the given period of time. Second, it influences the size of the "incoming" and consequently of the outgoing "waves" of flights, and thus the overall airline hub-and-spoke network performance.

Under such circumstances, many theoretical and practical endeavors have been made to understand, analyze, calculate, increase, and diminish vulnerability of the runway landing capacity to bad weather by identifying the most influential factors and then finding the corresponding solutions.

6.5.2 Analyzing Performances

6.5.2.1 Technologies

The theoretical and practical endeavors to increase the runway landing capacity are contained in the current U.S. NexGen and the European-EUROCONTROL SESAR research and development programs (Thompson 2002; http://www.faa.gov/nextgen; http://www.sesarju.eu). They are focused on developing advanced technologies aiming at supporting more effective, efficient, and safe operations of the future air transport system and those of the airport runways as well. Specifically, the technologies to support the advanced landing procedures and consequently contribute to increasing the runway landing capacity are given in Table 6.10.

The contribution of SESAR technologies is expected to be as follows (http://www.sesarju.eu):

- Reducing the current distance-based separation rules and complementing them with time-based separation rules during descent and final approach and landing, dynamic wake vortex detection and prediction in addition to recategorization of the current aircraft wake vortex categories (ASAS and WVDS onboard the aircraft and as the ATC automated supportive functions (tools));
- Developing the ATC decision support tools for efficient and effective sequencing of aircraft during their initial and final approach and landing (i.e., upgrading existing Integrated Arrival/Departure Manager);
- Introducing the initial and final approach procedures with Vertical Guidance (AVP), which can also be part of the 4D P-RNAV trajectories whose incoming (the first) segment is CDA (Continuous Descent Approach) procedure;
- Developing ATC aids for monitoring aircraft along their 4D RNAV trajectories and new approach procedures, conflict advisory and resolution, and improvement of safety nets (by using aircraft-derived data); and
- Developing ground ATC data link processing and management functions through considering initial data link applications building upon LINK2000+ and future enhanced applications.

The NextGen technologies are expected to contribute as follows (FAA 2011a, b, c; http://www.faa.gov/nextgen):

- Enabling efficient and reliable approach and landing at airports not equipped with ILS under reduced visibility conditions; this particularly refers to airports operating closely spaced parallel runways (RNP);
- Enabling precise final approach and landing at secondary airports not equipped with ILS, thus relieving the complexity of procedures for the aircraft approaching the primary airports in the same metropolitan area (WAAS or RNP1 and/or RNP with Curved Paths);
- Mitigating uncertainty concerning the current and prospective position of aircraft along their 4D RNAV trajectories including the final approach and landing segment by double monitoring, on-board the aircraft and at the ATC (ADS-B);
- Implementing CDA as a common practice with the aircraft engine minimum power, thus enabling mitigation of the environmental impacts (fuel consumption and related emissions of greenhouse gases, and noise burden); at the same time the aircraft is precisely guided along the descent trajectory, thus enabling their vertical separation in the vicinity of the FAG (Final Approach Gate) (CDA, RNP, WAAS as a supplemental system to GNSS); and
- Improving surveillance at airports which currently do not have radar surveillance.

Table 6.10 Advanced technologies for advanced airport landing procedures (EC 2005, 2007a; FAA 2010; http://www.sesarju.eu; http://www.faa.gov/nextgen)

Function/User	Technology	Availability
Air traffic flow management tools/ATC	• CTAS (Center/TRACON Automation System) assists in optimizing the arrival flow and runway assignment;	Now
	• Updated Integrated Arrival/Departure Manager enabling the advance planning and updating the arrival sequences according to the selected separation rules and service discipline	Now but still to be Improved in the medium term (NextGen)
Air traffic surveillance/ATC	• RADAR of improved precision enables reduction of the minimal separation between aircraft from 3 to 2.5 nm;	Now with additional improvements in the medium term (NextGen)
	• PRM (Precision Runway Monitor) consisting of a beacon radar and computer predictive displays, thus enabling the independent use of dual- and triple-dependent parallel runways spaced at less than 4300 ft (ft = feet).	Now
	• WVDS (Terminal Wake Vortex Detection System) providing information about the current and prospective wake vortex behavior during landings and take-offs	Medium to long term (NextGen)
Avionics/aircraft	• FMS (Flight Management System) in combination with RNP (Required Navigational Performance), thus enabling more precise following of 4D P-RNAV trajectories, thus reducing aircraft position error including that of arriving at the final approach gate;	Now (NextGen)
	• ADS-B (Automatic Dependent Surveillance Broadcasting) improving situation awareness onboard and with respect to other surrounding air traffic. is used independently, but in addition to TCAS and enhanced CDTI (Cockpit Display of Traffic Information);	Medium to long term (NextGen)
	• CDTI (Cockpit Display of Traffic Information) provides integrated traffic data onboard the aircraft, which may reduce the separation rules between aircraft;	Medium term (NextGen)
	• ASAS (Airborne Separation Assistance (Assurance) System) enabling airborne surveillance, display of traffic information, and consequently sequencing and merging at the final approach gate using data from ADS-B;	Medium term (NextGen; SESAR)

(continued)

Table 6.10 (continued)

Function/User	Technology	Availability
	• ACAS (Airborne Collision Avoidance System) improving the quality of information for ASAS, thus enabling reducing position error along 3D RNAV trajectories TCAS (Traffic Alert and Collision Avoidance System) indicating the spatial relation of two aircraft and thus providing instructions to avoid potential conflict(s);	Medium term (NextGen)
	• TCAS–Traffic Alert and Collision Avoidance System shows the spatial relation of two aircraft and provides instructions to avoid potential conflict(s);	Now
	• WVDS (Wake Vortex Detector System) onboard the aircraft enables the collection and display of information on the existing wake vortex to both pilots and ATC controllers;	Medium to long term (NextGen)
	• WAAS (Wide Area Augmentation System) supplementing and thus improving basic GNSS position accuracy;	Medium to long term (NextGen)
	• LVLASO (Low Visibility Landing and Surface Operating Program) reduces controls and predicts the runway occupancy time;	Medium to long term (NextGen, SESAR)
	• New data link infrastructures (improved LINK 2000+) enabling exchanging of the constantly updated aircraft current and expected position obtained from FMS to external recipients—other aircraft-pilots and ATC	Medium to long term (SESAR)
"Mixed" traffic surveillance and conflict alert/ATC and aircraft	• Distributed Air Ground Solution combining ADS-B, TCAS, and ATC tools, thus enabling simultaneous aircraft-ATC air traffic surveillance, alerting and resolution of potential conflicts	Medium term (NextGen, SESAR)

6.5.2.2 Landing Procedures, ATC Separation Rules, and Service Disciplines

Landing procedures

The advanced procedures for approach and landing on a single runway to be supported by the above-mentioned advanced technologies are classified into three categories: (a) Single segment with nominal GS (Glide Slope) angles; (b) Double segment with nominal and steeper GS angle(s); and (c) Single segment with ultimately arbitrary GS angle(s). Figure 6.10a–c shows a simplified scheme.

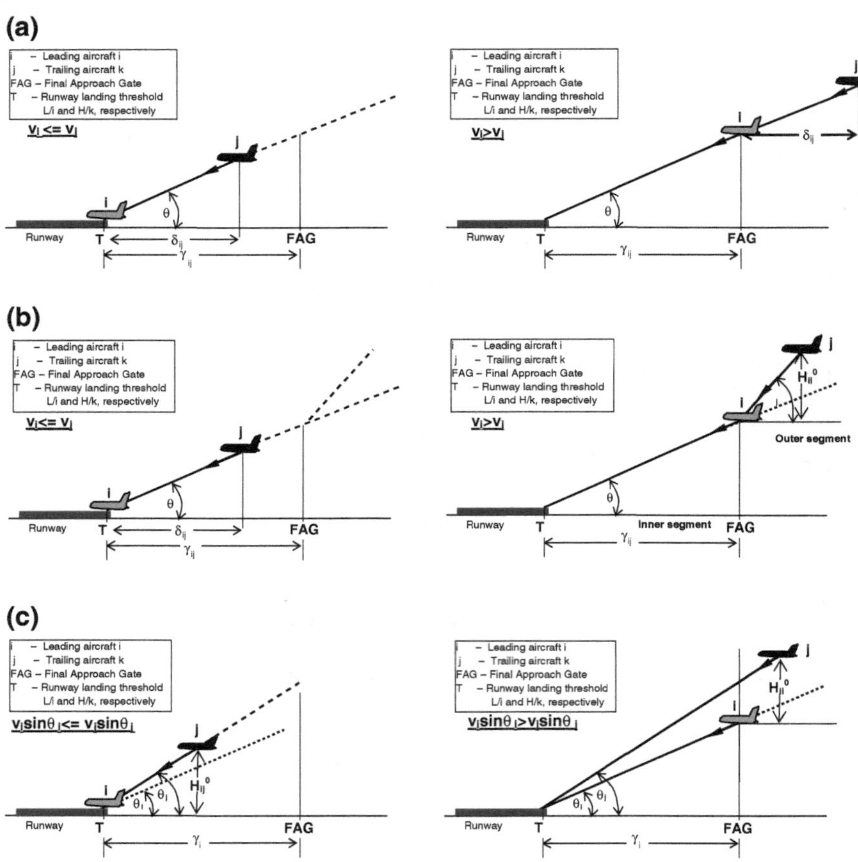

Fig. 6.10 Schemes of conventional and advanced approach procedures to a single runway (Janic 2012), **a** Single segment with nominal GS angles, **b** Double segment with nominal and steeper GS angles, **c** Single segment with ultimately arbitrary GS angles

Single segment—nominal GS angles

This final approach procedure is supported by the ILS (Instrument Landing System) or MLS (Microwave Landing System) where all aircraft use the same GS angle along the entire final approach path connecting the FAG (Final Approach Gate) and the landing threshold *T*. Figure 6.10a shows a simplified scheme in the vertical plane. Currently, at most airports, the ILS provides the standardized-nominal GS (Glide Slope) angle of 3°. In some specific cases such as when mitigating noise or avoiding obstacles in the final approach plane, the GS angle can be steeper, i.e., greater than the nominal GS. Consequently, all commercial aircrafts are certified for a nominal GS angle of 2.5–3°. Some smaller regional aircraft are certified for both nominal and steeper GS angles of 3–5.5°, respectively.[9]

The MLS (Microwave Landing System) has been designed as an alternative to the ILS offering higher flexibility of GS angles from 2.5 to 7–8°, but again only for certified aircraft (Kelly and La Berge 1990). However, except in some specific cases, it has not been widely applied. Furthermore, the above-mentioned research and development programs—US NextGen and European-EUROCONTROL SESAR—seemingly do not particularly consider MLS as a future alternative to ILS. Thus, at least presently, GS angle flexibility provided by MLS is not considered.

The ATC exclusively applies the minimum horizontal distance-based separation rules to all types of sequences of landing aircraft differentiated in terms of their approach speeds as follows: "Slow–Slow," "Slow-Fast," "Fast-Slow," and "Fast–Fast" (Tosic and Horonjeff 1976). Some research also suggests the application of ATC time-based separation rules under the same conditions, which could eventually increase the level and stability of the landing capacity with respect to weather (Janic 2008).

Double segment with nominal and steeper GS angles

This final approach procedure is based on the concept of "Individual Flight Corridor(s)" (i.e., 4D RNAV trajectories connecting WP (Way Point) of beginning the intermediate approach, FAG, and the landing threshold *T*). Figure 6.10b shows a simplified scheme in the vertical plane. This concept was originally designed for landing on closely-spaced parallel runways (Rossow 2003). 4D final approach trajectories consist of outer and inner segments. As applied to a single runway, these 4D trajectories overlap in the horizontal and vertical plane along the Inner segment. They can overlap and/or differ in both horizontal and vertical planes along the outer segment. This implies that the Inner segment applies a common (usually nominal) GS angle for all aircraft categories. The outer segment can have different (flexible) and/or the same GS angles for particular aircraft categories. Thus, the prime distinction from case (*a*) is that all aircraft use the common

[9] The earliest, De Havilland DHC-6 and DHC-8, were certified as STOL (Short Take-Off and Landing) aircraft. Later, regional aircraft Cessna Citation, BAe RJ 85/100, Fokker 50, Dornier 328, Embraer ERJ 135/170, and recently the larger Airbus A318 were certified for the steeper GS angle of 5.5^0 (EC 2005; TC 2004).

nominal GS angle (3°) along the Inner segment and different GS angles (3–6°) along the outer segment. Under such circumstances, the ATC mixed horizontal and vertical distance-based separation rules can be applied to particular types of landing sequences. For example, in sequences "Slow–Slow," "Fast–Fast," and "Slow-Fast," ATC horizontal distance-based separation rules and in the sequence "Fast-Slow," vertical distance-based separation rules can be applied.

Single segment with ultimately arbitrary GS angles

This final approach procedure is based on 4D RNAV trajectories consisting of a single straight and/or curved line segment connecting the FAG and the runway landing threshold *T*. For the same and/or different aircraft categories, this segment can differ in the horizontal plane except from the very short common portion in the vicinity of the runway landing threshold *T*. This segment is used for achieving full landing configuration and speed. In the vertical plane, the approach trajectories of particular aircraft categories can be entirely different due to distinctive GS angles. Figure 6.10c shows a simplified scheme in the vertical plane. In addition, each of these trajectories can be considered as the end part of a CDA (Continuous Descent Approach) trajectory aimed at reducing flying time, fuel consumption, related emissions of greenhouse gases, and noise around particular airports (Helmke et al. 2009). Thus, a range of GS angles (3–6°) can be assigned to particular aircraft. Consequently, contrary to cases (*a*) and (*b*), the trajectories of the leading aircraft in all sequences can be higher, the same, and/or lower than those of the trailing aircraft, and vice versa. At the same time, as in case (*b*), 4D RNAV trajectories are supported by multiple ILS GP or MLS, as well as other above-mentioned innovative and new technologies. The diversity of the actually small GS angles of 4D RNAV final approach trajectories offers an opportunity for the ATC to exclusively apply vertical distance-based separation rules to all sequences of landing aircraft. In this manner, depending on the location where this minimum vertical separation is established (the runway landing threshold *T* or FAG), the trailing aircraft in almost all sequences can come closer to the leading aircraft while always being above and thus fully protected from the wake vortices moving behind and below the flight paths of the leading aircraft. Consequently, the horizontal distances in particular aircraft sequences have become generally shorter than in the cases (*a*) and (*b*), i.e., when the current ATC horizontal distance and/or innovative mixed horizontal/vertical distance-based separation rules are exclusively applied, respectively.

ATC separation rules

Currently, landing aircraft is separated by the ATC (Air traffic Control) minimum horizontal distance-based separation rules, which for given aircraft approach speeds generate the minimum interarrival times at the runway threshold as the "reference location" for counting operations, and consequently the maximum flow, i.e., the landing capacity (Tosic and Horonjeff 1976). At US (United States) airports, two categories of ATC horizontal distance-based separation rules are applied depending on the weather: IFR (Instrument Flight Rules) under IMC

(Instrument Metrological Conditions) and VFR (Visual Flight Rules) under VMC (Visual Meteorological Conditions) (FAA 2010; NASA 1999). The former rules being stricter than the later are the sole responsibility of ATC; the latter are in sole responsibility of the aircraft/pilots. As being shorter, these rules imply higher landing capacity. Consequently, at the U.S. airports where both rules are applied, the ultimate runway landing capacity appears to be inherently unstable with respect to the weather conditions. At European airports, IFR are applied under both IMC and VMC, thus providing relatively higher stability of the runway landing capacity with respect to the weather (ICAO 2001, 2008).

Supported by the above-mentioned technologies, the ATC is supposed to use four types of separation rules between landing aircraft exclusively and/or in combination.

Horizontal distance-based separation rules

Thees rules are set up to protect trailing aircraft from the wake vortices of the leading aircraft in the particular landing sequences. In the USA, they are specified for both IMC and VMC as IFR and VFR, as given in Table 6.11 (FAA 2010, 2011). In Europe, they are only specified for the IMC as IFR.

Table 6.11 indicates that the current FAA IFR applied under IMC are about 40 % stricter than the VFR applied under VMC, which on average makes the landing capacity under IMC about 30 % lower than that under VMC (FAA 2010, 2011). Since the Super Heavy A380 still represents a rarity in the dense arrival streams at many airports, it is not given in Table 6.11. Nevertheless, it is worth mentioning that the ATC IFR separation rules vary from 4 nm between two A380 s to 6, 7, and 8 nm when an A380 is leading and a Small, Large/B757, and Heavy are trailing aircraft in the landing sequence, respectively (ICAO 2008). Furthermore, according to the ICAO rules, aircraft is categorized into four wake vortex categories: light, medium, heavy, and super heavy (A380) (EC 2005; ICAO 2007, 2008). The separation rules in Table 6.11 and their ICAO modalities can be applied to the above-mentioned operational procedure (a).

Mixed horizontal/vertical distance based separation rules

These rules can be applied similarly as horizontal distance-based separation rules, but by replacing the longest horizontal separation intervals in Table 6.11 with the vertical ones, thus being applicable to the above-mentioned operational procedure (b).

Vertical distance-based separation rules

These rules are supposed to be 1,000 ft (ft—feet) for all landing sequences if being exclusively applied, mainly to the above-mentioned operational procedure (c) (FAA 2010; NASA 1999). However, they can also be made dependent on the type of landing sequence similarly as their horizontal distance-based and time-based counterparts.

Table 6.11 FAA horizontal distance-based separation rules between landing aircraft (nm—nautical miles) (FAA 2010, 2011; NASA 1999)

i/j	VFR				IFR			
	Small	Large	B757	Heavy	Small	Large	B757	Heavy
Small	1.9	1.9	1.9	1.9	2.5	2.5	2.5	2.5
Large	2.7	1.9	1.9	1.9	4	2.5	2.5	2.5
B757	3.5	3	3	2.7	5	4	4	4
Heavy	4.5	3.6	3.6	2.7	5	5	5	4

Table 6.12 Derived ATC minimum time-based separation rules for landing aircraft (minutes) (Janic 2012)

i/j	Small	Large	B757	Heavy
Small	1.0	1.0	1.0	1.0
Large	1.5	1.0	1.0	1.0
B757	2.0	1.3	1.5	1.2
Heavy	2.5	1.5	1.5	1.2

Time-based separation rules

These rules can be applied to the above-mentioned operational procedure (a) instead of horizontal distance-based separation rules. An example of these rules is given in Table 6.12.

The time-nbased separation rules are actually based on the VFR horizontal distance-based separation rules in Table 6.12, and typical aircraft final approach speeds (Small—120 kts; Large—130 kts; B757—and Heavy—155 kts). They ensure protection of the trailing aircraft from the wake vortices of the leading aircraft in particular landing sequences on the one hand, and provide a higher stability of the landing capacity to the weather conditions on the other (Janic 2008).

The above-mentioned ATC separation rules are under permanent revision aiming at incorporating advanced knowledge about wake the vortex behavior of existing and particularly new aircraft. For example, this includes possible recategorization of particular aircraft wake vortex categories and setting up modified separation rules. Thus, the currently medium aircraft can be segregated into Upper Medium (A330, A300, A310, B767, B757-300, B757-200—MTOW >= 120 tons and <250 tons), Medium (A320, B737-[2|3|4|5|7]00, A321, B737- [8|9], MD 80—MTOW >= 40 tons <120 tons), and Lower medium (CRJ, ERJ, ATR—MTOW >0 7 tons and <40 tons). The remaining categories are Heavy (A380, B747/787, C5A, C17, MD11, A340, B777, DC 10—MTOW >250 tons) and Small (MTOW <7 tons). The corresponding ATC separation rules can be a combination of the rules given in Table 6.12 (Freville 2008). In addition, the most recent example are the separation rules of 10 nm set up for Heavy B787-8 as the leading aircraft in all landing sequences (see Chap. 2).

Service disciplines

ATC is assumed to be able to apply two service disciplines to the aircraft landing on a single runway. The first commonly applied at most airports is the FCFS (First Come—First Served) service discipline, where aircraft is accommodated on the runway according to the actual time of their arrival or request for service (landing). The other is the PR (Priority) service discipline, which gives priority to particular aircraft over others irrespective on their actual request for landings. In this case, ATC creates rather homogenous clusters of aircraft in terms of their wake vortex categories and landing speeds and then serves them according to the specified order. Under such circumstances, except between different clusters, ATC separation rules are uniform within the cluster. Usually, clusters of larger (faster) aircraft are given priority over smaller (slower) aircraft. In order to avoid unnecessary delays due to clustering, ATC can apply this service discipline most efficiently if the aircraft are already in the queue requesting service ultimately at the same time. In some cases, the airlines' schedule can favor application of this discipline (the airlines' flight scheduling at their hub airports can consist of "waves" of heavy aircraft followed by "wave" of large and "wave" of small aircraft, both during landings and take-offs (Janic 2009)). In any case, the eventual implementation of this (PR) service discipline, in addition to be manageable by ATC, would need to be accepted by the airlines. The former implies safety, efficiency, and effectiveness of execution, while the latter implies making the main prospective advantages and disadvantages transparent and clear.

6.5.2.3 Implementation of Advanced Landing Procedures

Both the SESAR and NexGen program aim at trajectory-based control implying performance-based navigation and aircraft position information (FAA 2011a; http://www.sesarju.eu).

The properly equipped aircraft with SESAR and NextGen technologies will be able to follow 4D P-RNAV (Precision Area Navigation) trajectories from the departure to the arrival airport gates. ATC will be able to rather accurately predict the aircraft position over time, i.e., where they will be as compared to the current system, which informs where they currently are. Aircraft will be able to follow these trajectories by useing FMS (Flight Management System) and fulfillment of RNP (Required Navigational Performance) (FAA 2011a). The latter will be possible thanks to continuously improving onboard systems, whose accuracy depends on GNSS (Global Navigation Satellite System) (an additional supplement to increase accuracy of GNSS is WAAS (Wide Area Augmentation System) (Table 6.10).

The aircraft position will be continuously and precisely monitored by using ADS-B both at ATC and onboard of other equipped aircraft (ADS-B uses GNSS to determine the aircraft position in real time, its transponder to provide its identity and altitude, and a data link to broadcast and receive positioning information (FAA 2011a)).

The ADS-B will enable carrying out the initial and intermediate descent of 4D P-RNAV trajectories as CDA. In particular, while performing the advanced final approach procedures, the aircraft will be continuously self-guided and monitored; thanks to 3D view surveillance monitors onboard in combination with CDTI. The ATC will monitor them similarly. Such double monitoring will enable ATC to systematically apply the current but reduced horizontal, mixed horizontal/vertical, exclusively vertical distance-based, or time-based separation rules between aircraft approaching and landing at the same and/or different (in advance assigned) GS angles provided by the multiple-angle ILS system. Choice and safe use of these separation rules will be supported by ASAS in addition to both on-board and ground WVDS (FAA 2011a; http://www.sesarju.eu).

The ATC will sequence the approaching and landing aircraft according to the selected service discipline using the updated Integrated Arrival/Departure Manager. The other onboard (existing TCAS) and forthcoming ATC conflict detection and resolving tool(s) will function constantly. Consequently, the arriving aircraft from en-route to the landing threshold will be automatically separated in order to provide minimum spacing and sequencing achieving the runway landing capacity under given conditions.

6.5.2.4 Operationalization of Advanced Landing Procedures

Operationalization of the above/mentioned advanced landing procedures in combination with applying diverse ATC separation rules and service disciplines requires the fulfillment of three sets of conditions as follows:

- Supportive advanced technologies should be available;
- The aircraft of particular wake vortex categories should be certified for different (including steeper) GS angles; and
- Both the ATC controllers and pilots should be appropriately trained.
  - Some existing and prospective supportive technologies to be in place are given in the above-mentioned Table 6.10.
  - Since GS angles greater than 2.5–3° are considered steeper, the aircraft must be certified. In the given context, these are assumed to be almost all Small and the majority of Large/B757 aircraft. The range of GS angles for each of these aircraft will depend on their aerodynamic characteristics, which implies that different aircraft of the same categories can be certified for different ranges of GS angles. In principle, smaller aircraft seem to be able to be certified for steeper GS angles than the larger ones. This certification will be a precondition for more frequent, systematic, and safe use of ATC mixed and particularly vertical distance-based separation rules. The distinctions of approaching at a steeper instead of a conventional GS angle(s) are as follows:

If DH (Decision Height) at the landing threshold remains 50 ft (15 m) and the flare distance 35 ft (10.5 m), changing the GS angle from 3 to 5.5° will shorten the flare and touchdown distance from 300 m and 160 to 200 m and 110 m, respectively (EC 2005);

If the aircraft approach speed(s) remains the same along steeper as compared to conventional final approach trajectory, its horizontal component will be lower, thus together with the shorter touchdown distance(s) generally resulting in a shorter runway landing occupancy time; under such circumstances, this time will certainly be shorter than the inter-arrival times at the landing threshold of particular aircraft sequences, thus not causing turn-around affecting the landing capacity.

- Training of ATC controllers will need to relate to modifying if not completely changing their existing mental model(s) used in sequencing landing aircraft. In addition, it will be necessary to develop a convenient tool for converting the ATC horizontal shown on the ADS-B and/or radar monitors into the vertical separation rules to be shown on the 3D view surveillance monitor(s). The pilots will be trained for approaching at a range of GS angles and monitoring the angle on the cockpit's 3D monitor or on vertical profile view surveillance and CDTI monitor(s).

Numerous barriers to fulfilling the above-mentioned conditions may emerge. The simplest one could be the lack of necessity for incremental increase in the landing capacity by advanced landing procedures, separation rules, and service disciplines at many airports despite the expected traffic growth and consequent increasing of congestion and delays. This can be due to the availability of other more easily implemented measures for matching the constrained capacity to demand such as slot constraints, spreading peaks, stimulating airlines to use the available slots by larger and fuller aircraft, and charging for congestion (Morrison and Winston 2007).

Additional barriers could be the rather time-consuming and costly operationalization of the above-mentioned advancements including their full acceptance by the system, i.e., airlines, airports, pilots, and ATC controllers, as compared to the expected air traffic growth. This could be because of maintaining the required (target) level of safety and reliability of these inherently more complex system components including still unsolved dilemmas such as, for example, division of the control responsibilities between pilots and ATC controllers for maintaining the given (prescribed) separation rules. Therefore, these and other still hidden barriers in combination with an inherent inertia of the system to more radically accept such rather revolutionary changes could postpone operationalization of the proposed advancements to the long-term future. A factor to eventually speed up this process a bit could be continuous increase in congestion at airports with strengthening constraints to provide additional capacity by building additional runway(s) as well with other means.

6.5.3 Modeling Performances

6.5.3.1 Background

The objectives of modeling the airport runway landing capacity are usually to develop suitable/convenient analytical and simulation models for estimating the ultimate capacity based on the advanced operational procedures, ATC separation rules, and service disciplines. In this context, the analytical models are based on the following assumptions (Janic 2006, 2007a, 2009, 2012; Tosic and Horonjeff 1976):

- The three-dimensional approach and landing trajectories of particular aircraft categories are known in advance; assignment of conventional and/or steeper approach GS angles depends on the type of arrival sequence(s) in terms of the aircraft wake vortex categories, approach speeds, and the capability to perform either approach procedure;
- Certified aircraft can safely use both conventional and steeper approach trajectories with different GS angles;
- The assigned GS angles do not influence the runway landing occupancy times;
- The system is error-free, implying that aircraft in all landing sequences appear on the particular segments of approach trajectories at the time when the ATC expects them; this implies that the time and space deviations of the actual from the prescribed aircraft positions in both horizontal and vertical planes are considered negligible mainly thanks to the above-mentioned new technologies assumed to be in place at that time (see Table 6.1);
- ATC applies only one type of minimum separation rules and service discipline at a time; and
- The runway landing threshold is considered as the "reference location" for counting the capacity.

6.5.3.2 The Runway Capacity Models

Landing capacity

The models for estimating the ultimate landing capacity of a single runway based on application of the advanced operational procedures, ATC separation rules, and service disciplines have been categorized as follows (Janic 2008, 2009):

FCFS (First Come-First Served) service discipline

When the FCFS (First Come-First Served) service discipline is applied, the ultimate landing capacity of a single runway can usually be estimated as (Janic 2007a, 2008; Tosic and Horonjeff 1976):

$$\lambda_1 = T/\bar{t} = T/\sum_{ij} p_i * t_{ij} * p_j \quad i,j \in N \tag{6.9a}$$

where

T is the period of time;

$\bar{t}$ is the minimum average inter-arrival time between successive aircraft passing through the "reference location", i.e., the runway landing threshold, where they are counted;

i, j is the index of the leading and trailing aircraft (i) and (j), respectively, in the landing sequence (ij);

p_i, p_j is the proportion of aircraft of wake vortex category (i) and (j), respectively, in the landing fleet mix;

t_{ij} is the minimum inter-arrival time between an aircraft of wake category (i) followed by an aircraft of wake vortex category (j) at the "reference location" for counting landings, i.e., the runway landing threshold; and

N is the number of different aircraft wake vortex categories in the landing fleet mix.

PR (Priority) service discipline

When the PR (Priority) service discipline is applied, Eq. 6.9a is transformed as follows (Janic 2009):

$$\lambda_2 = T/\bar{t} = T/\sum_{i=1}^{N} p_i * t_{ii} \tag{6.9b}$$

where

p_i is the proportion of aircraft of the wake vortex category (i) in the landing fleet mix; and

t_{ii} is the minimum inter-arrival time of the sequence of aircraft of the same wake vortex category (i) at the "reference location", i.e., the landing threshold.

The other symbols are analogous to those in Eq. 6.9a. This case implies that the arriving aircraft is clustered into rather homogeneous groups respecting their wake vortex and approach speed characteristics, and then landed sequentially, one after the other. The order of serving particular groups does not influence the ultimate landing capacity. This is because the average inter-arrival time at the landing threshold remains independent of the order of their serving. Carrying out the above-mentioned clustering seems feasible only if a relatively large number of aircraft of different categories requests landing on the given runway almost at the same time.

Inter-arrival times at the "reference location(s)"

Horizontal distance-based separation rules

Currently, ATC applies the minimum horizontal distance-based separation rules between the landing aircraft on a single runway exclusively (Table 6.11). In such case, the minimum inter-arrival time at landing threshold T of two aircraft in landing sequence (ij) independently of their GS angles is determined as follows (Janic 2007a, 2008; Tosic and Horonjeff 1976):

$$t_{ij/h} = \begin{bmatrix} \delta_{ij}/v_j,\ for\, v_i \le v_j \\ \delta_{ij}/v_j + \gamma_{ij}(1/v_j - 1/v_i)\ \ for\, v_i > v_j \end{bmatrix} \tag{6.10a}$$

where

δ_{ij} is the minimum ATC horizontal distance-based separation rule between leading aircraft (i) and trailing aircraft (j) in landing sequence (ij), and

γ_{ij} is the length of the final approach trajectory of aircraft (i) and (j) connecting the FAG and landing threshold T; and

v_i, v_j is the final approach speed of leading aircraft (i) and trailing aircraft (j), respectively.

The other symbols are analogous to those in previous equations.

The first condition in Eq. 6.10a implies that ATC minimum horizontal separation rules between aircraft (i) and (j) in landing sequence (ij) are applied at the moment when leading aircraft (i) arrives at landing threshold T (Fig. 6.10a—case $v_i <= v_j$). The second condition implies that these rules are applied at the moment when leading aircraft (i) is just at the FAG (Fig. 6.10a—case $v_i > v_j$).

Mixed horizontal/vertical distance-based separation rules

ATC mixed horizontal/vertical distance-based separation rules can be applied between particular sequences of landing aircraft using the above-mentioned Individual Flight Corridors with two segments, each with different GS angles. In such case, the minimum inter-arrival time at landing threshold T of aircraft sequence (ij) can be estimated as follows:

$$\begin{aligned} t_{ij/m} = [&\min(\delta_{ij}/v_j;\ H_{ij}^0/v_j \sin\theta),\ for\, v_i \le v_j;\ H_{ij}^0/v_j \sin\theta_j + \gamma_{ij} * (1/v_j \\ &- 1/v_i),\ for\, v_i > v_j] \end{aligned} \tag{6.10b}$$

where

H_{ij}^0 is the minimum ATC vertical distance-based separation rule between leading aircraft (i) and trailing aircraft (j) in landing sequence (ij);

θ, θ_j is the nominal and steeper GS angle on the Inner and Outer segment of the approach trajectory of the trailing aircraft (j), respectively;

The other symbols are analogous to those in Eq. 6.10a.

The first condition in Eq. 6.10b implies that the minimum ATC horizontal or vertical distance-based separation rules are applied at the moment when leading aircraft (i) arrives at landing threshold T (Fig. 6.10b—case $v_i <= v_j$). The second condition in Eq. 6.10b implies that the minimum ATC vertical separation rules are applied at the moment when leading aircraft (i) is just at the FAG (Fig. 6.10b—case $v_i > v_j$). In this case, the vertical distance-based separation rules can be also dependent on the type of landing sequence similarly as the horizontal ones

Vertical distance-based separation rules

ATC vertical distance-based separation rules can be applied to particular landing sequences in which the aircraft use different GS angles along their entire final approach trajectories. Depending on the approach speeds and GS angles, there can be 12 different combinations of landing sequences for which the minimum inter-arrival times at the reference location, i.e., the landing threshold, calculated as follows:

$$t_{ij/m} = \begin{bmatrix} H_{ij}^{0}/v_j \sin\theta_j, \mathit{for}\, v_i \leq v_j \sin\theta_j/\sin\theta_i \\ H_{ij}^{0}/v_j \sin\theta_j + \gamma_{ij} * tg\theta_i * (1/v_j \sin\theta_j - 1/v_i \sin\theta_i), \mathit{for}\, v_i > v_j \sin\theta_j/\sin\theta_i \end{bmatrix} \tag{6.10c}$$

where

γ_i is the length of the final approach path of leading aircraft (*i*);

θ_i, θ_j is the GS angle of leading aircraft (*i*) and trailing aircraft (*j*), respectively, in landing sequence (*ij*).

The other symbols are analogous to those in the previous equations.

The first condition in Eq. 6.10c implies that ATC minimum vertical separation rules are applied at the moment when the leading aircraft (*i*) arrives at the landing threshold *T* (see Fig. 6.10c—case: $v_i \sin\theta_i <= v_j \sin\theta_j$). The second condition of Eq. 6.10c implies that ATC minimum vertical separation rules for sequence (*ij*) are applied at the moment when the leading aircraft (*i*) is just at the FAG (Fig. 6.10c—case: $v_i \sin\theta_i > v_j \sin\theta_j$). In all cases, the minimum vertical distance-based separation rules can depend of type of aircraft landing sequence (*ij*).

Time-based separation rules

In order to anticipate application of ATC time-based separation rules, the term (δ_{ij}/v_j) in Eqs. 6.10a, 6.10b should be replaced with the term $\tau_{ij/\min}$,—the ATC minimum time interval applicable to aircraft sequence (*ij*) either at runway threshold *T* ($v_i <= v_j$) or at the FAG ($v_i > v_j$) (Janic 2008).

In all above-mentioned cases, the minimum inter-arrival time at the runway landing threshold *T*, t_{ij} must be at least equal or greater than the runway landing occupancy time t_{ai} of the leading aircraft (*i*) in the landing sequence (*ij*), i.e., $t_{ij} >= t_{ai}$. This ensures safe separation on the runway, namely that only one aircraft can occupy the runway at any time.

6.5.3.3 Application of the Models

The above-mentioned analytical models are applied to the generic case of calculating the landing capacity of a single runway according to the "what-if" scenario approach,

Input

Scenarios

The "what-if" scenario approach is used because at present, the prospective operationalization of proposed advanced operational procedures, separation rules, and service disciplines appears at most airports quite uncertain and thus hypothetical. In such context, three attributes characterize each scenario: (a) the aircraft fleet mix; (b) the ATC separation rules and associated operational procedure; and (c) the service discipline. As a result, eight different scenarios are defined in Table 6.13.

In each scenario, the ATC separation rules are exclusively attributed to the individual operational procedure. The FCFS service discipline is used in Scenarios 1–4 and the PR service discipline in Scenarios 5–8. The aircraft fleet mix is common for all Scenarios 1–8. This is specified to reflect the typical situation at most major airports during peak hours (negligible proportion of Small and B757, and substantive proportions of large and heavy aircraft) on the one hand, and enable consistent comparison of the landing capacity across Scenarios on the other. Scenario 1 in Table 6.13 adopted as the benchmark is characterized by using the ATC horizontal distance-based separation rules (IFR) given in Table 6.11, the FCFS service discipline, and the conventional (existing) final operational procedure (Fig. 6.10a) (FAA 2010, 2011). In addition, the ATC time-based separation rules in Table 6.12 derived from the VFR distance-based separation rules in Table 6.11 and the aircraft average approach speeds in Table 6.14 are used in Scenario 4. The other types of ATC advanced separation rules are used accordingly in the remaining Scenarios 2, 3, 5, 6, 7, 8 (Table 6.13).

The length of the common approach path connecting the FAG and the runway landing threshold is adopted to be: $\gamma = 6.2$ nm in all scenarios (Janic 2008) (nm—nautical mile; 1 nm = 1.852 km).

Table 6.13 Characterization of particular "what-if" scenarios for calculating the runway landing capacity (Janic 2012)

Scenario	ATC separation rules	Operational procedure	Priority discipline	Aircraft fleet mix
1	H	SS-NGS	FCFS	Small—5 %
2	MH/V	DS-NGS SGS	FCFS	B757—5 %
3	V	SS—UAGS	FCFS	Large/Heavy as complements
4	TB	SS-NGS	FCFS	(0–90 %)
5	H	SS-NGS	PR	Small—5 %
6	MH/V	DS-NGS SGS	PR	B757—5 %
7	V	SS—UAGS	PR	Large/Heavy as complements
8	TB	SS-NGS	PR	(0–90 %)

H Horizontal, *MH/V* Mixed Horizontal/Vertical, *V* Vertical, *TB* Time-based, *SS-NGS* Single segment-nominal GS angles, *DS-NGS SGS* Double segment—nominal and steeper GS angles, *SS-UAGS* Single segment—ultimately arbitrary GS angles
FCFS First Come First Served, *PR* Priority

Table 6.14 Characteristics of the aircraft wake vortex categories (averages) (EC 2005; FAA 2010; Thompson 2002; Janic 2012)

Aircraft type	Weight W (tons)	Weight v (kts)	GS angle θ (°)	Runway landing occupancy time t_a (s)
Small	20	120	3/4/5.5	30–40
Large	55	130	3/4/-	40–50
B757	117	155	3/4/-	40–45
Heavy	206	155	3/-/-	50–60

Aircraft fleet and assignment of GS angles

The technical/operational characteristics of the particular aircraft wake vortex categories including the assigned GS angles are given in Table 6.14.

According to these assignments, small, large, and B757 aircraft are assumed to approach and land at different (steeper) GS angles while heavy aircraft exclusively use the nominal GS angle of 3°.

Based on Table 6.14 and respecting the two operational procedures when the ATC mixed horizontal/vertical and exclusively the vertical distance–based separation rules are applied, the combinations of GS angles for particular aircraft landing sequences are given in Table 6.15a, b.

Results

The results from applying the models by using the above-mentioned inputs are shown in Figs. 6.11, 6.12, 6.13, and 6.14.

Figure 6.11 shows the runway landing capacity for Scenarios (1–4) in Table 6.13, i.e., their dependence on the proportion of Heavy aircraft in the fleet mix, type of ATC separation rules, and FCFS service discipline.

As can be seen, the capacity is the highest in Scenario 3 when ATC vertical distance-based separation rules are applied, followed by the capacity in Scenario 4 (ATC time-based separation rules), Scenario 2 (ATC mixed horizontal/vertical distance-based separation rules), and the benchmarking Scenario 1 (ATC horizontal distance-based separation rules).

The landing capacity in Scenario 4 also decreases as the proportion of Heavy aircraft in the fleet mix increases. It is lower than the landing capacity in Scenario 3 and the gap increases with the proportion of heavy aircraft in the fleet mix up to about 6 %.

In the benchmarking Scenario 1, the capacity decreases with increasing of the proportion of heavy aircraft in the mix, up to about 10 % for the given range of changes of this proportion. This capacity similarly decreases in Scenarios 3 and 4, up to about 8 and 12 %, respectively, for the given range of changes of the proportion of heavy aircraft (0–90 %). In addition, the landing capacity in Scenario 1 is lower than the capacity in Scenario 3 (between 20 and 30 %) and the capacity in Scenario 4 (between 18 and 28 %). This is as intuitively expected due to the shortening of distances and consequently the average inter-arrival times

Table 6.15 Combinations of GS angles for different ATC separation rules and related operational procedures (°)-degrees) (Janic 2012)

i/j	Small	Large	B757	Heavy
(a) Mixed horizontal/vertical distance-based separation rules				
Small	3/3	3/3	3/3	3/3
Large	3/5.5	3/3	3/3	3/3
B757	3/5.5	3/4	3/3	3/3
Heavy	3/5.5	3/4	3/4	3/3
(b) Vertical distance-based separation rules				
Small	3/5.5	3/4	3/4	3/3
Large	3/5.5	3/4	3/4	3/3
B757	3/5.5	3/4	3/4	3/3
Heavy	3/5.5	3/4	3/4	3/3

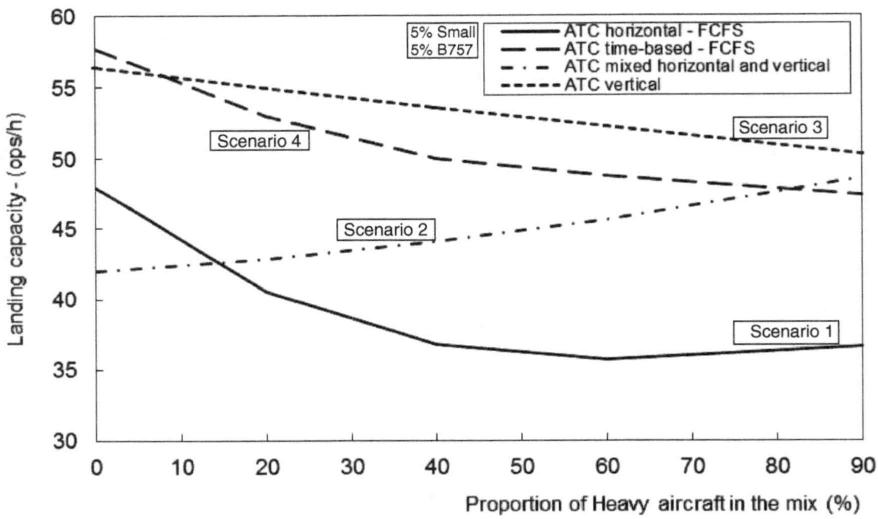

Fig. 6.11 Relationship between the runway landing capacity and the proportion of heavy aircraft in the mix and different ATC separation rules—FCFS service discipline (Janic 2012)

between landing aircraft thanks to the application of vertical distance-based and time-based separation rules. Decreasing capacity in Scenarios 1, 3, and 4 as the proportion of heavy aircraft in the fleet mix increases is mainly due to increasing the frequency of longer distances between these and other aircraft which could not be compensated by increasing the average speed of landing flow just thanks to the more frequent presence of these (Heavy) aircraft in the fleet mix.

Contrary to Scenarios 1, 3, and 4, the landing capacity in Scenario 2 increases with increasing of the proportion of heavy aircraft in the fleet mix. The difference between this and the capacity in Scenario 3 decreases from about 28 to 4 %, and the capacity in Scenario 4 from about 38 to 2.5 %. At the same time, the positive

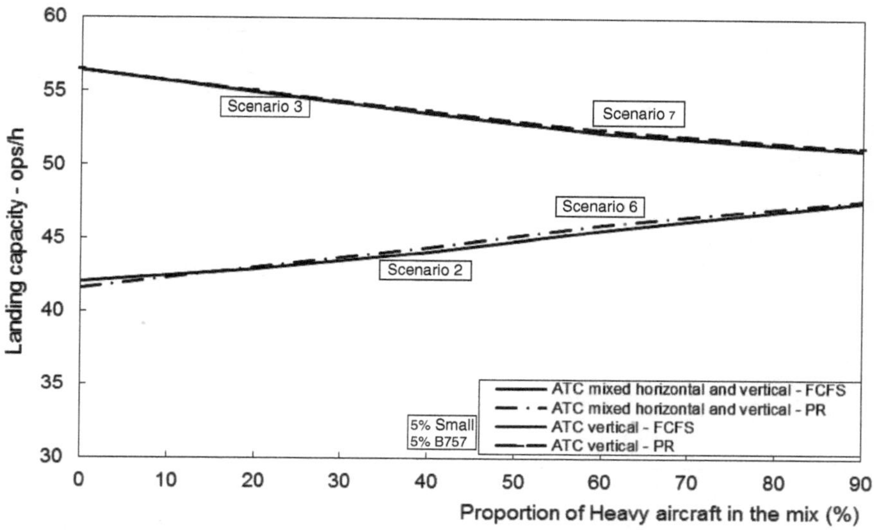

Fig. 6.12 Relationship between the runway landing capacity and the proportion of Heavy aircraft in the mix and ATC mixed and vertical distance-based separation rules—the FCFS and PR service discipline (Janic 2012)

difference between this and the capacity in Scenario 1 increases by 6–33 % for a given range of changes in the aircraft fleet mix. This increase of the capacity with increasing of the proportion of heavy aircraft in the fleet mix in Scenario 2 is mainly due to applying vertical separation rules to the FS ("Fast-Slow") sequences at the FAG, which shortens the horizontal distances between aircraft, and consequently the corresponding inter-arrival times at landing threshold *T*, which in turn contribute to increasing the landing capacity.

Figure 6.12 shows the influence of the ATC service disciplines (FCFS and PR) on the landing capacity given ATC mixed horizontal/vertical and vertical distance-based separation rules (Scenarios 2 and 6, and Scenarios 3 and 7 in Table 6.13, respectively), and the proportion of heavy aircraft in the mix.

As can be seen, the PR service discipline (Scenarios 6 and 7) increases the capacity only negligibly as compared to the FCFS service discipline (Scenarios 2 and 3). At the same time, the influence of heavy aircraft on this capacity appears to be of the same character for both service disciplines in each pair of the considered Scenarios (3, 7 and 2, 6).

Figure 6.13 shows the dependence of the runway landing capacity on the proportion of heavy aircraft in the mix, ATC horizontal distance-based and time-based separation rules, and service disciplines, i.e., the capacities in Scenarios 1 and 5, and Scenarios 4 and 8, respectively.

As can be seen, when ATC horizontal distance-based separation rules are used, the PR service discipline (Scenario 5) contributes to increasing the landing capacity

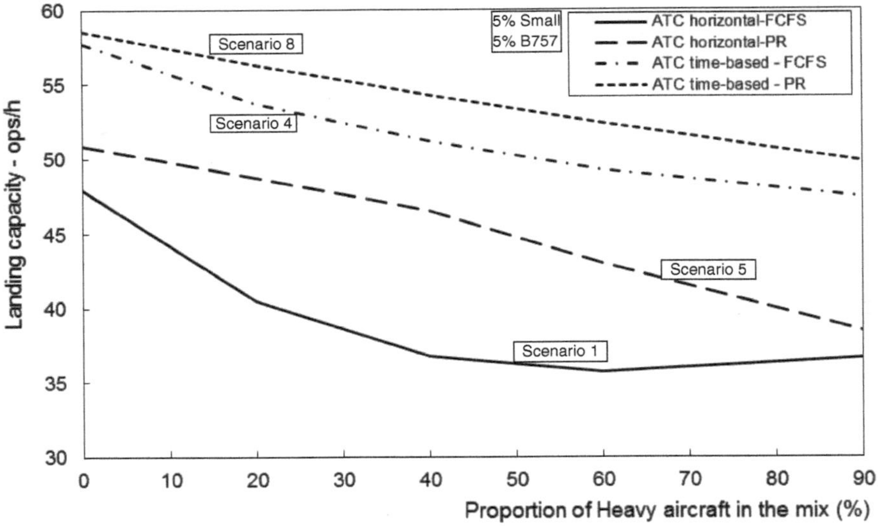

Fig. 6.13 Relationship between the runway landing capacity and the proportion of Heavy aircraft in the mix and ATC horizontal distance-based and time-based separation rules—the FCFS and PR service discipline (Janic 2012)

by about 6 to 7 % as compared to the FCFS service discipline (Scenario 1). This difference is the greatest (about 30 %) when the proportion of heavy aircraft in the mix is about 40–50 %. When ATC time-based separation rules are used, the PR service discipline (Scenario 5) contributes to increasing the runway landing capacity by between 2 and 5 % as compared to the FCFS service discipline (Scenario 4) for the specified range of proportions of heavy aircraft in the fleet mix (0–90 %).

Figure 6.14 shows the runway capacity envelopes when different ATC minimum separation rules are applied to the landing aircraft. The landing and departing aircraft fleet mix consists of 5 % small, 5 % B757, 40 % heavy, and 50 % large aircraft, all served according to the FCFS service discipline. In addition, ATC minimum time-based separation rules are exclusively used between successive departures (FAA 2010, 2011). As can be seen, when the landings predominate at the runway, different ATC minimum separation rules provide different landing capacities, namely the highest when ATC vertical and the lowest when ATC horizontal distance-based separation rules are used. When departures are inserted more frequently between successive landings, the effects of different ATC separation rules between landings diminish and almost disappear when the proportions of both types of operations equalize. After departures start to dominate, the effects of different ATC separation rules between landings continue to diminish further.

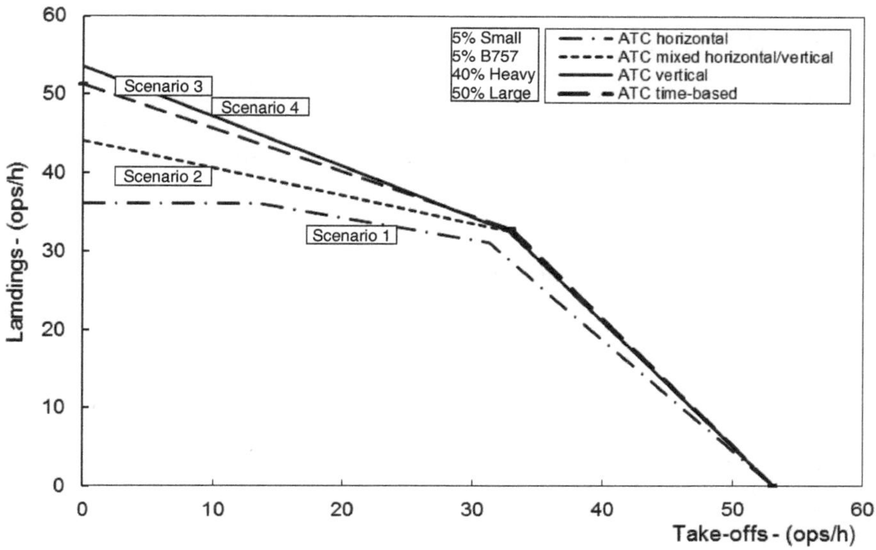

Fig. 6.14 Runway capacity envelopes for cases of application of different ATC minimum separation rules for landings (Janic 2012)

6.5.4 Evaluation

Advanced ATC (Air Traffic Control) technologies and procedures for increasing airport runway landing capacity possess some advantages (Strengths and Opportunities) and disadvantages (Weaknesses and Threats).

Advantages

- Contributing to a substantial increase in the runway landing capacity when the FCFS service discipline and ATC vertical distance-based separation rules are applied. This capacity would be the greatest as compared to that obtained by using ATC time-based, mixed horizontal/vertical and particularly horizontal distance-based separation rules, respectively; and
- Contributing to a substantial increase in the runway landing capacity when the PR service discipline is applied instead of the FCFS service discipline in combination with ATC horizontal distance-based and time-based separation rules.

Disadvantages

- Demanding a relatively long time for full operationalization both at the ATC and aircraft;
- Diminishing effects with increasing proportion of heavy aircraft in the landing mix except in the cases of applying the PR service discipline and ATC mixed horizontal/vertical distance-based separation rules; and

- Diminishing effects with increasing of the heterogeneity of operations on a single runway (more take-offs on the account of landings).

Finally, these technologies and procedures can be considered as advanced mainly due to their contribution to the rather substantial increase in the airport runway landing capacity as compared to their currently used counterparts.

6.6 Advanced Supersonic Transport Aircraft

1947	The rocket-powered Bell X-1 flown by C.Yeager exceeds the speed of sound during the first controlled level flight (U.S.)
1967	The air speed record of 3,940 kts (7,297 km/h Mach 6.1) is set by the X-15 aircraft (U.S.)
1975	Aeroflot begins regular services with the supersonic TY144 aircraft (USSR)
1976	British Airways and Air France begin commercial servicers with the French-British Concorde aircraft (UK, France)

6.6.1 Definition, Development, and Use

The STA (Supersonic Transport Aircraft) has already been commercialized as the French-British Concorde and the Soviet Union's TY144 supersonic aircraft. Their main characteristics are given in Table 6.16.

Both aircraft had very similar performances. However, during their commercial use over the limited time span,[10] they were unable to fulfill the requirements and expectations in terms of effectiveness, efficiency, environmental and social acceptance. This is illustrated by comparing some of the characteristics of Concorde and the Boeing 747 aircraft given in Table 6.17.

These characteristics indicate that, except as concerns its cruising speed, Concorde was inferior to the B747-400 in all considered aspects. As such, it was not so attractive for airlines faced with increased pressure to manage their profitability and take care about the environment, both under unstable economic conditions at the time when both aircraft were operating. Figure 6.15a, b additionally supports this perception of airlines by showing the quite substantial differences in the payload-range and technical productivity-range between Concorde (and TY144D) and the B747-400 aircraft.

In addition, the accident that occurred in 2000 strongly undermined Concorde's previous reputation as the safest aircraft. Consequently, the aircraft was retired in

[10] Concorde commercially operated during the 1976–2003 period and TY144D during the 1977–1983 period (http://en.wikipedia.org/wiki/Concorde; http://en.wikipedia.org/wiki/Tupolev_Tu-144).

Table 6.16 Characteristics of Concorde and the TY144 supersonic commercial aircraft (http://www.concordesst.com/home.html; http://www.tu144sst.com/index.html)

Characteristic	Aircraft type	
	Concorde	Ty144
Number built/used	20/16	15/3
Length (m)	61.70	65.50
Wingspan (m)	25.60	28.80
Height (m)	12.20	12.55
Cockpit crew	3	3
Seating capacity (seats)	92–120	120–140
MZFW[a] (tons)	92	92.2
Fuel (tons)	74	99.8
Payload (tons)	13	15
MTOW[b] (tons)	185	207
MLW[c] (tons)	111	120
Engines (number/thrust (kN))	4/140–170	4/200
Range (km)	7220	6000–6600
Rate of climb (ft/min)	5000	3000
Cruising speed (Mach/kts)	2.02/1157 at 60,000 ft/18300 m	2.15/1232 at 65,600 ft/20000 m
Average take-off speed (kts)	217	192
Average landing speed (kts)	161	146
Take-off runway length (m)	3599	2930
Landing runway length (m)	2169	2560
Fuel consumption (kg/s-km)	0.120–0.143	0.080–0.093
Maximum nose tip temperature (°C)	127	N/A

[a] Maximum Zero Fuel Weight
[b] Maximum Take-Off Weight
[c] Maximum Landing Weight

Table 6.17 Characteristics of Concorde and the Boeing 747-400 aircraft (Smith 1989)

Characteristic	Aircraft type	
	Concorde	B747-400
Seating capacity (seats)	120	413
Cruising speed (Mach/km/h)	2.02/2145	0.85/903
Max. range (km)	7,220	14,353
Technical productivity (s-km/h)	261480	377069
Operating costs (€/s-km)	0.0374	0.0202
Fuel efficiency (kg/s-km)	0.120–0.143[a]	0.025[b]
Emissions of GHG (kgCO_{2e}/s-km)	0.715	0.133
Noise[c] (arrivals/departures)[d] (EPNdB)	119/119/116	100.4/96/101.7

[a] Estimated from the total costs of 1.25 billion €, 50,000 flights and length of route of 5,564 km (London-New York)
[b] Estimated from the expression (2.26) for the same route
[c] EPNdB—Effective Perceived Noise Level (dBA); Sideline (450 m)/Take-off (6.5 km)/Approach (2 km)
[d] Smith (1989)

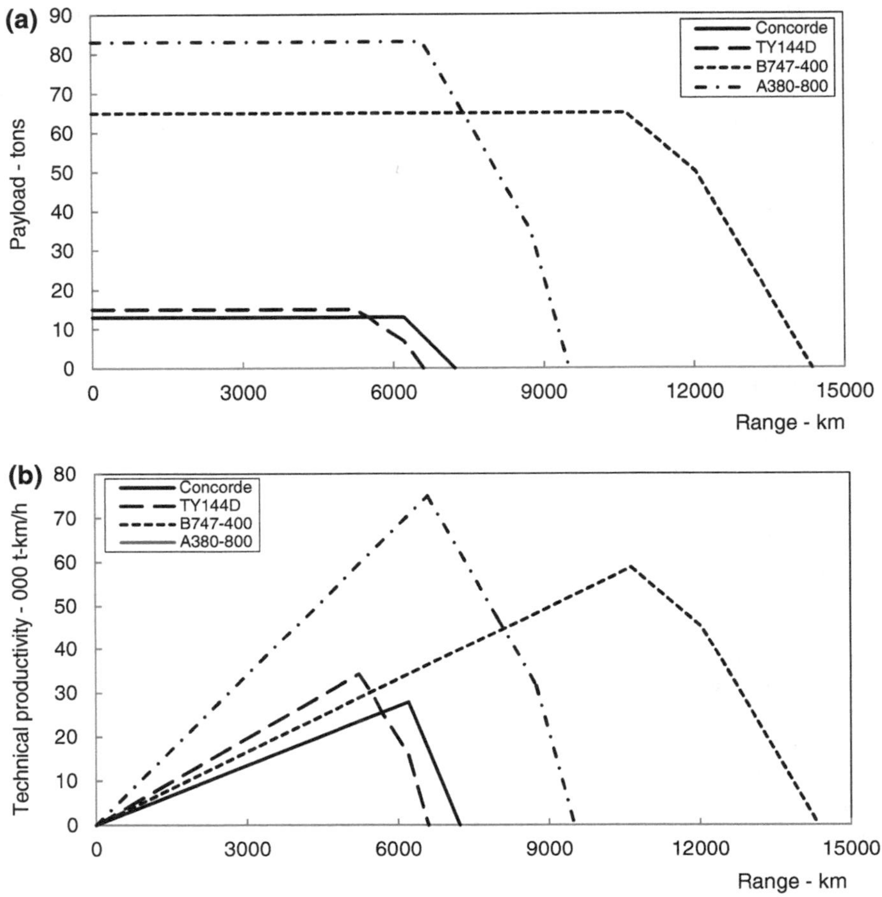

Fig. 6.15 Operational performances of the selected aircraft types (Airbus 2005; Boeing 2002; http://en.wikipedia.org/wiki/Concorde; http://en.wikipedia.org/wiki/Tupolev_Tu-144), **a** Payload-range, **b** Technical productivity-range

2003. After crashing in Paris, the TY 144 aircraft was retired in 1978 after just one year in commercial service.

6.6.2 Analyzing and Modeling Performances

6.6.2.1 Background

After retirement of the supersonic Concorde and TY144 aircraft, the question remained whether there might be sufficient passenger demand for an advanced generation of STA (Supersonic Transport Aircraft), which, in addition to speed, would be also close if not superior to their subsonic conventional counterparts particularly in terms of the economic, environmental, and social performances.

The full development, implementation, and operationalization, i.e., commercialization of advanced STA seems likely to be a matter of the forthcoming decades. This is because, at the moment, these currently main aircraft manufacturers Airbus and Boeing, and other agencies and institutions dealing with the visions of development of the air transport system, do not and will not take, at least publically, such development into account before 2030 (ACARE 2010; Airbus 2012; Boeing 2012). Indirectly, this reflects developments in the airline industry, which, at least in the short- to medium-term future, does not expect substantial (mass) increase in the passenger and freight demand for supersonic speed, and consequently remains uninterested in acquiring and operating such aircraft. In turn, the manufacturers consider in investing in the development of such aircraft too risky.

Nevertheless, despite such circumstances, some initiatives of developing advanced STA as successors of Concorde and TY144 have sustained within the aircraft manufacturers and other related agencies. Some concepts that have emerged after the retirement of Concorde are as follows:

- Toward the end of 2003, EADS (European Aeronautic Defense and Space Company)—the parent company of Airbus—announced its cooperation with Japanese companies to develop a larger and faster STA as a replacement for Concorde. As a result, in mid-2011, EADS unveiled the ZEHST (Zero Emission High-Speed Transport) aircraft concept at the Paris Air Show. This actually hypersonic aircraft would be able to fly at a cruising speed of Mach 4.0 (4,248 km/h).
- In the last quarter of 2005, JAXA (the Japan Aerospace eXploration Agency) carried out the aerodynamic testing of a scale model of an aircraft supposed to have the seating capacity of 300 passengers and speed of about Mach 2 (label NEXST). If shown commercially feasible, it has been expected to start commercial operations in around 2020/2025.
- In addition, the LAPCAT (Long-Term Advanced Propulsion Concepts and Technologies) research program half-funded by the EC (European Commission) examined a design of LH_2 (Liquid Hydrogen)-fuelled aircraft with the seating capacity of about 300 passengers (called the A2), range of about 20,000 km, and cruising speed of Mach 5.0 (5,110 km/h) (EC 2006; http://en.wikipedia.org/wiki/Supersonic_transport).
- Furthermore, in 2001, the U.S. NASA (National Aeronautics and Space Administration) announced the development, among others, of an Efficient Multi-Mach STA whose sonic boom would be low enough to enable operations over both land and water. Some specifications of the new EC and US NASA STA concept are given in Table 6.18 (Coen 2011; EC 2006; NAS 2001).

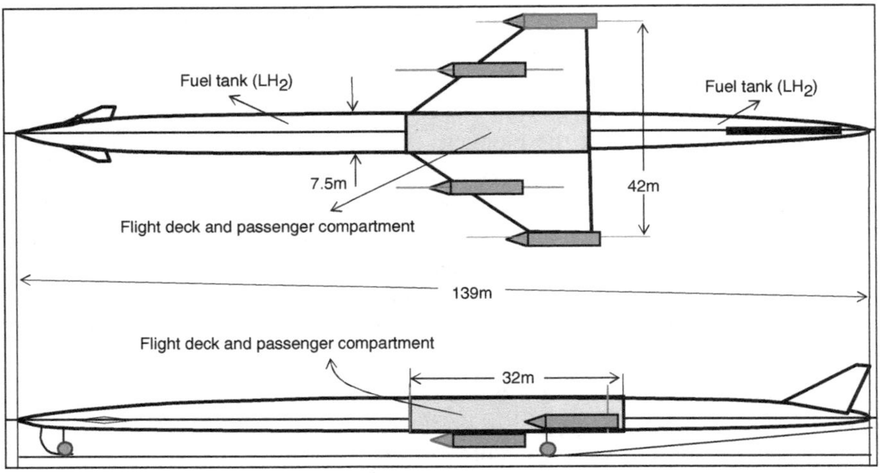

Fig. 6.16 Scheme of possible shape of the advanced STA—EC Hydrogen Mach 5 Cruiser A2 (EC 2008)

Figure 6.16 shows an example of possible design of the advanced STA—EC Hydrogen Mach 5 Cruiser A2 (Coen 2011; EC 2008).

The above-mentioned concepts indicate that the main driving force for developing new STA remains to be increasing technical productivity as the product of cruising speed and payload. For airlines, this is simple: if this productivity is higher, more services (passenger- or ton-km) can be carried out during a shorter period of time, thus requiring a smaller fleet with consequently lower costs. For passengers, flying at higher speeds would save time and related travel costs, which also would be generally beneficial for society. What remains is making the environmental impacts in terms of the energy consumption and related emissions of GHG, and noise around airports and wider area due to sonic boom, acceptable.

In general, the technical productivity could be increased exclusively by increasing speed, increasing payload, or both simultaneously. In the above cases of STA, the last option has been considered. But how much would this cost?

Operationalization of some of the above-mentioned concepts will likely depend of their overall costs, reflected through prices and operational costs for airlines, which in turn will be reflected through fares for passengers and freight shippers. These costs per unit could be reduced by increasing the number of units (STA) manufactured, which depends on the number of units sold, which itself depends on the prices per unit, and vice versa. For example, the number of units sold, in addition to price, depends on the level of satisfaction of airline requirements in terms of efficiency, effectiveness, and environmental and social acceptance on the one side, and the volumes of expected passenger and freight demand on the other. In addition, overall cost reduction could be achieved by improving the five-stage process of commercialization (Chap. 1), ensuring greater efficiency of the manufacturing process through the learning curve, and maintaining the optimal manufacturing pace (usage of resources—labor, material, and energy), and facilities and equipment).

6.6.2.2 Infrastructural Performances

The infrastructural performances of new STA generally relate to their fitting with exiting airports in terms of the footprint and other operational performances. For example, the EC Hydrogen Mach 5 Cruiser STA in Table 6.18 would fit with the ICAO Aerodrome Code 4 and Letter D, and IATA (International Air Transport Association) Aircraft Size Categorization Scheme 4. Regarding MTOW, the aircraft could be categorized as a Heavy (400 tons) aircraft of the approach category E (approach speed $\geq$ 166 kts (kts—knots (nm/h)). In addition, its MTOW should meet the current maximum allowable ground loads, i.e., wheel number, size, load, and distribution, thus preventing damage to airport runways, taxiways, and aprons. Furthermore, its dimensions (length, wing span, and ground footprint) should cope with the existing layout and size of airport maneuvering spaces (runways, taxiways, and aprons). In this context, the length of new STA (139 m in Table 6.18 and Fig. 6.16) could become a critical issue.

Ground handling in terms of embarking and disembarking passengers and cargo, fuel supply (either compatible with the existing (kerosene) or new system (LH_2)), etc., should enable a reasonable turnaround time of new STA comparable to that of its subsonic counterparts.

In addition, new STA will have to be able to follow the current and future ATC (Air Traffic Control) rules and procedures. This may eventually apply to steeper final approach angles, which could be requested at some airports in order to increase the airport runway capacity on the one hand and reduce noise around the airport on the other (Chap. 3).

6.6.2.3 Technical/Technological Performances

New STA would operate under rather extreme conditions generally characterized as follows (EC 2008):

- Higher cruise altitudes (FL500-600, i.e., 50,000–60,000 ft) than its subsonic counterparts (FL300-350, i.e., 30,000–35,000 ft) would require higher in-cabin pressure by approximately 25 %,
- Exposure of the structure to a wider range of temperatures, from about −50 °C to more than +150–200 °C, including thermal cycling under moisture and radiation impacts, and

Table 6.18 Specifications of advanced STA concepts (Coen 2011; EC 2008; NAS 2001)

Specification	The STA concept	
	NASA high-speed civil transport—beyond the year 2030	EC Hydrogen Mach 5 Cruiser A2
Design		
Length (m)[a]	89–95	139
Wing span (m)[a]	42–43	42
Wing Aspect Ratio	–	1.868
L/D (Lift-to-Drag) Ratio	10–11	8–9
Take-off weight (tons)	276	400
Number of engines	4	4
Cruise speed (Mach)	2.0–2.4	5.0
Range (km)	9000–11000	20000
Fuel/weight fraction	0.40–0.45	0.13
Payload/weight fraction	~0.20	~0.15
Payload (passengers)	300	300
Fuel efficiency		
SFC (kg of fuel/kg of thrust/h)[b]		
Kerosene	2.325	–
Liquid Hydrogen (LH_2)[c]	–	2.208
Environmental/social efficiency		
Emissions of GHG (gNO_x/kg of fuel)	<15 (lower speeds) ≤5 (higher speeds)	– ≤1
Noise		
Sonic boom—Low Boom Flight (EPNdB)	65–70	–
Sonic boom—Overwater flight (EPNdB)	75–80	–
Airport noise (cumulative below stage 3/4) (EPNdB)	10–20	–

[a] Approximate values
[b] TSFC—Thrust Specific Fuel Consumption
[c] The ratio of heating content: LH_2/Kerosene = 2.745

- Comparability to its subsonic counterparts in terms of technical productivity, which would require high utilization and reliability during the life cycle—about 25,000 flight cycles and 60,000 h over 20 years.

The above-mentioned conditions will require establishing an optimal balance between the required strength of structure and weight of the new STA, which will in turn influence all other technical/technological, operational, economic, environmental, and social performances.

Airframe

Supersonic aerodynamics

Supersonic aerodynamics are quite different from subsonic aerodynamics because of the significant increase of the dynamic pressure and fundamental change in the character of the fluid flow at very high cruising speeds (above Mach 1.6–2.0). This makes the aerodynamic performances of the new STA particularly important because: (i) the typical (L/D) (Lift-to-Drag) of new STA is generally lower than that of the equivalent subsonic aircraft by about 50 %, (ii) the engine size, fuel consumption and related emissions of GHG are directly influenced by drag, and (iii) the weight of new STA is increased due to the above-mentioned two factors, which in turn compromise all other performances. Consequently, the design of new STA generally requires simultaneous balancing of many conflicting issues. For example, supersonic aircraft such as Concorde had relatively low supersonic efficiency measured by the L/D (Lift-to-Drag) ratio, which was not the result of a weak design, but a generic consequence of its supersonic aerodynamics.

Regarding aircraft design, the (L/D) is the most important aerodynamic parameter influencing the most important attributes of economic and environmental performances such as the range, payload, fuel consumption and related emissions of GHG. Therefore, one of the main design objectives of new STA would be to increase this ratio at cruising speeds beyond Mach 1.6–2.0. Specifically, the supersonic lift force (*L*) can generally be expressed as follows:

$$L = (1/2)\rho * v^2 * S \tag{6.11a}$$

where

ρ is air density (kg/m^3);
v is the aircraft speed (km/h);
S is the reference wing area (m^2); and
C_L is the lift coefficient.

In Eq. 6.11a, the lift coefficient C_L depends on the wing characteristics, i.e., its airfoil or section (a cross section of the wing parallel to the plane of symmetry of the aircraft). This is shaped to generate lift without excessive drag. The supersonic drag (*D*) can be expressed as follows (NAS 2001):

$$D = qSC_{D0} + W/q * \pi * b^2 + 128q * V_{ol}^2 + \left[(M^2 - 1) * W^2\right]/(q * \pi * l^2) \tag{6.11b}$$

where

b is the wingspan (m);
C_{D0} is the coefficient of parasite drag at zero lift (i.e., skin drag and all other drag except induced drag);
l is the effective length of the aircraft (m);
M is the Mach number;

q is the dynamic pressure (N/m^2);
S is the reference wing area (m^2);
V_{ol} is the total volume (m^3); and
W is the total weight of the aircraft (tons).

If the optimal cruising altitude for a given Mach number is selected, the (*L*/*D*) ratio can be expressed as follows (NAS 2001; Seebass 1998):

$$(L/D)_{\max} = \left[\left(\frac{4}{\pi * A * R} + \frac{2(M^2 - 1)}{\pi * A * R_1}\right) * (C_{D0} + C_{D0w})\right]^{-1/2} \quad (6.11c)$$

where
AR is the aspect ratio given by b^2/S;
AR_l is the length aspect ratio given by l^2/S; and
C_{D0w} is the coefficient of zero lift wave drag depending on the aircraft length, volume, and wing area.

The other symbols are analogous to those in the previous equations.

Using Eq. 6.11c, the ratio (*L*/*D*) can be obtained for the specified values of the particular variables. Alternatively, the maximum achievable ratio (*L*/*D*) can be empirically estimated as follows (EC 2006, 2008):

$$(L/D) = 6 * (M + 2)/M \quad (6.11d)$$

where all symbols are analogous to those in the previous equations. Figure 6.17 shows a generic relationship between (*L*/*D*) and (*M*).

The (L/D) ratio increases approximately more than proportionally with increasing of the Mach number up to about $M = 1.5$; beyond $M = 1.5$, the ratio decreases more than proportionally with increasing of the Mach number. This implies a decrease in the aerodynamic performances at higher Mach numbers. Therefore, in order to achieve the required high (L/D) ratio, the drag (D) during the cruise phase of flight should be as low as possible, which could be achieved by increasing the length of the advanced STA. Such increase would make the aircraft much longer than its subsonic counterpart(s) with an equivalent payload. This would create disparity between the design for the best cruise and the design for low speed, thus causing degradation of the STA aerodynamic performances at low speed(s), increasing its weight, and consequently diminishing fuel efficiency and increasing related emissions of GHG, than otherwise. In order to overcome these imbalances, some supersonic features such as supersonic laminar flow, methods for modifying the flow around the aircraft, unconventional aircraft configuration (preferably long and slender), and computational systems and methodologies would need to be revolutionized.

Aerodynamic heating

The advanced STA would be exposed to aerodynamic heating during flying at supersonic speeds. This is caused by heating of the air near the surface of the

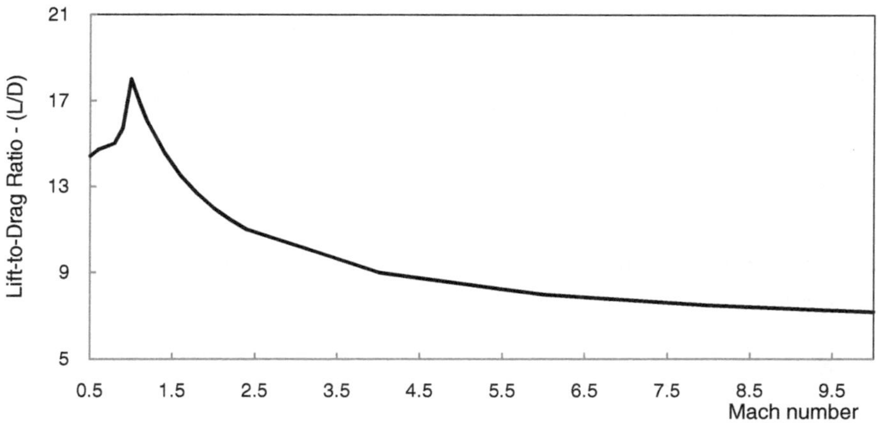

Fig. 6.17 Relationship between Lift-to-Drag (L/D) ratio and Mach number for STA

aircraft whose temperature can reach the so-called stagnation temperature expressed as follows:

$$T = T_0 + v^2/2c_p \tag{6.11e}$$

where

T_0 is the temperature of the impacting air (°C);

v is the aircraft speed (m/s); and

c_n is the specific heat capacity[11] of air at constant pressure (kJ/kg°K).

For example, if the new STA was cruising at the speed of $M = 2.0$ at the altitude FL600 (60,000 ft) where the temperature of the impacting air is: $T_0 = -56.5$ °C, the stagnation temperature would be about: $T = 112$ °C. If the speed was $M = 2.5$, the stagnation temperature would reach about $T = 200$ °C under the same conditions.

Materials

If the aircraft stayed under the above-mentioned operating conditions sufficiently long, the rather high air temperature would pass to its structure as the convective heating would dominate over the radiative heating. In order to sustain such temperature, the new advanced lightweight high temperature materials have been tested such as mainly aluminum and titanium alloys PMC (Polymer Matrix Composites) and MMC/IMC (Metal Intermetalic Matrix Composites/Intermediate Modulus Carbon) fiber, the latter considered as a compromise between the

[11] The specific heat capacity is the amount of heat required to change the temperature of 1 kg of a given material/substance by 1 °C. This capacity of air for temperatures (−50–400 °C) ranges from 1.005 to 1.068 kJ/kg°K).

required air-thermodynamic performances and costs. In addition, toughened Epoxy resins and thermoplastic materials have also been considered. In parallel, both the EC and NASA in the US have studied different lightweight advanced materials such as sandwich, honeycomb, and superelastic formed/diffusion bonded concepts (EC 2006, 2008; NAS 2001; Saounatsos 1998).

Range

The most important aerodynamic and operational characteristics of the new STA is its range. In general, this range can be expressed as follows (Breguet range equation):

$$R = \frac{H(L/D)}{g} * \eta * \ln\left(\frac{1}{1 - W_F/W}\right) = \frac{v * (L/D)}{g * SFC} \ln\left(\frac{1}{1 - W_F/W}\right) \quad (6.11f)$$

where

H is the fuel energy content (MJ/kg);
η is the overall installed engine efficiency;
SFC is Specific Fuel Consumption (kg/kN/h);
g is the gravitational acceleration (9.81 m/s^2);
W is the aircraft take-off weight (tons); and
W_F is the available trip fuel (tons).

The other symbols are analogous to those in the previous equations. Eq. 6.11f indicates that the range R increases linearly with increasing of the energy content H of the fuel and the overall installed engine efficiency η. By changing the fuel from kerosene to LH_2 (Liquid Hydrogen), this content and consequently the range R can be increased by the factor 2.745 (Chap. 4). For a given type of fuel, the range R of the advanced STA could be increased, on the one hand by increasing the cruising speed v, improving (*L/D*), and the amount of fuel carried W_F, and on the other by decreasing the engine SFC. However, at the same time, the SFC increases with cruising speed and decreases with increasing of the overall engine propulsion efficiency η as follows:

$$SFC = M/4\eta \quad (6.11g)$$

Figure 6.18 shows some relationships.

Specifying the values of particular variables in the above-mentioned expressions, while particularly intending to increase (L/D) as a vital issue for enabling economically viable STA, enables setting up some initial specifications for the STA design similarly to those in Table 6.18.

Propulsion systems

The propulsion systems-engines of the new STA should be designed to fulfill the following requirements (EC 2008):

- Sufficient thrust to enable operating at the designed supersonic cruising speed(s),

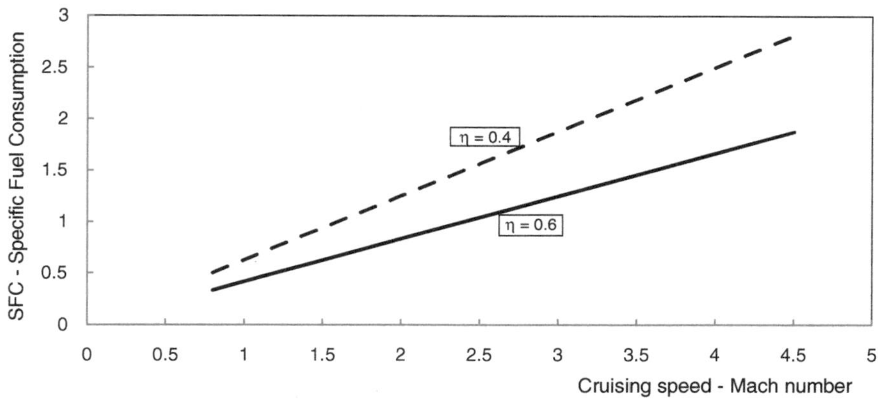

Fig. 6.18 Relationship between SFC (Specific Fuel Consumption) and the aircraft cruising speed, and the engine overall propulsion efficiency

- Relatively high propulsion efficiency implying lower fuel consumption and related emissions of GHG, particularly NO_x (Nitrogen Oxides), as compared to that of supersonic predecessors Concorde and TY144, as well as advanced subsonic aircraft (Chap. 2).
- Lightweight reliable construction materials resistant to high temperatures, and
- Meeting the noise requirements of Stage-3/4 during LTO cycle (Landing and Take-off Cycle) (ICAO 2001).

Thrust

The total thrust, i.e., power, of engines (T) for the new STA should generally be sufficient to balance the aerodynamic drag (D) during cruising phase of flight, i.e., $T = D$ (at the same time the lift force (L) should be in balance with aircraft weight (W), i.e., $L = W$). Thus, based on Eq. 6.11b, the thrust per engine can be estimated as follows:

$$T = \frac{1}{N}[D * (M/a)] \tag{6.11h}$$

where

N is the number of aircraft engines, and

a is the speed of sound at the cruising altitude (m/s).

The other symbols are analogous to those in the previous equations. In combination with Eqs. 6.11b, h for aerodynamic drag (D) implies that the required thrust for the new STA increases with the third degree of cruising speed.

Propulsion efficiency

The propulsion efficiency of engines powering the advanced STA generally depends on its cruising speed as follows (empirical relationship):

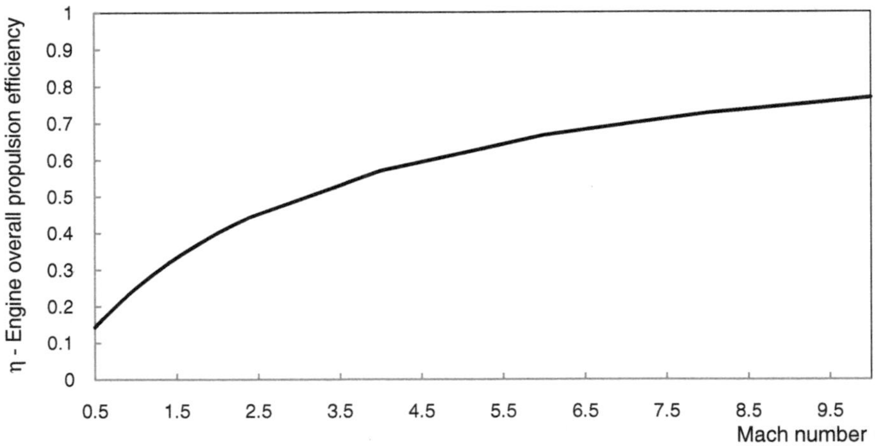

Fig. 6.19 Relationship between propulsion efficiency and Mach number for advanced STA

$$\eta = M/(M + 3) \tag{6.11i}$$

Figure 6.19 shows an example of this relationship indicating that the overall propulsion efficiency increases at decreasing rate with increasing of the Mach number.

This again indicates an inherent conflict with the above-mentioned aerodynamic design of new STA characterized by the substantial drop of the (*L*/*D*) ratio as the cruising speed increases.

Materials

In order to reduce the weight and improve performances of new STA, the materials the aircraft engines are made out of must at the same time provide low density, high strength, and long life at high temperatures. In particular, high temperatures refer to combustion temperatures that which would likely increase to about 3,000–3,500 °F (1,650–1,930 °C) and/or nozzle temperatures to about 2,300–2,400 °F (1,260–1,370 °C). Critical engine components such as disks, turbine airfoils, and nozzles would be exposed to these temperatures for a relatively long time, i.e., during the cruising phase of flight, which could last 4–5 h (EC 2008; NAS 2001). This is quite different compared to subsonic engines exposed to extreme temperatures for a shorter time-only during the take-off and climb lasting about 15–20 min. In addition, more extensive use of such actually advanced materials could contribute to reducing the overall aircraft weight, and consequently improve the fuel consumption and related emissions of GHG, and noise efficiency.

Ceramic Matrix Composites (CMCs) could be a promising solution to simultaneously reduce the engine weight and increase the above-mentioned engine aerothermodynamics performances. The materials considered have been polymers, intermetalics, metal matrix, and ceramics matrix composites. In addition, new fibers and fiber coatings have also been tested as materials to eventually withstand the high operational temperatures and long operational times of particular engine components. One additional material considered for fabricating the critical engine nozzles has been titanium aluminized and other PETI/5 (Phenylethynul/Terminated Imide), the latter particularly considered for high temperatures composite matrices (EC 2008; NAS 2001).

The particular issue would be creating materials for maintaining a safe fuel temperature in the fuel tanks during flying at supersonic speeds. Some preliminary investigations have shown that this could be achieved by advanced coatings with the desired emissivity, absorptivity, and corrosion ability to maintain the fuel temperatures below the acceptable levels (about 150 °F) reached at the speeds of about Mach 2.4 (NAS 2001).

Aircraft systems

Advanced STA would be long and slender lightweight high-temperature durability constructions preventing water absorption during the aircraft life cycle. As such, their construction would be able to fulfill the required aerodynamic performances. This particularly refers to a long pointed nose for drag reduction and delta wings for flight at high supersonic speeds on the one hand and high attack angles at lower speeds on the other. Such design at low speeds actually compromises the forward visibility of the crew from the cockpit. Consequently, this would require either a drop-nose design (similarly to Concorde) or advanced sensors and cockpit displays for restoring the required crew visibility. The latter alternative being under development in the scope of the US NexGen and European SESAR program could be promising (Chap. 2). Nevertheless, in order to relieve the crew from only a computer-generated view of the outside environment, the cockpit windows could be located at the lower instead of the traditional upper front location of the aircraft fuselage. This would enable visual contact to the outside during the aircraft approach, landing, and take-off. In other phases, computer displays would be used for controlling and managing the aircraft flight. In addition, a sonic boom shadow tracking system would be added as unique to new STA. Using standard, predicted, or real-time atmospheric conditions, this system would be able to depict location, coverage, predicted path, and strength of the STA sonic boom.

6.6.2.4 Operational Performances

The operational performances of new STA include their payload-range characteristics, service quality, and technical productivity, which all should be at least comparable to those of the predecessors Concorde and the TY144 and some of the subsonic counterparts.

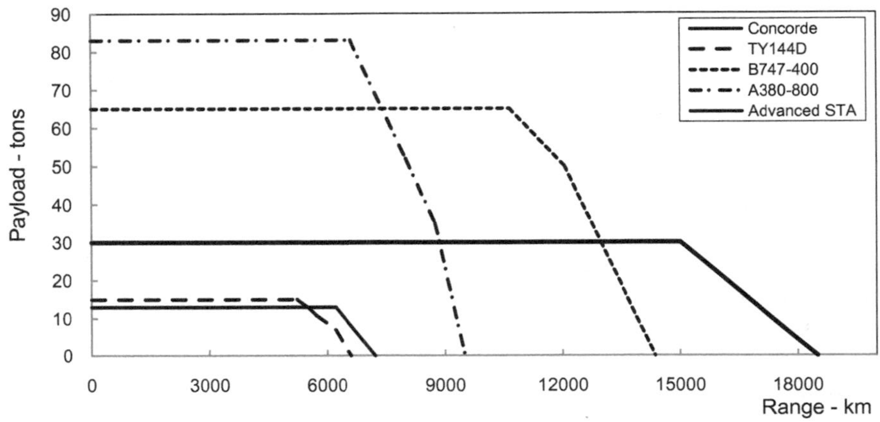

Fig. 6.20 Payload-range characteristics of selected supersonic and subsonic aircraft (*different websites*)

Payload-range characteristics

Possible payload-range characteristics of particular supersonic aircraft and the B747-400 subsonic aircraft are given in Fig. 6.20.

The payload-range characteristics of the Mach 5.0 STA are superior to those of both Concorde and the TY144 aircraft. This could be considered as some kind of increased capability to penetrate the market, which both predecessors were unable to achieve, and thus provide the necessary economic viability. At the same time, new STA have a longer range than the subsonic B747-400 aircraft has for about half of its payload.

Quality of service

The main attribute of the quality of service offered by STA is travel time. Subsonic commercial flights usually take place along the shortest—Great Circle routes connecting particular origins and destinations. However, due to passing through the sound barrier, the routes of the new STA are expected to very often deviate from the shortest great circle distances in order to avoid the impacts of sonic boom on populated areas. In terms of design, this will require extension of the range. In an operational sense, this would extend the travel (flight) time thus partially mitigating the effects of the supersonic speeds as compared to those of its subsonic counterparts. Figure 6.21 shows the relationship between the flight time and route length when flown by subsonic aircraft and advanced STA.

As can be seen, the time differences, which can be deemed as savings for both airlines and users/passengers, appear to be substantial and almost proportional to the differences in the cruising speeds of both categories of aircraft/flights. In the case of airlines, this contributes to decreasing the size of the required fleet for operating a given route network and schedule. In the case of users/passengers, this

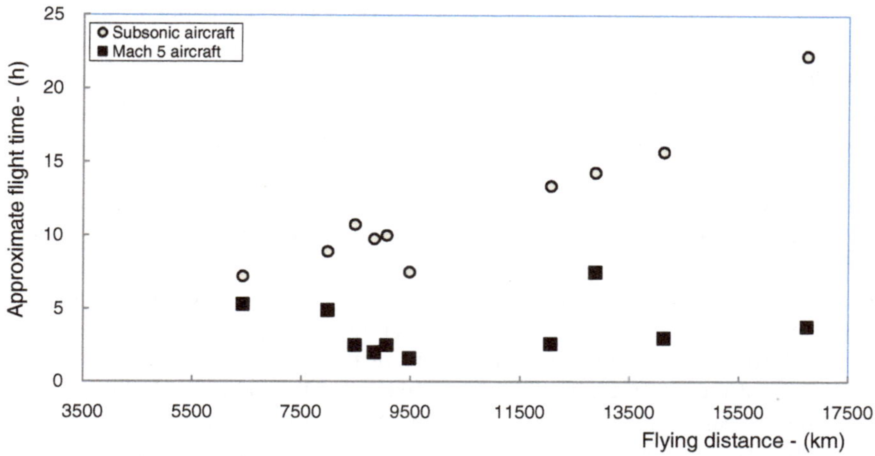

Fig. 6.21 Relationship between approximate flight time and length of route for subsonic aircraft and STA (Mach 5) (EC 2006)

contributes to decreasing their generalized travel costs shown to be particularly relevant for business passengers.

Technical productivity

The technical productivity of new STA during the cruising phase of flight can be estimated as follows:

$$TP = v * PL = M * a * PL \tag{6.12}$$

where

v is the aircraft speed (km/h);
a is the speed of sound (km/h); and
PL is the payload (tons).

Based on Eq. 6.12, the technical productivity for an advanced STA cruising at FL600 with the payload $PL = 30$ tons (300 passengers each weighting together with their baggage about 100 kg) at the speed of $M = 5.0$, where the speed of sound is: $a = 1{,}062$ km/h, would be: TP = 159,300 t-km/h.

6.6.2.5 Economic Performances

The economic performances of new STA in terms of aircraft operating costs per seat and passenger mile/kilometer need to be comparable to those of their subsonic counterparts particularly while operating on long distance continental and inter-continental routes for which they are primary planned and designed in order to maximize savings in the travel time, i.e., to shorten this time from about 7.2–22.5 h to about 1.7–3.8 h for a range of routes (6,000–18,000 km). For example, the Hydrogen (LH_2) M5 Cruiser STA in Table 6.18 has been estimated to cost about 639 M€. Its annual operating costs have been estimated to be about 554 M€. The former is based on the costs of developing and manufacturing 100 aircraft, while the latter is based on the average annual utilization—flight frequency of 2 flights/day on the route of length of about 18,000 km, availability 0.90, load factor 0.75, and the number of passengers on the route 148000 (Steelant, 2008). This gives the average unit operating costs of about 0.1054 €/s-km or 0.208€/p-km, which appears to be quite comparable to the costs of their subsonic counterparts under equivalent conditions. If LH_2 was derived from electrolysis of water, the share of the fuel costs in the total STA operating cost would be about 83 %. If LH_2 was derived from the steam reforming of hydrocarbons, the share of fuel costs would be about 30 %, which in turn reduces the total STA operating costs by about 50 %. In this latter case, the average unit operating cost would be 0.139 €/s-km and 0.104 €/p-km under the above-mentioned traffic scenario, which is again competitive to the subsonic counterparts of STA. If the passenger generalized costs were considered for long distance travel, taking into account the savings in the costs of travel time for all categories of passengers (business, leisure) would indicate an additional advantage of new STA over their conventional subsonic counterparts under the given conditions. Similar analysis has been carried out for the new STA—Kerosene Mach 4.5/5.0 Cruiser (EC 2006; Steelant 2008).

6.6.2.6 Environmental Performances

The environmental performances of the new STA would specifically include the energy/fuel consumption and emissions of GHG (Green House Gases), and land use/take.

Fuel consumption

Fuel consumption is crucial to the economic and environmental success of new STA. In terms of efficiency, it is usually expressed as SFC (Specific Fuel Consumption), which is closely related to the engine BP (Bypass Ratio). The latter is defined as the rate between the amount of air flowing around the engine core and the amount of air passing through the engine. In general, jet engines with a higher bypass ratio usually have lower SFC. Figure 6.22 shows the SFC-BR relationships for the cruise and take-off phase of flight for 20 jet engines powering subsonic commercial aircraft produced by different airspace manufacturers. These are: CFM Company (joint corporation of Snecma (France) and General Electric Company (USA)), Rolls-Royce (UK), Pratt and Whitney and General Electric (USA), and IAE (International Aero Engines AG made up of the engine manufacturers Pratt and Whitney, Rolls-Royce, MTU (Europe) and Aero Engine Corporation (Japan)). In addition, the SFC-BR relationships for the engines that powered supersonic Concorde and the TY144 are shown for comparison purposes (Janic 2007a).

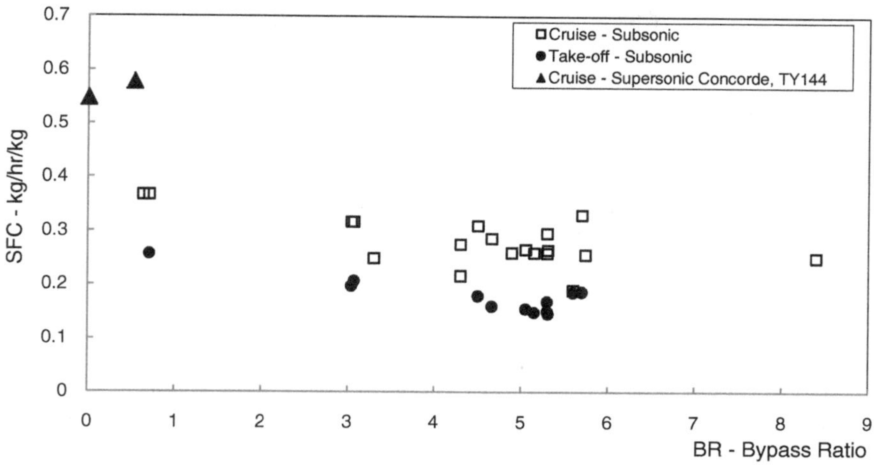

Fig. 6.22 Relationship between SFC (Specific Fuel Consumption) and BR (Bypass Ratio) for different commercial aircraft (Janic 2007b)

As can be seen, the SFC of most turbofan engines powering subsonic commercial aircraft during take-off varies between 0.25 and 0.40 and during cruising between 0.15 and 0.30 for the given range of BRs, 1–8. For the engines that powered the two retired above-mentioned supersonic aircraft, the SFC was greater than 0.5 for BRs ranging from 0 to 0.53. These relationships indicate that in order to make the new STA economically and environmentally viable, its engines would need to have a lower SFC (preferably 0.423 as specified in Table 6.18). This would be a challenge since advanced STA is expected to operate under different engine operating regimes and speeds (i.e., take-off, subsonic cruise, supersonic cruise, approach, and landing). In general, this implies operating at the subsonic regime requiring a high propulsive efficiency and the supersonic regime requiring a high specific thrust. In the former case the engine frontal area should be as wide as possible in order to provide a high propulsive efficiency at subsonic speeds, while in the latter it should be as small as possible in order to reduce drag at high supersonic speeds. One of the options using kerosene could be a variable cycle engine, which is similar to a subsonic mixed-flow turbofan engine, but with an additional secondary outer bypass duct aimed at increasing the overall bypass ratio and consequently the air flow handling capacity. As opposed to conventional turbofan engines, a variable cycle engine could change and adapt the bypass ratio to the particular above-mentioned flight conditions. For example, increasing BR during {XE "TO (Take-Off)"} take-off and landing decreases noise around airports, while reducing BR during cruise improves the fuel efficiency. Furthermore, for example, increasing BR from 0.5 to 1.0 or even higher for variable cycle engines would require application of more advanced cooling concepts for the turbine blades in combination with the use of lightweight heat resisting materials.

Last but not least, at present, the other alternative engine designs perceived as technologically, economically, and environmentally feasible include PDE (Pulse Detonated Engine), hydrogen-fueled engines, and the fuel-cell based engines (propulsive system).

PDEs are based on detonation waves aimed to initiate rapid combustion of a premixed fuel–oxidizer mixture contained in an array of tubes, which are opened at one end and closed on the other. These engines are expected to be even more efficient than the current turbofan engines, while enabling high (supersonic) speeds. Consequently, the US NASA has maintained PDE research efforts with the baseline being a Mach 5 STA (a PDE test was successfully carried out in 2008).

LH_2 (Liquid Hydrogen)-fuelled engines would seemingly be relatively easier to design if carried out by modifying existing turbofan engines (Chap. 4).

Fuel-cell based engines (propulsive systems) would be powered by electricity produced by fuel cells. The power density in terms of both weight and volume has already shown to be the critical in designing these systems for new STA (NAS 2001).

Emissions of GHG

The energy consumption of STA has already been discussed. The emissions of GHG (Green House Gases) would depend on the engines' injection and combustion performances, i.e., on the quality of air/fuel mixture, its distribution in the combustor primary zone, and the corresponding temperatures. In addition to other GHG with constant emission rates per unit of fuel burnt (CO_2, H_2O, etc.) (Chap. 5), the primary concern would be emissions of NO_x (Nitrogen Oxides) independently of the fuel used—kerosene or LH_2 (Liquid Hydrogen). This is because advanced STA would cruise at altitudes with the maximum concentration/density of O_3 (Ozone)−from 14 (FL450) to 26 km (FL850) as shown in Fig. 6.23.

O_3 protects the Earth from harmful solar UV radiation by absorbing all light with a wavelength less than 295 nm* (nanometer). However, O_3 is sensitive to free radicals such as atomic chlorine CI, Nitric Oxide NO, and hydroxyl radicals OH. They are formed from the water vapor (H_2O) and chlorofluorocarbons (CFCs), products of the fuel burned by the new STA, directly affecting the ozone layer formed in the stratosphere. Near the ground, emissions of NO_x and other hydrocarbons contribute to increasing the concentrations of O_3, which is considered as a pollutant harmful to inhale and a contributor to the formation of smog (Janic 2007b).

Thus, in contrast to its regional impact near airports, the NO_x injected into the stratosphere affects the ozone layer globally. In any case, the increased concentration of NO_x and other GHG might generally cause depletion of the ozone layer with inevitable impacts. For example, a 10 % depletion of this layer could result in an increase in UV radiation by about 45 %, which certainly could damage almost all biological cells and in particular cause skin cancer in the exposed population.

Currently, in the case of the existing subsonic turbofan engines operating at given fuel combustion temperatures, the highest rate of formation of NO_x appears to be when the fuel/air equivalence rate is equal to one (i.e., to the stoichiometric air-to-fuel ratio). Furthermore, in the absolute sense, $EINO_x$ (Emission Index of

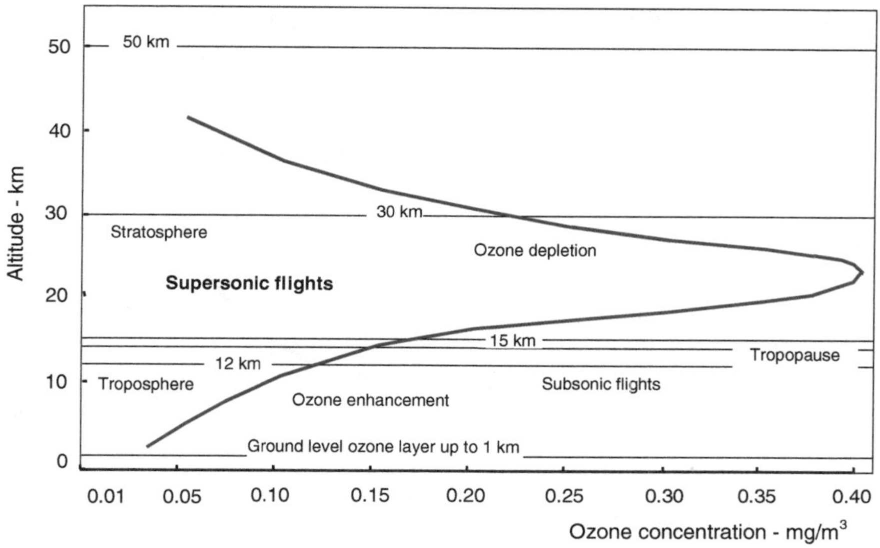

Fig. 6.23 Scheme of distribution of the concentration of ozone and impact of commercial air transport (mid-latitude) (Janic 2007b)

Nitrogen Oxides) increases with increasing of the fuel combustion temperature as shown in Fig. 6.24.

This is an exponential increase up to the temperature barrier of the aircraft constructive materials. Thus, if the above relationship was applied to engines powering advanced STA without any technological improvements and with the compressor temperature of about 1,900 °C, $EINO_x$ would reach about 95,367 g/kg of fuel, which is far above the specified target of less than 5 g/kg of fuel in Table 6.18. Consequently, at least in this respect, new technological solutions, for example, in the variable cycle engine concept, need to be developed in order to achieve, among the other requirements, also the prescribed targets of NO_x and other GHG emissions. The solutions should achieve either the fuel-rich or lean burning processes, reduce the compressor exit temperature, or both. Specifically, in the first case, two different combustion processes have been under investigation: RQL (the two stage Rich-burn/Quick-quench/Lean-burn), and LPP (Lean Pre-mixed Prevaporized concept). The former operates at the fuel/air ratios greater than one, thus affecting the formation of NO_x due to the lack of oxygen. The latter concept operates at a very low fuel/air ratio, which would be sufficient to maintain stability and efficiency of the combustion process. In the second case, the very high temperature in both processes (supposed to be around 1,900 °C) would require liner materials and cooling by air, which in turn could affect the emission rates of GHG achieved by the combustion process. Nevertheless, experiments so far with two above-mentioned burning processes and related materials have shown that it could be possible to achieve $EINO_x$, of about 45, 15, 5, and even lower, while not

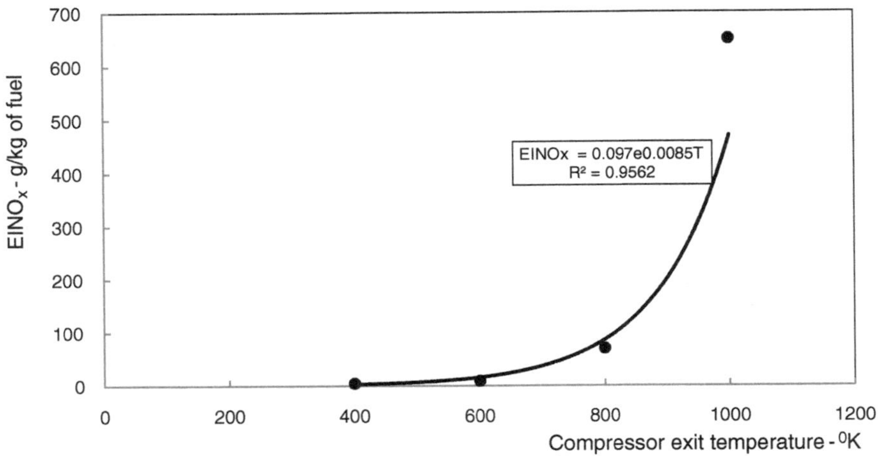

Fig. 6.24 Relationship between $EINO_x$ (Emission Index of NO_x) and the compressor exit temperature for subsonic turbofan engines (Coen 2011)

critically compromising the other required performances of the advanced STA propulsion system (engines) (Coen 2011; Saounatsos 1998).

Nevertheless, new STA is expected to contribute to global warming occurring as a result of radiative forcing, i.e., changes in the global balance between the incoming solar radiation and the outgoing infrared radiation. Some estimates have indicated that presence of advanced STA in the commercial aircraft fleet could increase the radiative forcing by commercial aviation by about 42 % in 2050, i.e., from about 0.2 without to about 0.284 W/m^2 with advanced STA (IPCC 1999).

Land use

As mentioned above, advanced STA should be designed to smoothly operate within the existing airport clearances. This implies that the aircraft will be provided with the maneuvering and parking space without the need for using additional land.

6.6.2.7 Social Performances

The social performances of STA relate to noise, congestion, and traffic incidents and accidents, i.e., safety.

Noise

The main attributes of the noise performances of new STA remain to be the airport landing, sideline, and take-off noise, and sonic boom while operating at supersonic speeds.

Landing, sideline, and take-off noise

In order to obtain high specific thrust at supersonic speeds, STA engines would need to have a rather high jet exhaust speed at subsonic cruising speeds. In general, this speed could range from about 300–900 m/s (Concorde), but has already been established at around 400 m/s with the average temperature of 600–800 °K. This could also guarantee an acceptable noise level during the aircraft landing, take-off, and sideline, i.e., within the ICAO Chap. 3 Noise Regulation for airports (ICAO 2001; Smith 1989).

Sonic boom

Another aspect of the noise of new STA is sonic boom. Sonic boom can be defined as an acoustic phenomenon associated with supersonic flight. A STA flying at supersonic speeds creates an acoustic wave, which propagates through the atmosphere and either strikes the ground or dissipates in the upper atmosphere. The observers (people) on the ground usually hear a sonic boom as a sharp double report (Hayes 1971). The signal of sonic boom is usually very short lasting typically from 200 to 300 ms. In addition, it is characterized by the sharp head and tail shock, which are actually caused by the sudden jumps in the air pressure of typically 1–3 psi (psi—pounds/in^2) or 60–120 P_a (P_a—Pascale). The loudness of sonic boom could reach 136–140 dBA. A typical impact of such loudness is its psychoacoustic effect on people and animals while covering a ground area of about 30–40 km on both sides of the aircraft track projected on the ground (NAS 2001). For example, Concorde generated an overpressure of 93 P_a, which consequently restricted its supersonic flying to over water areas and thus substantively diminished its economic feasibility. Some research indicates that advanced STA could achieve an overpressure of 85 P_a, but again they would need to be restricted from flying over populated areas, which would extend the routes and related flying times and furthermore result in additional fuel consumption and related emissions of GHG (EC 2006). Therefore, in the given context, the principal task for designers of advanced STA is to make the sonic boom acceptable for the affected population on the ground. This would imply including the sonic boom aspect as one of the criteria for aircraft design, which was not the case at both Concorde and TY144 aircraft.

The sonic boom can generally decrease by decreasing the aircraft weight and increasing: (i) L/D (Lift–to–Drag) ratio, (ii) length, and (iii) innovative propulsion integration (NAS 2001). The aircraft weight could be decreased, in addition to selecting a lightweight construction based on lighter materials, by reducing the amount of required trip fuel, which could be achieved by reducing the engine SFC (Specific Fuel Consumption) without compromising the range. The oblique/wing in combination to the relatively long fuselage and innovative integration of the propulsion systems (engines) have shown promising in providing both satisfactory aerodynamic (L/D) performances and acceptable sonic boom, particularly for Mach numbers of 1.4–2.0 (Kroo 2005; NAS 2001). Thus, because the required operational and economic performances need to be simultaneously fulfilled in order to make the advanced STA commercially and environmentally/socially viable, their design will very likely remain to be highly influenced by the sonic boom reduction requirements. Use of LH_2 as a fuel instead of the conventional jet fuel-kerosene could additionally require compromise in its design (Chap. 4).

Congestion

Advanced STAs would operate within the approaching and departing traffic flows at airports exposed to congestion and delays similarly as other subsonic aircraft. In some cases, they could be given the ultimate priority in service but the related ATC (Air Traffic Control) criteria, if any, still need to be elaborated. The length of the advanced STA (Table 6.18) could affect smooth traffic at the airport apron/gate complex, but this would be resolved by parking and maneuvering on of dedicated parking positions.

Safety

In addition to the absence of traffic incidents and accidents due to the already know reasons, the safety of advanced STA would also imply protecting the flight crew and passengers from high-altitude radiation and cabin decompression.

High-altitude radiation refers to primary and secondary cosmic radiation. The Earth is protected from this radiation by the atmospheric shield which is over 1,000 g/cm^3 thick at sea level, 200–300 g/cm^3 thick at typical cruising altitudes of subsonic aircraft (FL 300–400), and only 60 g/cm^3 thick at the cruising altitudes of new STA (FL600–650). This implies that advanced STA would preferably have some kind of the very light shield in addition to more careful selection of 4D (four dimensional) routes based also on the perceived (expected) cosmic radiation. Furthermore, monitoring the exposure of flight crews and frequent fliers to this radiation would be introduced as a part of STA flight (safety) regulations.

The crew and passengers of advanced STA would be exposed to a high risk of loss of life in cases of explosive and rapid cabin decompression or sudden air leakage from the STA fuselage occurring at altitudes above FL500 (50,000 ft). Since current regulations do not allow exposure to pressure at altitudes above FL 400 (40,000 ft) for any time (as compared to 2 min of allowable exposure to the pressure at an altitude of FL 250 (25,000 ft)), this requirement would be very difficult and complex to fulfill for the advanced STA. Nevertheless, the possible solutions could be using innovative technologies such as self-sealing materials to contain fuselage pressure leaks, thus providing sufficient time for an emergency descent initiated by an automatic emergency descent mode in the scope of the aircraft flight control system (NAS 2001).

6.6.3 Evaluation

Advanced STA possess both advantages (Strengths and Opportunities) and disadvantages (Weaknesses and Threats) as follows:

Advantages

- Superiority as compared to their supersonic (retired) predecessors and competitive to the future subsonic aircraft in terms of the overall efficiency and effectiveness from the viewpoint of both airlines and users/passengers;
- Complementing to the subsonic and supersonic aircraft fleet operating within the future APT (Air Passenger Transport) system;
- Contributing to the creation of an integrated and very high-speed transport system on both the national and particularly the international-global scale;
- Compatibility with the existing airport layout enabling smooth, efficient, and effective maneuvering, parking, and ground handling;
- Environmentally and socially acceptable in terms of the emissions of GHG and noise, the former globally and the latter both locally around the airports and globally (sonic boom); and
- Contributing to development of innovative and new knowledge, processes, and materials for both airframe and engines, which could be disseminated to other non-APT (Air Passenger Transport) fields.

Disadvantages

- Uncertainty in the final design (i.e., the most feasible supersonic speed and size, namely seating capacity) that would guarantee economic feasibility for both airlines and their users/passengers (similarly as at the ETT system: who would really demand to travel at the offered supersonic speed at the given price and under what conditions);
- Unknown technical/technological barriers, which would eventually need to be overcome to develop the components—fuselage, propulsion systems, and aircraft control systems;
- Diminishing advantages of flying at supersonic speeds due to the prohibition of passing over populated areas (creating sonic boom), which may prolong flying distance(s) (not always the shortest—Great Cycle ones) and related travel time(s); and
- The inherent risk of increasing the overall environmental and social impacts in terms of global emissions of GHG and noise (sonic boom), the former due to increasing emissions of NO_x and the latter due to some unavoidable passing over populated areas at supersonic speeds.

Finally, any further elaboration of the overall feasibility of advanced STA should include the above-mentioned and additional (depending on the actors/stakeholders involved) advantages and disadvantages. Some facts provided in this section indicate that advanced STA might be, under given conditions, a promising supersonic complement to the future APT (Air Passenger Transport) system and other long-haul very high-speed systems such as ETT.

References

ACARE, (2010). *Beyond vision 2020 (Towards 2050)*. The Advisory Council for Aeronautic Research in Europe, Directorate-General for Research Aeronautics and Transport. Brussels, Belgium: European Commission.

Airbus, (2012). *Delivering the future: Global market forecast 2012–2030*. Blagnac, Cedex, France: AIRBUS S.A.S.

Anderson, J. E. (2000). A review of the state of the art of personal rapid transit. *Journal of Advanced Transportation, 34*(1), 3–29.

Anderson, J. E. (2005). *"The Future of High-Capacity PRT", Advanced Automated Transit Systems Conference, November 7–8*. Italy: Bologna.

Anderson, J. E. (2007). *High-capacity personal rapid transit: Rationale, attributes, status, economics, benefits, and courses of study for engineers and planners*. Minnesota, US: PRT International LLC.

Antaki, A. G. (2003). *Piping and pipeline engineering, design, construction, maintenance, integrity, and repair*. Boca Raton, FL: CRC Press.

Boeing, (2002). *747-400 Airplane characteristics for airport planning*. Seattle, WA, USA: Boeing Aircraft Company.

Boeing, (2012). *Current market outlook 2012–2031*. Market Analysis. Seattle, WA, USA: Boeing Commercial Airplanes.

CC, (1999). *Automated underground transportation of Cargo: The 5th transportation alternative for the transport of goods in congested urban areas*. Bochum, Germany: Brochure, CargoCap GmbH.

CIA, (2012). *The world factbook*. Washington D.C., USA: Central Intelligence Agency.

Coen, P. (2011). *Fundamental aeronautic program: Supersonic projects. 2011 Technical Conference* (p. 34). Cleveland, Ohio, USA: National Aeronautic and Aerospace Administration.

EC, (2005). *Optimal Procedures and Techniques for the Improvement of Approach and Landing—OPTIMA*. Sixth Framework Program, FP6-2002-Aero 1502880, Deliverable 1.2. Brussels, Belgium: European Commission.

EC, (2006). *ATLLAS (Aerodynamic and Thermal Load Interactions with Lightweight Advanced Materials for High Speed Flight)*. European Commissions Sixth Framework Programme, Final Public Report. Brussels, Belgium: Thematic Priority 1.4 Aeronautic and Space.

EC, (2007). *CityMobil*. European Commission, DG Research, 6th FMP (Framework Programme), Thematic priority 1.6, Sustainable Development, Global Change, and Ecosystems, Integrated Project, No. 031315, Brussels, Belgium.

EC, (2007a). *Gaps Identification Airport, TMA, En Route and TBS—Reduced Separation Minima—RESET*. Brussels, Belgium: Deliverable D4.2, EC 6th Framework Program, European Commission.

EC, (2008). *LAPCAT (Long/Term Advanced Propulsion Concepts And Technologies)*. Brussels, Belgium: European Commissions Sixth Framework Programme, Final Public Report, Thematic Priority 1.4 Aeronautic and Space.

Freville, E. (2008). *The RECAT project*. Brussels, Belgium: Presentation of the TAAM-based Simulation Tool to Estimate Re-categorisation Impact on Runway Capacity, EUROCONTROL.

FAA, (2010). *Air traffic control,- FAA Order JO 7110.65T*. Washington DC, USA: Federal Aviation Administration, US Department of Transportation.

FAA, (2011). *Change 2, order JO 7110.65T*. Washington DC, USA: Federal Aviation Administration, US Department of Transportation.

FAA (2011a). *NextGen implementation plan*. Washington DC, USA: US Department of Transport, Federal Aviation Administration.

FAA, (2011b). *NextGen for Airports*. Washington DC, USA: US Department of Transport, Federal Aviation Administration.

FAA, (2011c). *NextGen operator and Airport enablers*. Washington DC, USA: US Department of Transport, Federal Aviation Admininistraion.

GAO, (2009). *Aviation and climate change: Aircraft emissions expected to grow, but technological and operational improvements and government policies can help control emissions*. Report to Congressional Committees- GAO-09-554. Washington D.C., USA: United States Government Accountability Office.

Helmke, H. (Hartmut), Hann, R., Muller, D., Witkowski S., (2009). Time based arrival management for dual threshold operations and continuous descent approaches. *8th USA/ Europe Air Traffic Management Research and Development Seminar (ATM2009)*, Napa, California, USA.

IBRD, (2012). *Air travel and energy efficiency*. Transport Papers TP-38, The International Bank for Reconstruction and Development/The World Bank, Washington D.C., USA.

ICAO, (2001). *Aircraft engine noise—Environmental protection*, Annex 16, Chapters 3 and 4. Montreal, Canada: International Civil Aviation Organization.

ICAO, (2007). *Procedures for air navigation services: Air Traffic Management*. Doc. 4444/ ATM501, Fifteen Edition. Montreal, Canada: International Civil Aviation Organization.

ICAO, (2008). *Wake Turbulence Aspects of Airbus A380-8/000 Aircraft*. TEC/OPS/SEP-08-0294.SLG. Neuilly sur Siene, Cedex, France: International Civil Aviation Organization.

IPCC, (1999). *Aviation and the global atmosphere*. Intergovernmental panel on climate change. Cambridge, UK: Cambridge University Press.

IWW, INFRAS, (2004). *External cost of transport: Update study*. Summary. Karlsruhe, Zurich, Germany, Switzerland: Universitaet Karlsruhe, INFRAS Zurich.

Janic, M. (2006). A model of ultimate capacity of dual-dependent parallel runways. *Transportation Research Record, 1951*, 76–85.

Janic, M. (2007a). A steeper approach procedure for increasing the ultimate capacity of closely spaced parallel runways. *Transportation Research Record, 2007*, 81–90.

Janic, M. (2007b). *The sustainability of air transportation: Quantitative analysis and assessment*. UK: Ashgate Publishing Company.

Janic, M. (2008). Towards time-based separation rules for landing aircraft. *Transportation Research Record, 2052*, 79–89.

Janic, M. (2009). A concept for prioritizing the aircraft operations at congested airports. *Transportation Research Record, 2106*, 100–108.

Janic, M. (2012). Modeling Effects of Different Air Traffic Control Operational Procedures, Separation Rules, and Service Disciplines on Runway Landing Capacity. to appear in *Journal of Advanced Transportation* 24 Aug 2012. DOI: 10.1002/atr.1208

Kelly, R. J., La Berge, F. E. C. (1990). MLS: A total system approach. *IEEE AES Magazine* (May), p. 10.

Kersting, M., Draganinska, S. (2005). CargoCap: Economic freight transportation in congested areas. *Proceedings of the 4th International Symposium on Underground Freight Transport by Capsule Pipelines and Other Tube/Tunnel Technologies*, Shanghai, China.

Kroo, I., (2005). Unconventional configurations for efficient supersonic flight. *VKI Lecture Series on Innovative Configurations and Advanced Concepts for Future Civil Aircraft*. Stanford University, USA, p. 25.

LHR, (2010). *LHR FEU Annual Report 2010*. London, UK: Heathrow Airport.

Liu, H. (2004). *Feasibility of underground pneumatic freight transport in New York*. Final Report, prepared for The New York State Energy Research and Development Authority (NYSERDA). Columbia, Missouri, USA: Freight Pipeline Company.

Morrison, A. S., & Winston, C. (2007). Another look at airport congestion pricing. *The American Economic Review, 97*(5), 1970–1977.

NAS, (2001). *Commercial supersonic technology: The way ahead*. Washington D.C., USA: National Academy of Science, National Academy Press.

NASA, (1999). *Benefit estimates of terminal area productivity program technologies*. NASA/ CR—1999 208989. Virginia, USA: National Aeronautics and Space Administration.

Rijsenbrij, J. C., Pielage, B. A., & Visser, J. G. (2006). *State of the art on automated (Underground) freight transport systems for the EU TREND project*. Delft, The Netherlands: Delft University of Technology.

Rodrigue, J.-P., Comtois, C., & Slack, B. (2006). *The geography of transport systems*. New York, USA: Routledge.

Rossow, V. J. (2003). Use of individual flight corridors to avoid vortex wakes. *Journal of Aircraft, 40*(2), 225–231.

RUF, (2008). RUF international: Investment case. Frederiksberg C, Denmark: RUF International.

Salter, R.M. (1972). The very high speed transit system. *The rand corporation* (p. 4874). Santa Monica, California, USA, p. 18.

Saounatsos, G. (1998). Supersonic transport aircraft (SST): Technology readiness and development risk. *ASCE Journal of Aerospace Engineering 1998*, 1–16.

Seebass, R. (1998). Supersonic aerodynamics: Lift and drag. *Paper presented at the RTO AVT Course on "Fluid Dynamics Research on Supersonic Aircraft"* (p. 6). Rhode-Saint-Gendse, Belgium.

Sirohiwala Y. A., Tandon A., Vysetty R. (2007). *Feasibility and economic aspects of vactrains*. An Interactive Qualifying Project. Worcester, Massachusetts, USA: The Faculty of the Worcester Polytechnic Institute.

Smith, M. J. T., (1989). *Aircraft noise*. Cambridge Aerospace Series. Cambridge, UK: Cambridge University Press.

Steelant, J. (2008). LAPCAT: High speed propulsion technology, in *Advances on Propulsion Technology for High-Speed Aircraft*. Educational notes RTO-EN-AVT-150, Paper 12. Neuilly-sur-Siene, France, pp. 12-1-12-38.

TC, (2004). *Approval of steep approach landing capability of transport category aircraft*. Toronto, Canada: Aircraft Certification, Civil Aviation, Transport Canada.

Thompson, S. D. (2002). *Terminal area separation standards: Historical development and process for change*. Lexington, Massachusetts, USA: Lincoln Laboratory, Massachusetts Institute of Technology.

Tosic, V., & Horonjeff, R. (1976). Effects of multiple path approach procedures on runway landing capacity. *Transportation Research, 10*(5), 319–329.

Zhang, Y., Oster, D., Kumada, M., Yu, J., & Li, S. (2011). Key vacuum technology issues to be solved in evacuated tube transportation. *Journal of Modern Transportation, 19*(2), 110–113.

Visser, J. G. S. N. (2010). Underground freight transport: What is the role of the public sector ion developing a new transport mode. *Netherlands Institute for Transport Policy Analysis (KiM)*. Den Hague, The Netherlands: Ministry of Transportation, Public Works and Water Management.

Winkelmans, W., Notteboom, T. (2000). In search of strategic positioning of underground freight transport in the framework of a coherent transport policy. *Proceedings of 2nd International Symposium on Underground Freight Transportation by Capsule Pipelines and Other Tube/ Tunnel Systems*. 28–29 September 2000, Delft, The Netherlands.

Chapter 7
Advanced Transport Systems: Contribution to Sustainability

7.1 Introduction

An important current and prospective medium- to long-term objective of transport systems worldwide is to continuously improve sustainability. On the one hand, this implies handling growing passenger and freight transport demand efficiently, effectively, and safely, and on the other mitigating their direct and indirect impacts on the environment and society such as energy consumption and related emissions of GHG (Greenhouse Gases), local noise, congestion, traffic incidents/accidents, land use/take, and waste on the other. Under such conditions, the main actors/stakeholders involved in dealing with transport system development, operation, and use also need to benefit from the perspective of their specific objectives. These include manufacturers of transport means/vehicles and supporting facilities and equipment, suppliers of energy/fuel, infrastructure providers and transport operators, users of transport services—passengers and freight shippers/receivers, the local population, and associations and policy bodies at different institutional levels (local, regional, national, and international). Consequently, the question of the contribution of advanced transport systems and their components to the above-mentioned sustainable development of the global transport system can be raised.

7.2 Contribution

Advanced transport systems are generally characterized by advanced infrastructure, transport means/vehicles, energy/fuel used, facilities and equipment, all of which support operations at the operational, tactical, and strategic level, together with the related organization of service networks. These advancements include: (i) design and materials used for building transport infrastructure and manufacturing transport means/vehicles; (ii) technology, size/capacity, speed, and energy used by transport means/vehicles; and (iii) different configurations/concepts of transport service networks aimed at serving the expected passenger and freight transport

M. Janić, *Advanced Transport Systems*,
DOI: 10.1007/978-1-4471-6287-2_7, © Springer-Verlag London 2014

demand efficiently, effectively, and safely. In order to indicate their contribution to the sustainability of the transport sector, advanced transport systems have been scrutinized by analyzing, modeling, and evaluating their infrastructural, technical/ technological, operational, economic, environmental, social, and policy performances.

Most of the considered advanced systems have already been commercialized although some are less likely to be commercialized in the medium- to long-term future. As far as infrastructural performances are concerned, commercialized advanced systems can use the same (in some cases partially modified and adapted) infrastructure as their conventional counterparts. In the given context, these include: (i) BRT (Bus Rapid Transit) systems, roadmega trucks, and advanced passenger cars in the road transport mode; (ii) High-Speed Tilting passenger trains, LIFTs (Long Intermodal Freight Trains), and advanced freight collection/distribution network services in the rail and intermodal transport mode; (iii) advanced ATC (Air Traffic Control) technologies and operations, advanced conventionally and LH_2 (Liquid Hydrogen)-fuelled subsonic passenger and freight aircraft, and advanced STA (Supersonic Transport Aircraft) in the air transport mode; and (iv) large advanced container ships in the sea-water transport mode. Others, such as the commercialized HSR (High-Speed Rail), partially commercialized TRM (Trans-Rapid Maglev), and the non-commercialized UFT (Underground Freight Transport), and ETT (Evacuated Tube Transport) systems in the rail, and partially commercialized PRT (Personal Rapid Transit) systems in the road transport mode would require completely new dedicated (and very expensive) infrastructure.

The technical/technological performances of the above-mentioned systems are fully developed, while those of the non-commercialized systems require further elaboration. The main performances of these systems relate to the transport means/ vehicles and facilities and equipment for their control and management. The former includes being: (i) made of a greater proportion of lighter (composite) materials; (ii) powered by electric engines (supposedly obtaining energy from from renewable primary sources such as wind, water, and solar energy) and renewable fuels such as LH_2 (Liquid Hydrogen); (iii) having high to large carrying capacity; and (iv) operating at as high as possible speeds (large advanced container ships, which operate at reduced speed in order to improve both economic and environmental performances, the latter in terms of fuel consumption and related emissions of GHG, are the exception). The latter includes the rather wide application of ITS (Intelligent Transport Systems) in direct and indirect supporting operations. Technical/technological performances are expected to generally contribute to improving the operational, economic, environmental, social, and policy performances of the particular advanced transport systems.

Operational performances, which include the service networks and patterns of their operations, are set up as close as possible to the characteristics of user/ customer demand. They depend on conventional and/or advanced infrastructural and technical/technological performances. As such, they directly influence all economic and some environmental performances.

For example, BRT systems, road mega trucks, large advanced container ships, and freight aircraft, all using existing or slightly modified infrastructure, could improve the efficiency and effectiveness of related services. At the same time, however, the latter two can contribute to increasing congestion, particularly when operating in congested urban and suburban areas and spatially constrained ports and airports, respectively. Similarly, advanced (mainly electric) passenger cars using the present roads and streets in combination with modified supporting facilities and equipment mainly for energy/fuel supply can also contribute to congestion but not to increasing direct emissions of GHG. On the contrary: PRT and UFT systems, which require completely new dedicated and very expensive infrastructure if developed to the network level, can mitigate congestion and both direct and indirect emissions of GHG in given urban or suburban area(s). High-speed tilting, HSR, and TRM passenger trains operating on completely new infrastructure (railway lines) do not contribute to increasing congestion due to the nature of railway operations. In particular, high-speed tilting trains sharing infrastructure with conventional passenger and freight trains are always given the highest priority. By taking over medium- to long-distance trips from individual passenger cars, these trains could indirectly contribute to reducing road congestion and related emissions of GHG.

Advanced freight collecting/distributing networks and LIFTs in Europe operating on current rail/road/inland waterway and rail infrastructure, respectively, can contribute to improving particularly the economic and environmental performances of the supply chains served, mainly due to replacing road standard (and prospectively mega) trucks.

Advanced subsonic commercial aircraft and advanced STA remain neutral in terms of contributing to congestion as compared to their less advanced counterparts. They can use existing infrastructure with modified (or new) supporting facilities and equipment mainly for fuel supply in case of LH_2–fuelled models. The relative fuel consumption and type of fuel used (also LH_2) can contribute to improving the environmental performances in terms of energy/fuel consumption and related emissions of GHG. Aircraft engine technology in combination with advanced ATC technologies and operations for increasing airport runway capacity can contribute to reducing local noise. In addition, they can directly improve the technical/technological and operational performances, and indirectly the economic, environmental, and social performances of a given airport runway system(s). However, due to operating at supersonic speed, advanced STA create sonic booms, which, in parallel to the requirement for increased energy/fuel consumption and related emissions of GHG, can affect large (also populated) area(s) and consequently compromise social performances in general.

ETT (Evacuated Tube Transport) systems can apparently contribute to improving the operational and environmental performances of long-distance (intercontinental) passenger transport. At the same time, economic performances can be compromised due to the very high costs of completely new infrastructure that needs to be built and maintained under very specific conditions (underwater).

The level of maturing and consequently the dynamics of commercialization of particular advanced transport systems can influence their technical/technological, and consequently policy performances. In some cases, the influence is positive, while in others, it is negative. In general, the relevant system must be technically/technologically mature in order to be fully commercialized. However, very often, technical/technological maturity does not imply immediate commercialization. The positive influence of the technical/technological performances on commercialization is already present with respect to the fully and/or partially commercialized systems such as BRT and PRT systems, high-speed titling passenger trains and HSR, advanced commercial subsonic passenger and large advanced freight aircraft, advanced freight collection/distribution networks, LIFTs, road mega trucks, and large advanced container ships. The negative influence of these performances on commercialization is present in case of currently non-commercialized systems such as advanced passenger cars, TRM, UFT, and EET system, advanced ATC technologies and operations for increasing airport runway system capacity, and advanced STA.

7.3 Some Controversies

Consequently, many of the above-mentioned advanced transport systems appear controversial regarding their contribution to the overall medium- to long-term sustainability of the entire transport sector. This controversy becomes more obvious after considering their technical productivity, the energy/fuel consumption and related emissions of GHG, and safety (i.e., traffic incidents and accidents).

7.3.1 Technical Productivity

The evidence so far indicates that the above-mentioned advanced transport systems are characterized by gradually increasing technical productivity partially thanks to the increasing size/weight/capacity of transport means/vehicles and their number in the fleet to serve growing demand, and primarily thanks to increasing technical and operating speed(s).

7.3.1.1 Vehicle Size

The size/capacity/weight of particular transport means/vehicles operated by different transport modes has seemingly reached constructive/design limits. But is that really the case? Where are these limits? Are they contained in the sustainability and durability of the vehicle's structure—design and material? Maybe they lie in constraints in the existing transport infrastructure (roads, railway lines, inland waterways, airports, and river/sea ports) requiring substantial adjustment in order to enable efficient, effective, and safe maneuvering of these increasingly larger transport means/vehicles? Is the limit stricter regulation constraining the absolute impacts of these vehicles/means and the entire transport sector on the environment and society? Or is this the overall shortage of land for expansion of infrastructure in combination with an increasing shortage of (and more expensive) currently used energy/fuels? Or perhaps a combination of all the above-mentioned factors?

What is the general influence of these larger transport means/vehicles on the transport processes and related effects and impacts?

In general, the increased size/weight/capacity of transport means/vehicles implies less frequent services to satisfy a given volume of demand on a given route(s), and on the entire network(s). Such less frequent services contribute to increasing schedule delays of passengers at their origins and the inventory cost of freight shipments at both ends of the given supply chain. In addition, such more sizeable vehicles are designed to operate on medium to long distances, which in combination with the longer loading and unloading (ground handling time) generally require greater fleets to serve the given volumes of passenger and freight demand during a given period of time. On the other hand, the inherently lower service frequencies require a smaller number of these larger vehicles. At the same time, the investment and operating costs of these vehicles per service are higher, while the unit costs are lower compared to their smaller counterparts thanks to economies of scale and density. These larger vehicles also consume more energy/fuel per service, which depending on the type and primary sources of energy/fuel production generally create higher total emissions of GHG. However, in relative terms, these emissions can be lower than those of their smaller counterparts. The same applies to the local noise created by these vehicles.

7.3.1.2 Vehicle Speed

The dynamism of raising the technical and operating speed of the commercialized and the non-commercialized advanced transport systems is different. The former is an evolutionary gradual process, while the latter will likely be a revolutionary process. Take for example the case of evolutionary/gradual increase in the speed of the urban public passenger transport achieved through operational advancements in the BRT system. High-speed tilting passenger train(s), HSR (High-Speed Rail) and its modification Super HSR, and TRM are all examples of gradually increasing the speed by technical/technological modifications. In urban and suburban freight transport, a gradual increase in the freight/goods delivery speed can be achieved by as yet non-commercialized UFT systems. But, again, how big should such an increase be combined with the other advantages to justify the generally high investment cost in UFT infrastructure? In case of the ETT system and advanced STA as yet non-commercialized future systems, the increase in speed will likely be revolutionary. At the EET system, this implies an application of TRM technology through a vacuum tube enabling a very high (supersonic) operational speed. In case of STA, this implies development of a new supersonic configuration expected to be much faster, efficient, effective, safe, and particularly less noisy than the previous retired models—the Anglo-French Concorde and the Soviet Union's TU144.

Higher technical and operating speeds generally contribute to shortening the travel time and related costs for the users/passengers and freight shipments. In parallel, they enable shorter turnaround times, thus requiring engagement of smaller fleets to serve given volumes of demand during a given period of time. This generally diminishes the investment, maintenance, and operational costs of the transport means/vehicles. However, higher speeds require higher energy/fuel consumption, which contributes to increasing the cost per service. Depending on the type of primary sources for producing the energy/fuel, such increased energy/fuel consumption generates higher emissions of GHG. The noise level generated by the vehicles passing-by at higher speeds is generally also higher. The question of the maximum possible technical and operation speed of the above-mentioned systems also arises. Where are the technical/technological barriers and where are the commercial constraints? The latter implies who actually needs very high (supersonic) speeds—to what extent are such speeds beneficial and when do they become counterproductive? Do savings of the users' time justify setting up usually very expensive infrastructure, high energy/fuel consumption and related emissions of GHG, and increase in the local (and global) noise? Large advanced container ships can be considered as a case of such controversy: by reducing operating speeds and consequently fuel consumption, these ships can improve their economic and environmental performances, but only on account of an increased fleet size and longer delivery time of the freight/goods shipments, which need to be acceptable for both operators and users/customers.

7.3.2 Energy/Fuel Consumption and Emissions of GHG

The environmental performances of particular advanced transport systems are expected to contribute to sustainability through reducing energy/fuel consumption and related emissions of GHG (Green House Gases). These performances are generally influenced by the vehicle size/weight/capacity, operating speed, the scale of operations (size of the network and the service frequency), and type of the energy/fuel used. As mentioned above, the larger size, speed, and scale of operations by using a given fuel require a greater total fuel consumption, which consequently generates greater related emissions of GHG, and vice versa. Liquid fuels as derivatives of nonrenewable primary sources such as crude and LNG (Liquid Natural Gas), electric energy, and LH_2 obtained from different nonrenewable and renewable primary sources are used by these systems. While crude oil and LNG will seemingly remain the primary source of fuels used by large advanced container ships and aircraft, road mega trucks, the BRT (Bus Rapid Transit) System system, and advanced commercial subsonic aircraft. Electric energy obtained from nonrenewables such as coal, crude oil, LNG, and nuclear, and renewable sources such as water, wind, and solar energy is and will continue to be used by high-speed tilting, HSR, and TRM trains, advanced (electric) passenger cars, and the PRT, and UFT In addition, LH_2, which can be obtained from the electrolysis of water, appears to be under consideration as a future fuel for subsonic commercial APT including STA. Again, electric energy is needed for producing LH_2. Therefore, the question arises how to obtain sufficient quantities of electric energy for satisfying humanity's overall (generally growing) needs including those of advanced transport systems on the one hand while maintaining the related impacts on the environment within the prescribed targets, on the other? As well, the he question which kind of propellant/fuel would be used by the ETT system remains.

Nonrenewable primary sources will be exhausted sooner or later. Water, wind, and solar energy produced using dedicated plants installed on the Earth's surface will remain the most important sources of renewable electric energy. Alternatively, SSP (Space Solar Power) or SBSP (Space-Based Solar Power)[1] as a complement to existing sources/plants appear to be the most feasible long-term solution(s).

7.3.3 Safety

Advanced transport systems are expected to be safer, namely freer from traffic incidents and accidents than their conventional counterparts. This implies that, under given conditions, incidents and accidents should not occur due to already known reasons. In addition to the adequate design and construction of the infrastructure, transport means/vehicles, and supportive facilities and equipment, this will be achieved through the increased automation of operations, which will be established at the level of an individual vehicle and at the level of a route and/or the entire network. For example, advanced HS (High Speed) trains and LIFTs, advanced subsonic and supersonic commercial aircraft, and large advanced container ships and aircraft are guided automatically by autopilots mainly during the cruising phase of their trip. It is only a matter of time when advanced (electric) passenger cars, BRT buses, and mega trucks will start to be guided in a similar way. Analogously, advanced ATC technologies and operations will enable allocation of a part of the responsibility for aircraft separation from ATC controllers to the pilots. PRT and UFT are fully automated driverless systems at both the level of

---

[1] At present, solar energy is routinely used on nearly all spacecraft. On a larger scale, this technology combined with already demonstrated wireless power transmission could satisfy nearly all needs for electricity on Earth. Thus, the SSP system would consist of lower-cost environmentally friendly launch vehicles, large solar power satellites, and a power transmission system, technologies already known today at least at the conceptual level (NSS 2007).

an individual vehicle/capsule and the level of the route network. In semi-automated systems, automation helps drivers change their role from the previously more intensive controlling and monitoring to the present and future increasingly if not exclusively monitoring. In driverless systems, the controlling and monitoring role is and will be carried out fully automatically by central computer systems. Such developments will certainly bring advantages in terms of improving the efficiency, effectiveness, and inherent safety of these systems. On the other hand, disadvantages may include less employment and increased complexity requiring longer learning/training time and resulting in higher repair costs in case of technical failure. In addition, safety of operations could be significantly compromised when the drivers again take over a controlling and monitoring role. In light of the above-mentioned issues, the long-standing principal dilemma about the division of tasks between man and machine and consequently sharing the ultimate responsibility remains. In addition, incidents and accidents still remain possible. In such cases, due to operating at high speed and/or due to large size, the human causalities and property damage could be devastating. Therefore, in any case, safety will remain to be of the highest priority in designing and operating both existing and forthcoming advanced transport systems, as well as their conventional counterparts.

Reference

NSS. (2007). *Space solar power: An investment for today-an energy solution for tomorrow*. Washington, D.C.: National Space Society.

About the Author

Dr. Milan Janić is a transport and traffic engineer and planner. At present, he is a Senior Researcher at the Transport and Planning Department of the Faculty of Civil Engineering and Geosciences of Delft University of Technology (Delft, The Netherlands) and Research Professor at the Faculty of Traffic and Transport Engineering of the University of Belgrade (Belgrade, Serbia). Previously he was a Leader of the Research Program of the Section Transport and Infrastructure at the OTB Research Institute for the Built Environment of Delft University of Technology (Delft, The Netherlands), Senior Researcher at Manchester Metropolitan University (Manchester, UK) and Loughborough University (Transport Studies Group) (Loughborough, UK), and the Institute of Transport (Ljubljana, Slovenia).

He has been involved in transport-related research and planning projects on both the national and international scale for almost 30 years. He has also published many scientific and professional transport-related papers, and presented many of them at national and international transport conferences. In addition, he has published four books: "Greening Airports: Advanced Technology and Operations" (2011), "Airport Analysis, Planning and Design: Demand, Capacity and Congestion" (2009), "The Sustainability of Air Transportation: A Quantitative Analysis and Assessment" (2007), and "Air Transport System Analysis and Modeling: Capacity, Quality of Services and Economics" (2000). He has been a member of NECTAR (Network on European Communications and Transport Activity Research), Delft Aviation Centre (Delft University of Technology, Delft, and The Netherlands), GARS (German Aviation Research Society, Cologne, Germany), ATRS (Air Transport Research Society), and Airfield and Airspace Capacity and Delay Committee of TRB (Transportation Research Board) (Washington DC, USA).

DOI: 10.1007/978-1-4471-6287-2, © Springer-Verlag London 2014

Index

M. Janić, *Advanced Transport Systems*,
DOI: 10.1007/978-1-4471-6287-2, © Springer-Verlag London 2014

B

C

FSC
www.fsc.org
MIX
Papier | Fördert
gute Waldnutzung
FSC® C083411

Kolibri 360 GmbH
Ferdinand-Jühlke-Straße 7
99095 Erfurt, Deutschland
produktsicherheit@kolibri360.de